Improving Changeover Performance

Improving Changeover Performance

A strategy for becoming a lean, responsive manufacturer

R. I. McIntosh, S. J. Culley, A. R. Mileham
and G. W. Owen

OXFORD AUCKLAND BOSTON JOHANNESBURG MELBOURNE NEW DELHI

Butterworth-Heinemann
Linacre House, Jordan Hill, Oxford OX2 8DP
225 Wildwood Avenue, Woburn, MA 01801-2041
A division of Reed Educational and Professional Publishing Ltd

A member of the Reed Elsevier plc group

First published 2001

British Library Cataloguing in Publication Data
A catalogue record for this book is available from the British Library

Library of Congress Cataloguing in Publication Data
A catalogue record for this book is available from the Library of Congress

ISBN 0 7506 5087 7

Typeset in India at Integra Software Services Pvt Ltd, Pondicherry, India 605 005, www.integra-india.com
Printed by Antony Rowe Ltd, Chippenham, Wiltshire

Contents

Preface

Better changeover performance: design and organizational improvement options

We have been researching industrial changeover performance for nearly 10 years. Our work has taken us to different sites throughout the UK and mainland Europe. Diverse industries have been studied. *Improving Changeover Performance* is the result of this research.

Our book joins existing books on this topic. Previous texts have tended to emphasize refinement to work practice, where local shopfloor teams are usually encouraged to seek minimal cost retrospective improvement. This approach can lead to markedly better changeovers, but should not be viewed as an exclusive improvement route. Similarly, a few texts focus more specifically upon the process hardware, emphasizing design change to improve changeovers. In some cases, as well, businesses have decided not to pursue internal retrospective improvement: perhaps more changeover-proficient equipment has been installed, or, occasionally, the services of external design specialists have been engaged. Our text considers all these options. It concentrates, however, upon internal retrospective improvement, and proposes a number of tools that might be used for this purpose. In particular the book describes how organizational change and design change can be consciously adopted in unison, as part of a structured overall improvement programme in which the full strategic implications of seeking better changeover performance are also considered.

Ideas book

Changeover improvement is an on-going topic. The current book assesses changeover improvement methodologies, concepts and techniques. Space considerations, however, limit the number of improvement examples that can be included.

We believe that knowledge of previous changeover improvement successes – with emphasis on either method improvements or equipment modification – can greatly assist a practitioner. Just as a cook will be aided by an extensive knowledge of possible ingredients, understanding also the limits of what different ingredients can contribute and when they might be used, so too those seeking better changeovers will be assisted by an extensive knowledge of possible improvement options.

We propose to follow the current book with a further book that is devoted to changeover improvement examples. It is intended for use as a 'browse-through' catalogue. To this end we invite companies to contact us with examples that we might include. Improvement ideas should be submitted with drawings and/or photographs, alongside a concise explanation of what has been done. We would also like to know how the improvement was conceived, how much it cost to implement and what gains have been achieved. All entries that are accepted will receive a copy of the book upon publication.

Kindly visit http://www.altroconsulting.com or write to Dr G. W. Owen, Department of Mechanical Engineering, University of Bath, BA2 7AY, UK. Email: G.W.Owen@bath.ac.uk.

Indeed, we anticipate that this website will develop into rather more than just a simple contact point. Ultimately, we hope, it will be built into a comprehensive source of changeover information, including for example details of service providers, or details of useful devices and where they can be bought. Perhaps, in time, it will also become a forum where practitioners across the globe can exchange their experiences, and by doing so gain ever deeper understanding of this complex topic.

Bath
January 2001

R. I. McIntosh
S. J. Culley
A. R. Mileham
G. W. Owen

Acknowledgements

We would like to acknowledge the many companies that have participated with us. In particular we are grateful to different personnel within these companies who have given their time to us. We would like to single out the assistance of Peter Black, Mattias Buchmüller, Paul Davies, Martin Davis, Brian Graham, Bob Guyan, Mark Hickey, Charles Hill, Johann Illisie, Erik van Leeuwen, Fred Price and John Webbern. This book would be greatly diminished without their contribution.

Graham Gest was a fellow researcher at the University of Bath during the early 1990s, concentrating upon classifying changeover improvement options. He proposed the 'Reduction-In' strategy that is developed further in this book.

Much of the research that this book is based upon was funded by the EPSRC (Engineering and Physical Sciences Research Council). The authors also gratefully acknowledge this support.

Finally, the authors are indebted to the Royal Commission for the Exhibition of 1851 for the significant financial contribution made towards the writing of this text.

1

Introduction

The importance of good changeover performance in multi-product manufacturing environments is becoming ever more widely understood. Certainly, the academic community has devoted considerable attention to this topic. There is also a greater awareness of changeover issues within manufacturing industry. This applies to industries as diverse as presswork, plastic moulding, pharmaceutical, foodstuffs, printing and domestic consumables. Many other industries can also be nominated, including those that are concerned primarily with assembly operations. Whereas there was a particular early focus upon changeover improvement within the presswork industry, this is no longer the case. It is now recognized that any business where a range of products is to be manufactured on non-dedicated equipment should be conscious of its changeover capability.

Awareness of changeover issues should include an awareness of what better changeovers might contribute to business performance. The rewards can often be substantial. One option is simply to devote any time that is saved to increasing production volume. At one site, for just one line, we estimated that this could contribute an additional profit of nearly £1 million per annum. Yet in many respects this is a crude exploitation of an improved manufacturing capability. Often it is more advantageous to exploit this new capability instead by increasing the frequency with which changeovers are conducted. Products are manufactured in smaller batches. Many potential benefits are possible by manufacturing this way, not least of which are those that arise by being more flexible, or more responsive to customer requirements. At the same time, for example, significant inventory reductions might also be achieved. Using a very basic analysis, halving changeover time while at the same time doubling the number of changeovers that take place can potentially permit a 50 per cent inventory reduction.

Much of the current attention to changeover issues can be attributed to Shigeo Shingo's pioneering work. His changeover book *A Revolution in Manufacturing: The 'SMED' System*[1] is rightly recognized as a reference text. Yet it is now nearly 20 years since Shingo's changeover work became widely known. Our book aims to update Shingo's work. We argue that emphasizing procedural improvements to shopfloor activity, as largely described by Shingo's single minute exchange of die (SMED) methodology,[2] can be re-examined in the light of what has been learnt over these last two decades. We suggest that a wider, more comprehensive approach to changeover improvement can be adopted; an approach that addresses diverse issues that can otherwise significantly diminish the results that an improvement initiative achieves.

1.1 Changeover and modern manufacturing practice

It is now generally agreed that the philosophy of mass manufacture, which has long been at the heart of volume manufacturing practice, has become obsolete.[3–6] The limitations of mass manufacturing have become apparent as competitive performance criteria have broadened, where the prominence of price and cost efficiency, which have been a primary justification of mass manufacture, have declined.[7–9] Increasingly a business now has to compete simultaneously on price, product quality, product differentiation, delivery performance, and rapid product development.[10–12] These objectives must be achieved additionally without compromising overall productivity, while still retaining excellent control of costs and staying abreast of technological advance.[13,14]

Traditional mass manufacturing techniques, in which businesses typically employ a rigid process technology, are ill-matched to accommodate all these demands.[15] Instead a responsive, multi-product, small-batch, low-inventory, flexible manufacturing capability is required to enable an organization to compete successfully in today's volatile and congested markets. A change towards such a manufacturing capability has gradually come about since the early 1980s as JIT (Just-In-Time) manufacturing techniques have increasingly been adopted.[16,17] Subsequently, JIT manufacturing techniques have largely become embodied in 'world-class' manufacturing[18–20] and 'lean'/'agile' manufacturing.[21,22] A highly prominent component of these modern manufacturing philosophies is the need for better changeover performance.[23–27]

1.2 The aims of the book

This book is uniquely concerned with changeover performance – what it is, what it can contribute and how to improve it. We will assess what constitutes a changeover. We will consider the potential strategic impact of improved changeover performance. We will also consider in detail how to set about achieving significantly faster, higher-quality changeovers. Examples of the approach we advocate, drawn from case studies, will be provided. We will investigate the many pitfalls or difficulties that a changeover improvement initiative can face.

It has frequently been asserted in the past that better changeovers should be achieved principally by continuous, incremental improvement activity, conducted by a shopfloor team and concentrating upon changing existing work practices and procedures. Although this approach has prevailed when seeking retrospective improvement, this need not represent the only possible approach, nor indeed, in many cases, the best.[28] For example, there are factors that can influence the outcome of an initiative that are likely to be beyond the influence of a shopfloor team. Potential limitations to team activity, among many other considerations, should be understood. A comprehensive understanding of what an improvement initiative is likely to involve should be in place before the initiative gets underway. An initiative should involve more than simply selecting a production process to work upon, engaging shopfloor personnel, seeking to maximize 'external time' effort (see below) and setting arbitrary improvement targets.

A major theme of this book will be the potential contribution of design to achieve better changeovers, by physically changing the existing manufacturing system. We will investigate how design may be applied either as an alternative or complementary strategy to seeking improvement by organizational change. This is a topic that, to date, has received little attention. Design for changeover may occur retrospectively, or may be applied to new equipment as supplied by the original equipment manufacturer. This book is targeted primarily at those seeking to improve their manufacturing operations but, by setting out generic design rules, it can also help original equipment manufacturers to respond to customers who are increasingly demanding changeover-proficiency in the specification of their new equipment.[29–34]

1.3 The structure of the book

The book is structured for use by manufacturing industry. Different chapters will be of particular interest to managers and to production personnel. In general, the earlier chapters are aimed at senior business managers, where the strategic requirement for rapid changeover will be investigated, and our overall methodology for changeover improvement will be described. Later chapters are more focused on achieving improvement in detail, addressing topics that are of particular relevance to improvement practitioners – those who are responsible for effecting change on the shopfloor. Throughout the book we shall propose different tools and documentation (as later summarized by Figure 7.15) that might assist the improvement effort.

Apart from those directly concerned with manufacturing operations, our discussion of design issues should be of use to original equipment manufacturers. In addition, by highlighting how the disciplines of changeover and maintenance can significantly impinge upon one another, we believe that sections of our text will also be useful to maintenance staff.

Finally, we have also attempted to write a book that is relevant for academic purposes. To do so without unduly disrupting the text for industrial readers we have provided our references – and, often, a brief explanation of the point we are drawing from them – at the end of each chapter. Nevertheless, it is still sometimes necessary to describe our thinking in detail in the main body of the text, particularly concerning the development of the different tools that we propose. A summary of the topics of each chapter is presented in Figure 1.1.

1.4 Definitions

What constitutes a changeover needs to be defined if changeover activity is to be investigated. Some important definitions are presented below that will be applied throughout this book. Other definitions will be introduced where appropriate later in the text. A full glossary is presented in Appendix 1.

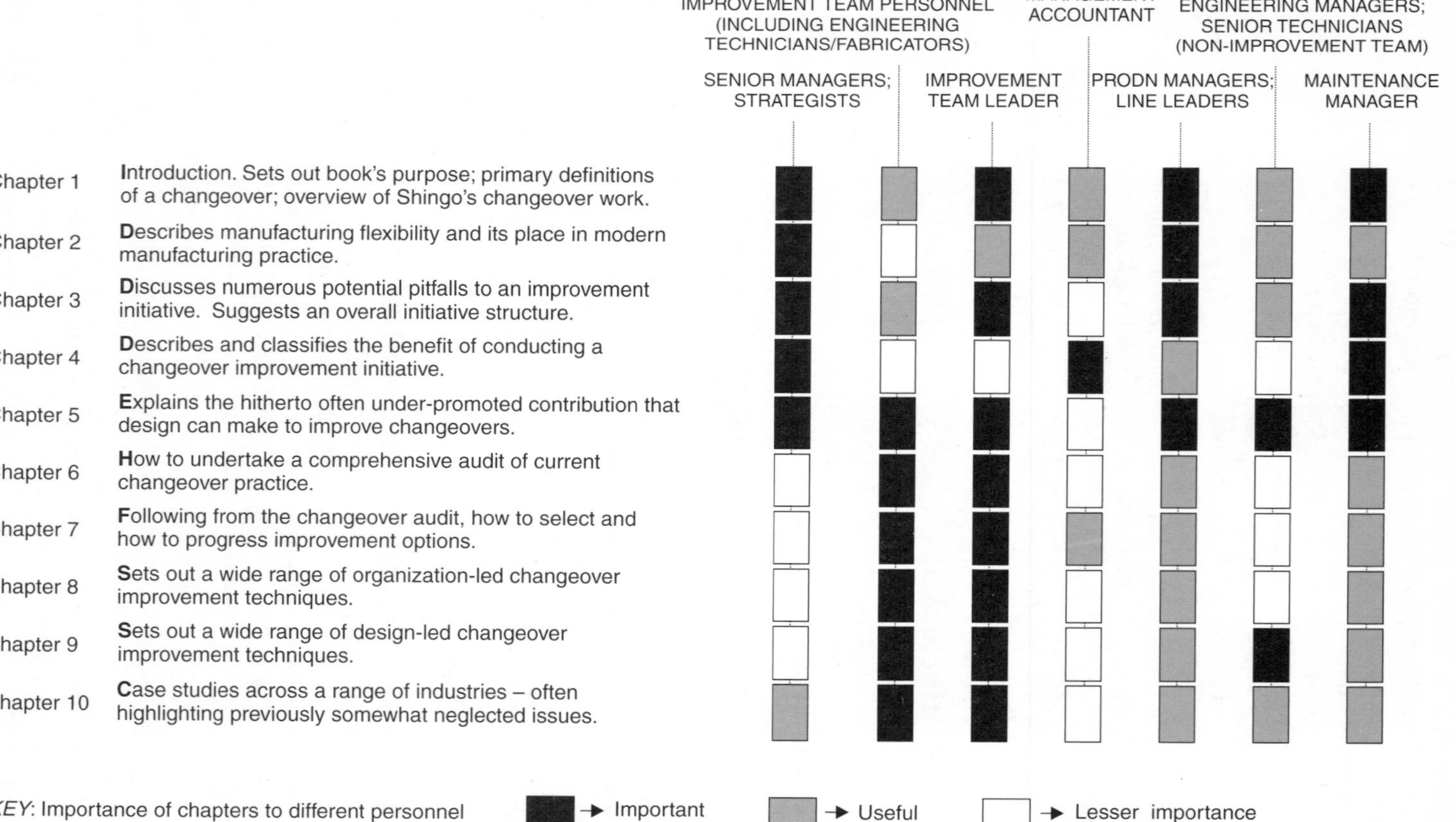

Figure 1.1 The content and likely relevance of different chapters

To date, changeover improvement has been very closely associated with Shigeo Shingo's 'SMED' methodology (which will be discussed in detail in later chapters). Apart from investigating the advantages and potential limitations of adopting the 'SMED' methodology, the term 'SMED' will not be used in our text. The term 'set-up reduction' (SUR) is also likely to be encountered in academic and industrial literature, as is the term 'make ready', which can be used when describing changeover activity in the printing industries. These terms similarly will not be used in this book. As noted below, the preferred terminology, which will be used throughout, will be variations of 'improving changeover performance'.

Improved changeover performance: faster and higher-quality changeovers

In general, we shall refer to 'changeover improvement', 'a better changeover' or 'improving changeover performance' in preference to 'a faster changeover'. Although the likely major thrust of any improvement initiative will be to reduce the changeover's duration, a practitioner may also seek to improve the changeover's quality. A higher-quality changeover will occur when line parameters at the completion of the changeover have been set more precisely, possibly allowing, for example, a higher production rate, reduced scrap, higher product quality or greater line reliability.[35–37] The term 'changeover improvement' embraces improvement of this nature. The term 'faster changeover' will be used, but only when the duration of the changeover alone is under consideration.

The total elapsed time for a changeover, T_c

A changeover is the complete process of changing between the manufacture of one product to the manufacture of an alternative product – to the point of meeting specified production and quality rates. The total elapsed time for a changeover, T_c, is shown in Figure 1.2.

Set-up period

What we have termed the 'set-up period' is the readily defined interval when no manufacture occurs. It is important to differentiate between activities which take place during the set-up period and those which are required to set a machine up (i.e. to adjust it for production): the two need not be the same. For example, presetting of tooling is an adjustment activity that occurs during external time. Similarly, for many changeovers, final adjustment of the machine only occurs once preliminary production has commenced – in the 'run-up period'.

Run-up period

The run-up period starts when manufacture of product B is commenced, and continues until steady production at full capacity occurs, at consistently acceptable product quality. In our experience the duration of the run-up period can be up to ten times the duration of the set-up period. Many commentators only define a changeover as the time from 'good piece to good piece', but this definition takes no

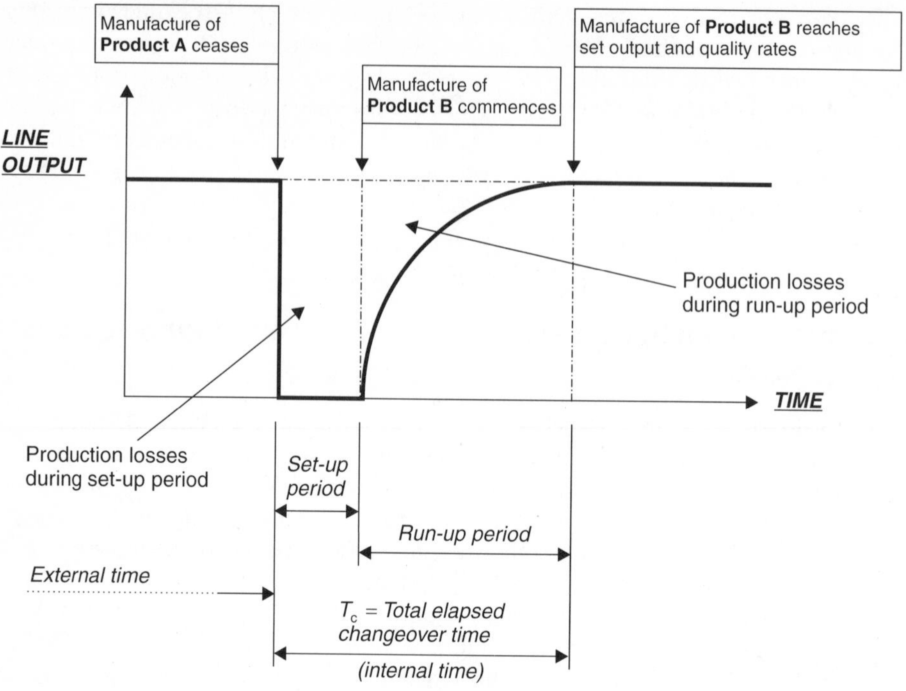

Figure 1.2 Representative line output during changeover

account of the run-up of the line or machine, where full production has still not been achieved.[38] This book will demonstrate how the set-up period and the run-up period are related to one another and why, therefore, it is wrong to consider the set-up period in isolation as representing the elapsed time for a changeover.

Figure 1.2 shows one possible way that run-up may occur. Other run-up regimes also exist, which are entirely different from the gradual and uninterrupted increase in production that Figure 1.2 presents. These will be discussed later in the book. The difficulty of establishing when the run-up period ends and when true volume production begins will also be discussed.

Internal time

Based on an original description by Shingo,[39] internal time is usually taken to refer to the period during changeover when no manufacture occurs. The term 'internal time' needs to be used with considerable caution, however, because it could be argued that it also includes the run-up period (which Shingo did not adequately distinguish).

Sparing and qualified use of the term 'internal time' will be made in this book, which includes both the set-up period and the run-up period – the period T_c. An 'internal time activity' is thus, by our interpretation, an activity that cannot occur during external time. A major thrust of improvement to date, based on the work of Shingo, has been to isolate and move tasks that are needlessly conducted in 'internal time' into external time (often without altering the tasks in any other way).

External time

Shingo's term 'external time' applies to the period before the line or machine ceases manufacture. It is frequently used in connection with preparatory activities prior to halting production.

Run-down period

Figure 1.3 shows that under some circumstances a manufacturing process may occasionally experience a run-down period, as the line or machine is slowed down and halted.[40] A linear, non-interrupted run-down period is illustrated.

Changeover activity

Changeover activity comprises each and every task that must occur for a changeover to be successfully completed (culminating in full production at a specified product quality). These tasks may be conducted before the line has halted, while the line is halted or during run-up. Some tasks required to complete a changeover might be undertaken by personnel who are not part of the immediate shopfloor environment.

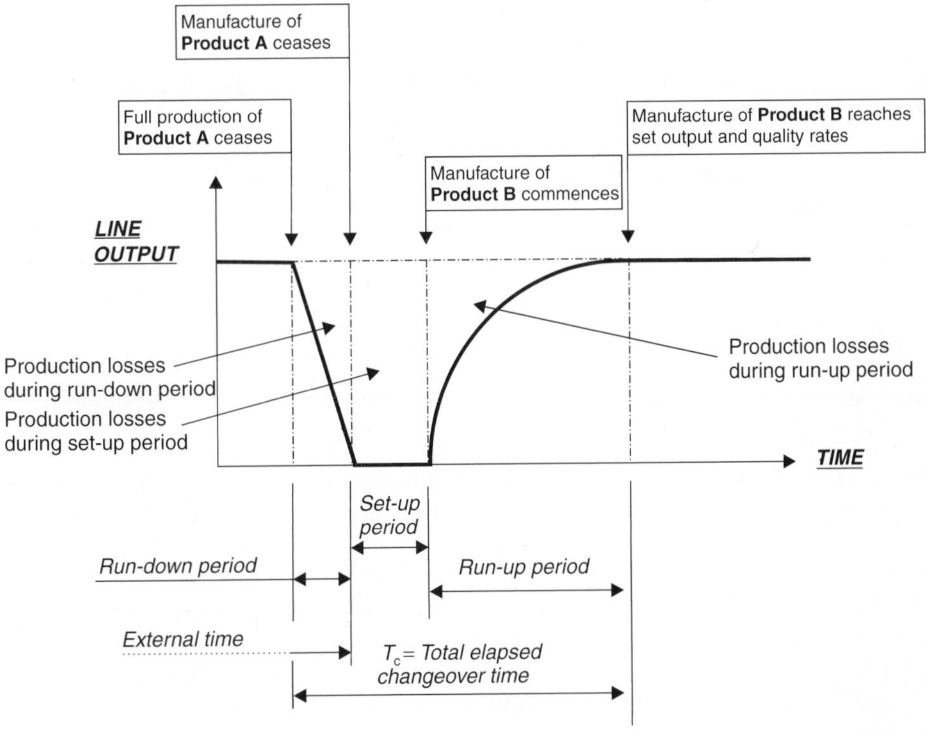

Figure 1.3 Line output during changeover when including a run-down period

Organization-biased improvement

Organization-biased improvement describes that emphasis is placed on changing the way that people work, without fundamentally altering the hardware that is worked upon. Organizational improvement occurs, for example, when people work in a more disciplined manner, or when more appropriate hand tools are used. It also takes place when the sequence in which tasks are conducted is altered, including arranging for parallel working to occur, or for tasks to be completed in external time.

Design-biased improvement

In contrast, design-biased improvement occurs when there is an emphasis on physically altering manufacturing equipment, thereby, typically, necessarily altering the changeover tasks that previously had to be completed. In some cases, also representing a design change, the product itself can beneficially be altered to improve the changeover.

1.5 Shigeo Shingo's 'SMED' methodology

Since its publication in the West in 1985, Shingo's 'SMED' methodology has had a very considerable influence upon retrospective changeover improvement initiatives.[41] The four-stage (stage 0 to stage 3) 'SMED' methodology, which is set out below, is based on the dominant rationale of separating/converting internal time tasks into external time. All tasks are then 'streamlined'.

- Stage 0 'Internal and external set-up conditions are not distinguished'
- Stage 1 'Separating internal and external set-up'
- Stage 2 'Converting internal to external set-up'
- Stage 3 'Streamlining all aspects of the set-up operation'

These are referred to by Shingo as 'conceptual stages'. Shingo supports the four-stage 'SMED' methodology by numerous improvement techniques, which are assigned to the individual stages. Based upon his own experience, Shingo also presents many examples of improvements that have been undertaken.

Such has been the impact of Shingo's work that the term 'SMED' is now common in both academic and industrial usage, to the extent that it has become almost synonymous with 'changeover improvement' itself.[42–44] By some it might even be regarded as the sole means by which better changeovers are to be achieved. The methodology has been widely adopted by changeover improvement consultants, which will have heightened this perception. This book will place Shingo's contribution into perspective. There can be little doubt that Shingo's work has had a substantial impact, but it should be recognized that the 'SMED' methodology is explicitly a shopfloor tool. Its use should be limited as such, while at the same time understanding that a comprehensive initiative typically will involve much that the 'SMED' methodology does not purport to address. In making these observations we do not mean to imply that the 'SMED' methodology is necessarily free of limitations, inhibiting its usefulness for the specific purpose for which it is intended, nor that it is always interpreted in a way that allows optimum results to be achieved. For example, we have often observed that comparatively minor consideration is given to the methodology's final 'streamlining' stage or concept yet, as we shall explain, some potentially highly important improvement techniques reside here, not least those that can promote design change to existing hardware. Or, considered another way, a low-cost, continuous incremental improvement, team-based approach[45] can inherently exclude substantive design change from taking place. Such an approach can sometimes generate improvements that are difficult to sustain[46] or, possibly in other respects as well, 'you will get what you pay for'.[47] These topics will be investigated in detail in later chapters. As previously noted, this book aims to update Shingo's work, and a retrospective improvement methodology (for use by a shopfloor improvement team) that could be used in place of the 'SMED' methodology will later be presented.

An introduction to the scope and complexity of the changeover improvement task

That meaningful, lasting improvement is more likely when conducting a wide-ranging, structured initiative is worth repeating. Too often we have observed patchy, *ad hoc* programmes; perhaps championed by an enthusiastic line manager, but at the same time both reluctantly undertaken by line operators and essentially unsupported by senior management. Appropriate understanding of how to identify and implement improvements might be lacking, or the initiative's purpose (specifically why better changeovers are being sought) might not have been adequately thought through. We have strong evidence that such programmes experience a high chance of failure (Chapter 3).

Initiatives rarely fail exclusively because line personnel do not work to their capabilities, using those improvement tools that they are provided with (for example the 'SMED' methodology). Many other factors are also likely to be influential. For example, staff should expect to have been trained. Also, it should be evaluated what resources warrant being committed, and, indeed, whether it is more appropriate to buy and install new equipment in preference to retrospectively improving what already exists. Neither should an initiative be closed once the desired level of improvement has been achieved: attention still needs to be given to sustaining the gains that have just been fought for.

Issues of sustaining and exploiting better changeover performance can themselves be complex. There is no point in being able to conduct a changeover, say, 60 per cent more quickly if this capability is not both maintained and used to business advantage. Thus awareness and discipline across the organization will be needed and, quite possibly, revision to internal business processes required. As with seeking improvement in the first place, staff should neither be expected to work unsupported nor be expected to be intrinsically aligned, at least initially, to changed working methods.

1.6 Changeover and maintenance

The way that the disciplines of changeover and maintenance can impinge upon one another has received little attention to date. First, changeover tasks can be closely correlated to maintenance tasks in that on-machine maintenance frequently involves removing and replacing, or adjusting, a part/module/assembly. During maintenance the original part can be re-used, or a new, identical part substituted in its place.[48] Maintenance attention is given to allow continued manufacture of the same product. Maintenance might thus be regarded as changeover from 'Product A' back to 'Product A', rather than a more conventionally perceived changeover from 'Product A' to 'Product B'. The majority of the techniques that will be described in this book will be identical in both situations, and therefore will be of direct relevance to industrial maintenance managers and technicians.

Maintenance can also impinge upon changeover performance in terms of the quality of the change parts that are supplied.[49–51] If maintenance is poorly conducted, where the change parts are used that do not conform to specification, this can have a very considerable impact on both changeover time and the changeover's quality. This too will be investigated.

1.7 Summary

It has been deliberately chosen not to dwell on the role of changeover performance as a cornerstone of JIT/Lean/Agile/World Class manufacturing: this topic has been covered in depth elsewhere. References have been provided for readers who may wish to pursue this topic further. A topic that will be investigated, however, is that of manufacturing flexibility (which these post-mass production manufacturing philosophies or paradigms seek to ensure). This is the subject of the next chapter.

This book aims to give a complete and thorough appraisal of changeover improvement activity. However, two topics are deliberately omitted. First, the book will not consider mathematical modelling methods (in which scheduling issues feature prominently).[52–55] Second, the book does not directly address issues of introducing a new product to an existing product range. In all other respects it seeks to provide the reader with a full understanding of the many facets of a successful changeover improvement initiative, coupled with an understanding of why such an initiative should be undertaken.

References

1 Shingo, S. (1985). *A Revolution in Manufacturing: The SMED system*. Productivity Press, Cambridge, MA.

2 Shingo, S. (1985). *Op. cit.* ref. 1. We shall make considerable reference to Shingo's book *A Revolution in Manufacturing: The SMED system*. This largely describes a particular approach for retrospective changeover improvement – which we shall refer to throughout as the 'SMED' methodology (see also Appendix 1).

3 Duguay, C. R., Landry, S. and Pasin, F. (1997). From mass production to flexible/agile production. *Int J. of Operations and Production Management*, **17**, No. 12, 1183–1195.

4 Jaikumar, R. (1986). Postindustrial manufacturing. *Harvard Business Review*, Nov.–Dec., 69–76. Jaikumar provides an early appraisal of changing manufacturing practice. US performance is contrasted with Japanese performance, particularly with regard to high-technology manufacturing systems. Emphasis is placed on the need for flexibility and responsiveness; aspiring to multi-lot, high-productivity, small-batch manufacture. Human/

management issues in using advanced technology are also described. Jaikumar concludes: 'The days of Taylor's immense, linear production systems are largely gone.'

5 Ayers, R. and Butcher, D. (1994). The flexible factory revisited. *American Scientist*, **81**, No. 5, 448–459.

6 Hayes, R. H. and Wheelwright, S. C. (1985). *Restoring our Competitive Edge: Competing through manufacturing*. Wiley, New York.

7 Schonberger, R. J. (1986). *World Class Manufacturing: The lessons of simplicity applied.* Free Press, New York.

8 Bolden, R., Waterson, P., Warr, P., Clegg, C. and Wall, T. (1997). A new taxonomy of modern manufacturing practices, I. *J. Operations and Production Management*, **17**, No. 11, 1112–1130. Bolden *et al.* present one, among many, of the more recent academic articles in which changed manufacturing priorities are discussed.

9 Mapes, J. (1996). Performance trade-offs in manufacturing plants. *Proc. 3rd. Int. Conf. European Operations Management Association*, London, pp. 403–408.

10 Dept of Trade and Industry (DTI) (1994). *Competitive Manufacturing – a practical approach to the development of a manufacturing strategy*. DTI, London.

11 Womack, J. P., Jones, D. T. and Roos, D. (1990). *The Machine that Changed the World.* Rawson Associates, New York.

12 Maskell, B. H. (1991). *Performance Measurement for World Class Manufacturers*. Productivity Press, Cambridge, MA.

13 Lamming, R. (1993). *Beyond Partnership, Strategies for Innovation and Lean Supply.* Prentice Hall, Hemel Hempstead.

14 Mair, A. (1994). Honda's global flexifactory network. *Int. J. of Operations and Production Management*, **14**, No. 3, 6–24. Chapter 2 assesses changeover performance in terms of its impact on manufacturing flexibility. Issues of flexibility, including rapid line changeover, are described for a motor vehicle manufacturer. The need to maintain high productivity is described.

15 Abernathy, W. J. (1978). *The Productivity Dilemma: Roadblock to innovation in the automobile industry*. Johns Hopkins University Press, Baltimore, MD. Abernathy's description of 'an ever more rigid process technology' was made of US automobile production plants that were slowly being rendered uncompetitive against Japanese auto manufacturers in the late 1970s. At this time JIT was not widely practised in Western manufacturing, and the inflexibility of existing mass production systems was becoming a major issue, as Western auto manufacturers sought to compete on cost efficiency – the primary area in which mass manufacture was perceived to offer manufacturing advantage.

16 Schonberger, R. J. (1982). *Japanese Manufacturing Techniques – Nine Hidden Lessons in Simplicity*. Free Press, New York. Schonberger's influential book was one of the first to articulate JIT manufacturing practice.

17 Slack, N., Chambers, S., Harland, C., Harrison, A. and Johnston, R. (1995). *Operations Management*. Pitman, London. Slack *et al.* write: 'JIT principles which were a radical departure from traditional operations practice in the early 1980s have now themselves become the accepted wisdom in operations management.'

18 Steudel, H. J. and Desruelle, P. (1992). *Manufacturing in the Nineties – How to Become a Lean, Mean World-class Competitor*. Van Nostrand Reinhold, New York.

19 Flynn, B. B., Schrpeder, R. G., Flynn, E. J., Sakakibara, S. and Bates, K. A. (1997). World class manufacturing project: overview and selected results. *Int. J. of Operations and Production Management*, **17**, No. 7, 671–685.

20 Sheridan, J. (1997). Lessons from the best. *Proc. Annual Int. Conf. American Production and Inventory Control Soc.*, Washington, pp. 126–131.

21 Gunneson, A. O. (1996). *Transitioning to Agility: Creating the 21st century enterprise.* Addison-Wesley, Reading, MA.

22 Quinn, R. D., Causey, G. C., Merat, F. L., Sargent, D. M., Barendt, N. A., Newman, W. S., Velasco, V. B., Podgurski, A., Jo, J. Y., Sterling, L. S. and Kim, Y. (1997). Agile manufacturing workcell design. *IIE Trans.*, **29**, No. 10, 901–909. Changeover can sometimes be singled out when defining 'agility': 'We define agile manufacturing as the ability to accomplish rapid changeover from the assembly of one product to the assembly of a different product.'

23 Karlsson, C. and Åhlstrom, P. (1996). Assessing changes towards lean production. *Int. J. of Operations and Production Management*, **16**, No. 2, 24–41.

24 Spencer, M. S. and Guide, V. D. (1995). An exploration of the components of JIT: case study and survey results. *Int. J. of Operations and Production Management*, **15**, No. 5, 72–83. By conducting a survey of previously published work, Spencer and Guide analysed the components of JIT. The requirement of a rapid changeover capability was common throughout all the publications that were surveyed.

25 Prasad, B. (1995). A structured methodology to implement judiciously the right JIT tactics. *Production Planning and Control*, **6**, No. 6, 564–577. Prasad also provides a review of JIT tactics, taken from a review of approximately 200 articles and books, but concentrating particularly on publications that are regarded as 'classics of JIT'. Rapid changeover features at the head of the list that Prasad provides.

26 Black, J. T. (1991). *The Design of the Factory with a Future*. McGraw-Hill, New York.

27 Koufteros, X. A. and Vonderembse, M. A. (1998). The impact of organizational structure on the level of JIT attainment: towards theory development. *Int. J. Production Res.*, **36**, No. 10, 2863–2879.

28 McIntosh, R. I., Culley, S. J., Mileham, A. R. and Owen, G. W. (2000). A critical evaluation of Shingo's 'SMED' methodology. *Int. J. Production Res.*, **38**, No. 11, 2377–2395. This paper discusses the impact of Shingo's 'SMED' methodology on changeover improvement practice in industry, describing how improvement typically is approached by emphasizing aspects organizational change. This is contrasted by the potential impact of using design which, it is argued, the 'SMED' methodology (also taking account of the way that it has been interpreted by changeover improvement consultants) does not adequately promote.

29 Emproto, R. (1994). Label with a cause. *Beverage World*, **113**, Dec., 78. There are many examples of OEMs advertising the changeover capability of their equipment in the trade press. This trend is particularly notable in the press tool industry, the printing industry and the plastic moulding industry. Emproto's article presents a brief review of labelling systems from a selection of manufacturers exhibiting at InterBev94, USA. One company, CMS Gilbreth, entitles its machine 'Trine Quick Change 4500', demonstrating the significance of changeover in this market. The machine has been designed to be changed over by one operator. Four parts only need to be changed, using just two different tools.

30 Rover Group Ltd (1991). *Triaxis Press* (promotional publication). Swindon, UK. Rover Group identify a few key performance parameters for this large vehicle body panel press, including changeover, which is specified to take 5 minutes or less.

31 Demetrakakes, P. (1997). Quick change artists. *Food Processing*, **58**, November, 80–83. Demetrakakes cites a machinery manufacturer's representative on the subject of packaging equipment changeover capability: 'Now customers are saying, "We want in the quote an estimate of what the changeover is going to be, and that'll be a big factor in our decision to buy."'

32 Harsch, E. and Rundel, P. (1994). Demands on present-day presses. *European Production Engineering*, **18**, No. 3–4, Oct., 15–18. Harsch and Rundel describe that changeover capability is an important feature of presses: 'Apart from the self evident requirements relating to quality and productivity, repeatability of technical parameters and rapid changeover are demanded nowadays.' Press systems represent one industrial sector that

has concentrated upon rapid changeover for many years. Others, for example injection moulding, also fall into this category, whereas other industries might be identified, hitherto, as having been less changeover conscious. These more mature 'changeover industries' can be studied to determine how changeover practice has evolved and, particularly, to determine how design change has been introduced to markedly improve changeover performance.

33 Matupi, N. and Kay, J. M. (1996). Defining manufacturing flexibility requirements, Advances in Manufacturing Technology. *Proc. 12th National Conference on Manufacturing Res.* (A. N. Bramley, A. R. Milcham and G. W. Owen, eds), pp. 186–190, Bath, UK. Enhanced equipment capability can also be perceived in terms of 'flexibility'. Matupi and Kay write: 'It has been shown that flexibility, as a concept, has enjoyed unprecedented support from all quarters of the manufacturing industry, including marketeers, equipment suppliers, academics, operations managers and customers.' It will be argued in Chapter 2 that better changeover performance is a tangible mechanism by which a desire for enhanced manufacturing flexibility can be satisfied.

34 Jones, J. and Griffin, D. (1993). Achieving quick changeover, I Mech E Total Productive Maintenance Workshop, 25–29 October 1993, Stratford-upon-Avon, UK. The incentive to specify changeover performance in new equipment is highlighted, following a changeover improvement exercise at Avon Cosmetics.

35 Smith, D. A. (1991). *Quick Die Change*. SME, Dearborn, MI. While there are many references to possible benefits of conducting faster changeovers, reference to benefit arising from conducting a better quality changeover is less common. Smith has extensive experience as a press tool consultant working in the USA. The benefits of adopting quick die change methods (which Smith largely advocates on the basis of precision, good working practice and standardization) are cited as: Increased Capacity; Scrap Reduction; Job Security; Safety; Die and Press Maintenance Costs; Reduced Inventory; Improved Quality. The topic of benefit arising from better changeover performance will be investigated in depth in Chapter 4. Smith's nomination of 'Safety' and 'Die and press maintenance costs' in particular, though valid, are unusual. In his text Smith emphasizes the importance of conducting a changeover properly as well as conducting it quickly, describing also that there is potentially a trade-off between these two objectives: 'Many stamping executives see the main benefit of quick die change as a process variability reduction tool. The ability to know with certainty that a die can be in a given press and producing good parts in 20 minutes is more important to them than having the die in the press in five minutes and producing junk.'

36 Schonberger, R. J. (1990). *Building a Chain of Customers*. Free Press, New York. The topic of higher-quality changeovers will be investigated again later in this book. Schonberger identifies that higher performance can be manifest in other ways, beyond experiencing simply a quicker changeover: 'A common by-product of quick set-up and fail-safing improvements is making equipment work better and break down less often.'

37 Souloglou, A. (1992). Scroll shear changeover. Bath University changeover group, internal case study report, July 1992. Souloglou – a Bath University research student working on-site at a UK plant – was tasked to assess in what ways improved changeover performance benefited the business. Changeover time reduction from 10 hours to 2 hours had been achieved when this attachment took place. Souloglou reports: 'It has been found that changeover improvement on the scroll shear has resulted in consistently higher quality production.' This improvement, for the most part, is attributed to precision setting using stops, thus eliminating previous extensive trial-and-error adjustment steps. Machine parameters are established by this means both swiftly and precisely (see also Chapter 5).

38 Trevino, J., Hurley, B. J. and Friedrich, W. (1993). A mathematical model for the economic justification of set-up time reduction. *Int. J. Prod. Res.*, **31**, No. 1, 191–202.

Some notable examples exist of changeover being defined as the period from 'good piece to good piece'. In situations where complex mathematical algorithms are used to model lot size and scheduling performance under differing constraints, it is possible that such a definition of changeover will have a significant impact on the results that are generated, particularly if the run-up period is considerably longer than the set-up period.

39 Shingo, S. (1985). *Op. cit.* ref. 1.

40 Konrad, G. (1997/8). Dynachange – 'on-the-fly' job changeover for newspaper production. *MAN Research Engineering Manufacturing*, 12–15.

41 Shingo, S. (1985). *Op. cit.* ref. 1. The extent to which the 'SMED' methodology is employed will be investigated later in our book.

42 Oliver, M. V. L. (1989). Back to the future – where is JIT going? *Proc. 4th Int. Conf. JIT Manuf.*, pp. 65–73, IFS Ltd, Bedford, UK. By 1989 the 'SMED' methodology had become well enough established for Oliver to proclaim 'SMED' as a buzzword: 'to reduce set-up times'. In our own research we have encountered the word 'SMED' applied very loosely both as a noun and as a verb: 'We must be aware of the benefits of using "SMED"' and even: 'We need to "SMED" our factory'.

43 Rizzo, K. (1993). Trimming press make-ready time. *Gatfworld*, **5**, Issue 4, 1–8. This is just one of many discussions of the application of the 'SMED' methodology to improve changeover performance – in this case for printing presses. Further examples will be cited and assessed in later chapters.

44 Black, J. T. (1991). *Op. cit.* ref. 26. Black's book provides an example of the 'SMED' methodology being reproduced in industrial management/operational texts.

45 Hay, E. J. (1988). *The Just-in-Time Breakthrough*. Wiley, New York. For example, Hay, who is a consultant, is emphatic that improving changeover performance should be a low-cost activity.

46 McIntosh, R. I., Culley, S. J., Gest, G., Mileham, A. R. and Owen, G. W. (1996). An assessment of the role of design in the improvement of changeover performance. *Int. J. of Operations and Production Management*, **16**, No. 9, 5–22. Sustainability is one aspect of a typical changeover improvement initiative that is discussed, where the role of design change to the existing manufacturing system is highlighted. Other possible benefits attributable to emphasizing improvement by design change – as will be set out in this book – are also argued.

47 Hughes, D. and Hobbs, S. (1994). SMED – Assessing the cost effectiveness of set-up reduction. *Proc. 27th int. Symp. on Automotive Tech. and Automation*, Aachen, 31 Oct.–4 Nov., pp. 601–607.

48 Souloglou, A. (1992). *Op. cit.* ref. 37. We shall give further attention to the topic of maintenance (and provide some further reference material) in Chapter 4. Some observations regarding changeover improvement and its interaction with maintenance are made by Souloglou. It is noted that the particular site in his study were planning to reduce the time allocated to maintenance activity on equipment on which changeover performance had been greatly improved:

1 '[Changeover improvement]... has led to many of the more complex mechanisms being replaced by simplified versions. This has reduced the number of breakdowns as well as reducing the time required to maintain the machine.'

2 'More frequent and consistent changeovers result in greater machine inspection and thus problems can be spotted earlier.'

49 Lee, D. (1986). Set-up reduction: making JIT work. *Management Services*, May, 8–13.

50 McIntosh, R. I., Culley, S. J., Mileham, A. R. and Owen, G. W. (2000). Changeover improvement: a maintenance perspective. *Int. J. Production Economics* (accepted for publication 13 November 2000). Our paper assesses changeover improvement and its relation-

ship with maintenance activity. It is argued both that focused maintenance activity can aid changeover performance and, conversely, that better changeover performance, including application of changeover improvement techniques, can assist maintenance. These topics are also assessed at different stages within the current book.

51 Smith, D. A. (1991). *Op. cit.* ref. 35. Smith similarly discusses the importance of preventive maintenance practice and its impact on changeover. One area that is singled out is the need to maintain alignment of the press, to maintain a common shut height. Smith goes on to describe 'press and die alignment maintenance' as the foundation of process repeatability, and argues the focused application of maintenance to minimize the need for special adjustments during changeover.

52 Allahverdi, A., Gupta, J. N. D. and Aldowaisan, T. (1999). A review of scheduling research involving set-up considerations. *Omega*, **27**, No. 2, 219–240. Provides a useful review of work in this area.

53 Gallego, G. and Moon, I. (1995). Strategic investment to reduce setup times in the economic lot scheduling problem. *Naval Res. Logistics*, **42**, No. 5, 773–790. Modelling scheduling in respect of production changeovers (specifically, the Economic Lot Scheduling Problem – ELSP) has been an important topic of academic research for at least 40 years. The modelling that is undertaken can be highly complex. We will occasionally refer to work and results that have been concluded, but we will not make any direct assessment of this specific area of changeover research.

54 Watson, E. F. (1997). An application of discrete-event simulation for batch-process chemical-plant design. *Interfaces*, **27**, No. 6, 35–51.

55 Kimms, A. and Drexl, A. (1998). Some insights into proportional lot sizing and scheduling. *J. Operational Res. Soc.*, **17**, No. 11, 1196–1206.

2

Manufacturing flexibility

Much has been written on configuring businesses to be competitive where markets experience change and uncertainty – conditions in which a business must be able to thrive. This chapter will investigate what improved changeover performance can contribute to manufacturing flexibility, particularly in this context of uncertain and volatile markets. The limits of what better changeover performance can contribute need to be known, for there are many aspects of manufacturing flexibility that better changeover performance cannot influence.[1] Manufacturing flexibility will also be assessed briefly as a component of overall business flexibility.

Better changeover performance may be used to improve manufacturing flexibility but, as will be described fully in Chapter 4, it may also be exploited by the business in other ways. Different potential uses of improved changeover performance may conflict with one another, and a decision will need to be taken as to how better changeovers are to be used to greatest advantage. This chapter will contrast the application of better changeovers as part of a rigidly conducted Total Productive Maintenance (TPM) programme with changeover improvement undertaken in its own right. Respectively, faster changeovers are used either to increase line availability for production or, typically, by conducting more changeovers and reducing batch sizes, to enhance manufacturing flexibility.

While better changeover performance can influence manufacturing flexibility, enhanced manufacturing flexibility can also be achieved by employing dedicated parallel equipment, which is only used as and when required to manufacture the product for which it is permanently set up. If this option is not wholly pursued, and changeovers are necessary, there are potentially many ways by which better changeover performance may be achieved. These range from low-cost, retrospective improvement of existing hardware and procedures, through to the wholesale replacement of existing facilities by alternative, more changeover-proficient equipment. The company must be able to evaluate all possible options. It must do this, for example, in terms of cost, time scale and implementation issues. Additionally, the business should evaluate each option's likely ability to achieve the improvement that is sought, and then to sustain it. Thus, in summary, a business needs to carefully consider how better changeover performance is both to be achieved and exploited, including possible issues of operational change beyond the shopfloor. To do so requires an understanding of the multifaceted nature of manufacturing flexibility.

2.1 Reassessing economic order quantity

Economic Order Quantity (EOQ) represents a pillar of traditional manufacturing theory, dating from 1915,[2] at the dawn of mass production.[3] The economic order quantity indicates that as the order quantity, or batch size, increases so too inventory costs increase, as shown in Figure 2.1.[4] Conversely, as the order quantity increases, so the order costs per unit decrease. For every batch there is a point at which the total cost for the batch may be minimized, representing the economic order quantity.

A formula may be used to estimate the value of the economic order quantity, which is derived by differentiating the sum of two simple expressions; respectively for order costs and inventory costs.[5] These expressions – and hence the formula – are based on a number of significant assumptions:

- Demand from inventory is at a constant (linear) rate.
- Full replenishment of inventory occurs both instantaneously and with no lead time.
- Inventory replenishment is triggered when inventory reaches zero.
- Replenishment is made at a constant batch size.
- Inventory costs increase linearly.

Nevertheless, as noted by Naddor,[6] '... this formula has had more applications than any other single result obtained from the analysis of inventory systems'.

As noted, the basic EOQ equation is simple to determine. Take the steady rate of demand as D (units/time); the repeated order quantity to inventory as Q (units); the cost of inventory as C_i (£/unit/time); the cost of ordering as C_o (£). The EOQ is a cost equation which seeks to determine the value of Q that minimizes costs overall:

$$\text{EOQ} = \sqrt{\frac{2 \times D \times C_o}{C_i}}.$$

The time between EOQ orders = EOQ $/D$ (see Figure 2.1).

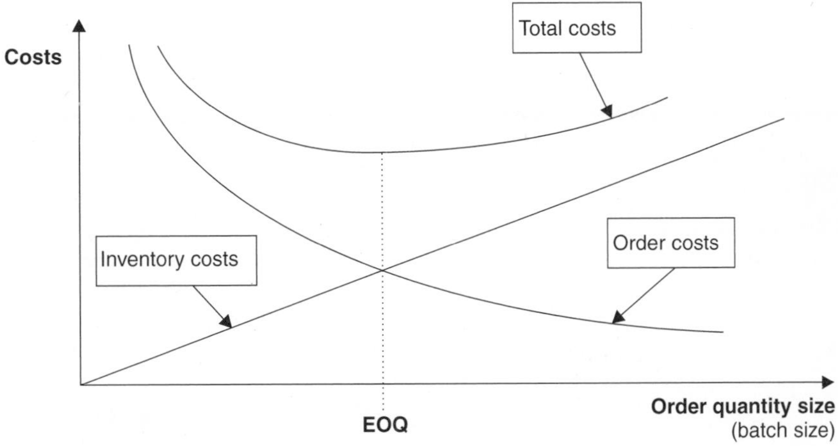

Figure 2.1 Traditional determination of the economic order quantity

The economic order quantity, especially because a mathematical formula can be applied to determine it, might lead the business to assume there exists a scientifically derived optimum order quantity for a situation of fixed operational constraints (inventory holding cost and order cost). This assumption is flawed because a business may undertake improvements that can impact upon either cost: neither is fixed.[7]

The EOQ equation above represents a simple analysis of the relationship between order and inventory costs. The EOQ analysis may be made significantly more complex, to overcome some of the basic underlying assumptions on which it is based.[8] Yet an EOQ analysis still should not dictate a subservient role for the operations manager – simply to accumulate the best possible cost data, then to use these data in the best available EOQ model equation and then, finally, to base batch scheduling decisions on the results that are generated. This rationale seeks to optimize the existing situation. By doing so, the business implicitly excludes seeking improvement to current practice.

This book focuses on order costs – altering the costs of putting an order into production. By improving changeover performance the order cost curve can be brought down markedly to the left, as shown in Figure 2.2. This can occur for many different reasons, to do with both quality and time improvements to the existing changeover. Where faster changeovers occur at the same frequency, order costs can be forced down by the profitable use of what was previously lost production time. Additional significant influences may also exist, for example employing fewer per-

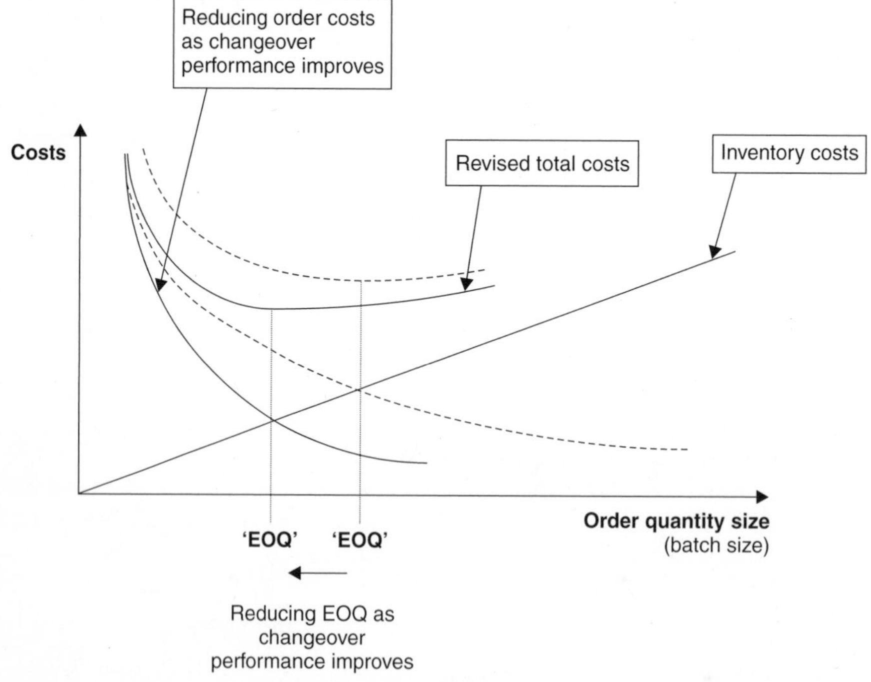

Figure 2.2 The impact of improved changeover performance on the order cost curve, and hence on the EOQ

sonnel to undertake a changeover, or conducting a changeover that has a lower skill requirement.

By using improved changeover performance to alter the order cost curve it is seen that the economic order quantity rapidly reduces from its previous level. Although the cost of holding inventory has not changed, Figure 2.2 predicts that as changeover performance improves it becomes necessary to hold significantly less inventory for the business to maintain a consistent delivery performance to its customers. Improved changeover performance in itself, therefore, is a powerful tool for inventory reduction,[9] and can impact upon many of the costs of carrying excess inventory.[10] Ultimately, if changeover performance was sufficiently good to allow the economical manufacture of single unit batches, potentially no inventory at all need be held (neither work-in-progress nor finished goods inventory), with the business instead producing and shipping immediately to order.[11] Any business that is able to operate in this way, assuming that this can be achieved with commensurate control of human and physical resources, will realize significant benefits, in line with those claimed of JIT manufacturing.[12] These benefits will include reduced cycle times, fast material throughput, and better production and quality problem identification – as well as a reduction or elimination of costs associated with inventory storage. The ideal of being able to manufacture in unit batch sizes is aspired to by modern manufacturing philosophies,[13] commencing with JIT and now embraced by the lean/agile manufacturing philosophy.

A blind spot that significant reductions in changeover time are possible

While it is straightforward to assert that improved changeover performance can reduce the economic order size, and while there is much evidence in the literature that considerable reductions in changeover times have been achieved,[14–17] our experience is that many businesses still find it difficult to accept that significant improvement to changeover performance is possible.

Other authors have referred to this inhibition as a 'blind spot'.[18,19] Often business managers and shopfloor personnel independently have expressed disbelief that a 50 per cent reduction in changeover time might be a relatively modest target. In one particular case a production manager who endured single shift (8-hour) changeovers for one piece of production equipment in his charge could not accept that near-identical equipment at another site in the same company was being changed over, following an improvement programme, in approximately 90 minutes. The production manager perceived that the circumstances at this other site were different whereas, in reality, the difference was only one of acceptance that improvement was possible. Until this mind set changed no improvement initiative could commence, or at least, if one did, it was not likely to be supported in such a way that it would be particularly successful.

At a higher level still, similar difficulties of awareness and acceptance of the probable improvement that can be wrought to changeover performance may still be found among the most senior personnel within manufacturing industry. Not all senior managers demonstrate a belief in the level of improvement that can be achieved nor, in some cases, can the purpose of conducting such improvement be identified. While these shortcomings prevail, at all levels, manufacturing industry will not be able to embrace completely the opportunities afforded by the new manufacturing philosophies.

The contribution of senior managers should not be limited simply to sanctioning a changeover improvement initiative. Senior managers have an important role as part of the overall initiative. This assertion is contrary to many authors' previously expressed opinions that senior managers should have no significant active role to play, and that an improvement initiative essentially should be wholly concentrated upon the shopfloor.

2.2 Business flexibility and manufacturing flexibility

The desirability of achieving a highly flexible business has frequently been discussed.[20–23] It has also been acknowledged that flexibility is 'a complex, multi-dimensional and hard to capture concept'.[24] Manufacturing flexibility is but one dimension, or component, of overall business flexibility and is itself also multi-faceted.

It is aimed here to gain a better understanding of what are the components of, or what contributes to, flexibility – both overall flexibility of the business and, more specifically, manufacturing flexibility. To do this, flexibility will be assessed from different perspectives,[25] with a focus where appropriate on changeover performance.[26,27] Our multi-perspective assessment of flexibility is presented with the understanding that flexibility is a significant ingredient of the modern manufacturing philosophies identified in Chapter 1.[28]

A global perspective of offensive and defensive uses of enhanced flexibility

Flexibility has been described as a 'fashionable virtue', being likened to a shock absorber to absorb the effects of uncertainty.[29] Flexibility similarly is often cited to deal with, or accommodate, uncertainty as an alternative defensive strategy to the use of inventory.[30,31] Uncertainty is often regarded as an external influence beyond the control of the organization, particularly through changes in customer demands,[32–34] although other uncertainties that originate externally may also prevail, for example in changing macro-economic conditions.[35] Additionally, uncertainties may be identified that arise internally, through the inability and ignorance of personnel within the organization,[36] or through the poor availability and use of systems that allow information to be located and processed.[37] By accommodating this internally generated uncertainty, flexibility can effectively buffer the organization from particularly adverse effects of poor long-term planning or incohesive strategy: it allows the business to operate more successfully using short-term planning, and under circumstances of changeable strategic direction.[38]

These defensive uses of flexibility may be contrasted with opportunities to exploit possible offensive uses of flexibility, for example in increasing efficiency or in allowing a business to introduce large numbers of new products within a reduced lead time.[39] Equally, there may be opportunities to use flexibility to be more responsive to customer needs, particularly in being able to win lucrative niche markets,[40] generating income with a particular emphasis on margin rather than production volume. Many of the offensive opportunities that JIT/lean manufacturing present to a business are dependent, at least in part, on a rapid changeover capability.

Ideally, all possible offensive and defensive opportunities need to be identified and compared. Flexibility is a wide-ranging and complex topic, and to do this in a way that addresses every facet of flexibility would be a highly complex undertaking.[41,42] The task becomes less daunting (though still complex) if only the potential contribution of better changeover performance to enhance flexibility is considered. This, though, still remains an exercise that should be treated with great caution. To undertake such a narrow assessment of the benefit of better changeover performance (only considering its impact on flexibility) ignores its wider potential contribution. Like flexibility, better changeover performance should not be sought as an end in itself, but should be regarded as a means to an end: it allows a business to become more flexible if it chooses, but equally it can be exploited to gain benefit in other ways. This topic will be picked up again in Chapter 4 where it will be argued that the sometimes conflicting potential applications of improved changeover performance need to be quantified if they are to be evaluated properly.

Defining manufacturing flexibility

It is useful to reflect on a possible definition of manufacturing flexibility, as this can also contribute to an understanding of what manufacturing flexibility is.

Physical changeover activity predominantly occurs within the production department of the business and, as has been indicated, may take place for both defensive (reactive) or offensive (proactive) reasons. Some definitions have been provided in the literature that concentrate, typically, on a defensive stance.[43] A more useful definition is one that accounts for both offensive and defensive positions, while at the same time also being more explicit than simply describing manufacturing flexibility as the ability of the production facility to 'change'. One such more comprehensive definition has been provided by Upton,[44] where manufacturing flexibility is succinctly described as '. . . the ability to change or react with little penalty in time, effort, cost or performance'.

This definition alludes to being able to achieve flexible manufacturing performance without recourse to substantial trade-offs in other areas of manufacturing performance.[45–47] The definition is also general in its scope, and does not single out the manufacturing function or manufacturing processes. By not restricting itself in this way the definition need not be confined to the immediate production environment, and might also be taken, for example, to include the work planning and performance reporting systems that the business employs. As will be argued later, many factors can influence process changeover times, including aspects of human attitude and performance, as well as factors that are to do with the physical nature of the manufacturing hardware.[48] The definition is inclusive of these factors.

In respect of changeover performance alone, one possible definition of manufacturing flexibility might be[49] 'An ability to commence economic manufacture of any quantity of a given product, from a specified product range, at any chosen time, by any chosen personnel'. This definition brings together aspects of flexibility that are investigated later in this chapter. The definition describes an ideal situation which, like being able to manufacture in unit batch sizes, represents a target rather than a situation that, typically, will ever be achieved.*

Importantly, this latter definition can be used to identify measurable criteria by which on-going changeover performance may be monitored. These criteria may be extended, where other measurable parameters may also be introduced. This topic of monitoring changeover performance will be covered in the next chapter.

Range and response

Two important determinants of manufacturing flexibility are 'range' and 'response'.[50] Range applies to the range of products that may be manufactured. Response applies to the time taken to commence manufacture of a desired product.

This distinction is important. In terms of overall business objectives it raises questions as to the degree to which both range and response need to be present within the manufacturing system. Range is one measure of the extent to which a manufacturing organization can change, but greater range – the ability to make a greater variety of products – will possibly entail both significant capital equipment purchase and significant business reorganization as the range is extended. It may also entail setting up collaborative partnerships with customers to develop new products, and/or collaborative partnerships with suppliers for product components.[51,52] It is possible that if products become more complex, and similarly if the range becomes more diverse and more extensive, the response time from one product to another will suffer. This is particularly likely where ever more radical adaptations of the existing production equipment are necessary for the manufacture of different products.

Improved changeover performance is discussed in this book in the context of changing over manufacturing processes between products from an existing, identified product range. Issues of expanding the product range will not be considered. The focus of this book in later chapters will be essentially on the determinant of response, where the existing product range is retained.

Assessing flexibility to accommodate changing customer requirements

One top-level distinction, or categorization, that can be made of flexibility is the ability of the business to react to changes in either environmental or market conditions.[53] Although this distinction is defensive in its emphasis, it does isolate the role of the customer – who *en masse* constitute the market – as an agent of change.

The distinction between environmental change and market change needs to be carefully drawn. Environmental change – change that is not directly to do with customer demands – may be induced by many different influences, for example by variable macro-economic conditions, labour disruption, step technological advance or the bankruptcy of a supplier. A task of the business should be to maintain an

---

* This typically becomes less true as higher levels of automation are used – whereby the number of personnel involved and the skill required of those personnel can be greatly diminished. These benefits are likely to be countered by cost considerations and, possibly, the effect of imposed automation on a culture of team-working and continuous improvement.

awareness of possible impending environmental changes, while at the same time being prepared to react to them.[54] Each of the circumstances noted above is likely to destabilize business operations – conditions in which the business still has to remain competitive.

One highly likely instigator of instability will be fluctuating customer requirements (changing market conditions), particularly through changes in product volume and mix, both for individual customers or across a range of customers.[55–57] Fluctuating customer requirements, as noted elsewhere, may be manifest as a demand for a new product (increasing the existing product range), or as changed order patterns for products from the existing product range. It is possible to be more specific than this, and identify individual customer requirements, for example:[58]

- Supply of the right product (supply of the correct product from the range)
- Supply at the right quality (supply of a product that meets all its functional and aesthetic specifications)
- Supply in the right format (supply of the product packaged and presented in an agreed format, for example 100 × 1 litre cans of oil, rather than 4 × 25 litre cans; plastic bottles boxed in the same orientation, rather than being presented in a random orientation)
- Supply at the right place (supply of the product to the agreed customer location)
- Supply at the right time (supply of the product within an agreed delivery time 'window')
- Supply in the right quantity (supply of the product in an agreed quantity)
- Supply at the right price (supply of the product at the price agreed with the customer)
- Supply in a professional manner (courteous supply and interaction with the customer, giving the perception of attentiveness, responsiveness and competence throughout the organization: in short, reinforcing a perception of a 'total quality' supplier)

As well as these requirements, comprising elements of both response and product range, the need for volume flexibility is also apparent, which may be regarded as an additional, separate, determinant of manufacturing flexibility.[59]

Volume flexibility

Volume flexibility may be assessed in terms of both production and delivery to the customer. Finished goods inventory may be used to decouple production from delivery. When assessed in terms of production, volume flexibility may be considered in relation to individual products or may be used to describe the total production capability (across a range of products) at a production facility. In addition, sub-contracting may take place to make use of external facilities, and thereby extend the possible options to enhance volume flexibility.

For a single specified product that is not sub-contracted, volume flexibility on the shopfloor might be termed 'production volume flexibility', which is the ability to change the quantity of the product that is manufactured in a given time interval. There are two components to this: the production rate and the chosen time interval. If the chosen time interval is fixed (say, a time interval of one minute) and if the manufacturing equipment is fully utilized, production volume flexibility can only be achieved by altering the rate of production (units/minute). On the other hand, a greater quantity of a specified product may also be manufactured by running production equipment at a constant production rate for a longer period of time. This might involve, for example, working a period of overtime whereby production volume flexibility will still be observed on a units/day basis. This assumes that the production equipment has available what would otherwise be 'idle' periods. One further method to achieve internal production volume flexibility will be to make use of additional internal production facilities for the manufacture of the specified product. Where there is a limited total production capability, a greater production volume for one product (the total production rate across all the manufacturing facilities that are dedicated to manufacturing the specified product) will necessitate a reduced production volume for the total of the other products that are being manufactured at the same time.[60]

In summary, production volume flexibility on the shopfloor may be achieved by increasing the rate of production of individual lines or machines, by running production facilities at a constant rate for a longer period of time or by bringing additional production facilities into use. Unencumbered production volume flexibility is the ability to manufacture at any chosen production volume at will. Realistically (and certainly in the short term) there will always be an upper bound to the volume production rate for any particular product. There is a likely cost dimension to production volume flexibility – which needs to be closely controlled if economic production volume change is to be undertaken.

Changeover performance can impact upon production volume flexibility. Consider a situation where a changeover takes 4 hours to complete on a line for which the average batch size is 75 000 units and the production rate is 4500 units/hour. In a 24-hour period, on average, a maximum 87 097 units could be manufactured. If the same average batch size is manufactured once the changeover time has been reduced to just 30 minutes, a 24-hour period will allow, on average, a maximum of 104 854 units to be manufactured. By improving changeover performance, while keeping a constant batch size, the line production capability has risen by just over 20 per cent.

Changeover performance can also have an impact on volume flexibility when it is used to bring additional production facilities into use. Consider a situation where a number of similar manufacturing lines exist which can all be used for the manufacture of any product from a product range. For three-piece food can manufacture, for example, a large number of different production lines may operate side by side within a factory. These lines may be dedicated to one product or may change over between different products (different size and specification food cans). An ability to change over quickly may be used to assign additional (or fewer) production lines for the manufacture of one particular product as its volume demands change, and by doing so achieve step changes to its production volume.

Volume flexibility may also be assessed in terms of delivery performance. Volume flexibility in the eyes of the customer might be termed 'delivery volume flexibility', which means taking delivery of a requested quantity of a product at a specified delivery time. The customer may change the requested delivery quantity, or the frequency at which delivery is requested; both changes possibly necessitating production volume flexibility by the supplier. Fluctuating delivery volume requirements are likely to be a major factor in a supplier's decision to hold high levels of finished goods inventory, where high inventory is used to decouple the customer's demands from the production environment. The ability to manufacture responsively (at a chosen time) in chosen batch sizes can greatly reduce the reliance on finished goods inventory to meet delivery schedules. Even so, finished goods inventory might still be important in the short term when the delivery requirements of the customer exceed the total production capability of the supplier.

The way that changeover performance is used by the business may differ depending on whether it is sought to enhance either production volume flexibility or delivery volume flexibility. In one case rapid changeover might be exploited to maximize the total production volume for a specified product. Alternatively, better changeover performance might be used to permit economic manufacture in smaller batch quantities.

Customer perspective of flexibility

It is customers who place orders with a business and who, ultimately, ensure a business's survival and growth. Flexibility to be able to accommodate changing customer requirements, both in the short term and in the long term,[61] and in whatever form such changing requirements are manifest, is therefore of fundamental importance. An understanding of customer requirements coupled with a desire to compete to meet these requirements to a 'world-class' standard[62] provides much of the impetus for strategic change in business operations.[63]

Customer–supplier relationships can occur internally within the same organization, or can apply between the organization and an external entity. Customer–supplier relationships are also abundant in our own personal transactions. One example of an everyday customer–supplier situation outside manufacturing industry is presented in Exhibit 2.1. The exhibit explores some of the issues that can apply to it. These issues are analogous to those that can apply to customer–supplier relationships in manufacturing industry. As is explained in Exhibit 2.1, the customer's perspective is highly influential in deciding how future transactions will be made. In the rail travel situation of Exhibit 2.1, as in a parallel industrial situation, the role of enhanced supplier flexibility may be identified.

Exhibit 2.1 A delayed rail journey

This exhibit concerns a passenger train operated by the company Wales and West.

A passenger waits at Bristol Temple Meads railway station for a Wales and West local service to Yatton. The passenger is a customer of Wales and West. The Yatton service is late departing because it is being held back, awaiting an overdue incoming Virgin Trains service from Glasgow, so that local connections may be made. The customer at Bristol is unaware of this situation. Also unbeknown to the customer, the Glasgow train is late because signalling problems near Cheltenham have arisen. Track and signal maintenance is the responsibility of a third company, Railtrack.

Exhibit 2.1 *(Continued)*

From a customer perspective, it is Wales and West who have failed to deliver the expected service. More specifically, the service criterion of departing on time has not been met. Other criteria such as the price of the ticket, train cleanliness, lighting and heating are all as expected (or as agreed in an unwritten contract).

The customer knows that the performance of Wales and West is sub-standard because lateness is an easily measured criterion. This sub-standard service may well be held against the company by the customer – even though the service is late for reasons that are beyond Wales and West's control. For example, were an additional local service available (run by an alternative train operator) that departed to Yatton on time, Wales and West might be penalized by losing some of its customers. That the 'fault' does not lie with Wales and West, from a customer's perspective, is of no consequence.

Wales and West's difficulties can be argued to have arisen because of circumstances that the company has not been flexible enough to accommodate. One possible option to respond in a more flexible manner would be to have a spare train available that could depart on time, while the original train continued to await the late-running incoming service. Alternatively, were it possible to do so, the original train might have been split and departed in two 'halves'. These options are dependent on having additional drivers and guards available. They are also dependent on having flexibility in the signalling network to accommodate new services and, if an extra train is to be run, platform space available – in other words, that there is scope within Bristol's congested rail network for additional services to operate. Moreover, the cost of enhanced flexibility needs to be assessed. Additional services will cost more money to lay on, particularly if additional rolling stock is employed. These costs need to be offset by the extra income from increased customer patronage. While a train operator might be able to operate in a more flexible manner, both the cost and the benefit of doing so need to be carefully evaluated.

In an industrial context, as in customer–supplier relationships that apply elsewhere, inadequate flexibility (particularly defensive flexibility) can lead a business to fail its customers in one or more of the criteria on which the customer–supplier transaction was originally agreed. These criteria are measurable, and the supplier may be penalized for this failure, either now or in the future, even though external circumstances might have been wholly beyond the supplier's control. The ultimate penalty would be to lose that customer's future business.

Exhibit 2.1 indicates that there may be different options to improve flexibility, but that these options may involve additional cost and are likely to require enhanced flexibility in more than one area of the operation if they are to be of use. Reliance on the use of inventory (analogous to additional rolling stock in the rail example) is typically a costly strategy. Flexibility that is provided by manufacturing in small lots (or splitting the existing rail service into smaller units) may well provide a more viable alternative solution, particularly if there is little or no cost penalty in being able to manufacture in this way. Whatever strategy is employed, the cost of becoming more flexible should be assessed in relation to the benefit that greater flexibility will provide.

As a business will almost certainly be competing simultaneously on many different criteria (supply of the right product; at the right quality; in the right format; at the right place; at the right time; in the right quantity; at the right price), care should be taken that any initiative that seeks to improve supply performance in one area should not lead to unacceptable downgrading of performance in other areas, thereby losing the competitive advantage that the company sought to enhance.

Strategic flexibility

It is reasonably straightforward for a customer to compare the relative performance of its suppliers. If it wishes to continue trading as profitably as possible in an existing market, a supplier's objective should be to maintain a high-ranking position in any customer assessment – ideally right at the very top.[64] Importantly, it is the customer that sets the criteria on which this comparison is made – not the businesses that supply it.

Inevitably different customers will emphasize different criteria. For the supply of 'commodity' products,* for example, a greater emphasis on price, product quality and service quality (including Just-In-Time, small-batch delivery) is likely than might otherwise have been the case.[65,66] An emphasis on different criteria will apply both between customers in the same market and between customers in alternative markets.

A business needs to be able to benchmark its own performance against those of its rivals.[67] By doing so it will know what the performance standards are that apply to its own market, and have data to assess what its competitive strengths and weaknesses are. By benchmarking its own performance the business will be able to determine how effectively, relative to its competitors, it can compete, one by one, against all the criteria that it is able to identify. Certainly benchmarking should be done within the markets that the business currently serves. If the business is able to understand criteria that it needs to meet in alternative markets, it will also be able to assess its relative strengths in these other markets (for example, for Exhibit 2.1, a coach operator might identify cost and punctuality as competitive strengths for carrying passengers between Bristol and Yatton).

On the basis of a comprehensive benchmarking exercise, strategic decisions for the development of the business may be made. Decisions relating to business strategy may be taken for either offensive or defensive purposes, and the ability both to make and to pursue well-judged strategic decisions may be termed strategic flexibility. Alternatively, strategic flexibility may be defined as the ability of the business to reposition itself in its own or new markets, so as to gain greatest long-term benefit.[68,69]

Strategic positioning of the business in the boardroom only has effect when it is translated into action. Better changeover performance is one candidate (among many) to effect the translation of strategy into deed. In respect of changeover performance, a possible strategic decision might be to exploit a niche market that larger, high-volume manufacturers currently ignore[70,71] (which can be opened up by being able to respond in small batches). Better changeover performance might also be considered strategically in other ways, for example to be quicker in responding to orders,[72] or to be more

* A commodity product exists when there is little likelihood of immediate product development and where an essentially identical product can be bought from different suppliers. A 33 cl. aluminium beverage can is an example.

dependable in supply.[73] As this book will indicate, other strategic repositioning may be effected, at least in part, by the use of better changeover performance.

Rapid changeover extends the degree of practical strategic manoeuverability available to senior management. Notwithstanding other potential strategic applications, the case for improved changeover performance can often be made alone by management's desire to respond better to fluctuating product mix demands.[74]

Internal considerations

A further perspective of manufacturing flexibility may be provided by consideration of where, when, how and by whom manufacture is to be undertaken. 'Where' means at which site (if the business has more than one site); 'when' means at what time, including weekends and nights; 'how' means deciding which systems to employ (including selection of which production hardware to use); 'by whom' means deciding what personnel to involve. These are all internal issues. Just as flexibility may be considered from an external or customer perspective, so too manufacture should be striven for that is not restricted internally. For example, issues of deficient operator skill, labour shortage, labour demarcation or ill-suited internal systems each might detrimentally impact upon business performance. Ultimately, ideally, manufacture of different products should be possible at any time within a 24-hour period, at any location and by anyone.[75] This determines that, as well as having flexible manufacturing hardware available, high levels of both operator mobility and skill typically will be needed.[76]

Flexible human performance contrasted with flexible internal systems

Consideration in the foregoing discussion of the internal means by which flexible manufacture occurs may be pursued further. Following on, a simple top-level perspective of flexibility may be proposed that distinguishes between the flexible performance capability of personnel, on the one hand, and the flexible performance capability of the systems employed by the organization, on the other.[77,78]

Flexible performance of people within the organization needs little elaboration: it may be regarded essentially as the ability and willingness of personnel to undertake a variety of tasks. Flexible performance of personnel should also include being able to develop or change existing working practices, including how and when tasks are conducted.

Internal systems (now often called business processes), under the above distinction, encompass all the systems that the personnel within the organization employ. All internal systems will be intended to give structure to day-to-day activities and help ensure the organization is able to function. Managers might boast: 'We have a system that allows us to identify when a defective product was made'; 'We have a system that allows us to forecast expected sheet metal usage over the next 4 months'; 'We have a system that prevents us from discharging any toxic waste'. The much-vaunted ISO 9000 standards are to do with having in place sound internal systems, and ensuring that those systems are used.

Ostensibly an organization's internal systems are fully under its own control. This means that ultimately each internal system can be adapted or replaced as the organization chooses. Thus changes may be made to one or more of, for example, an internal system's procedures, computer programs, documentation or its associated hardware. These constitute design changes to the system in question.[79] Design changes, by their nature, impact upon the way that people work and, typically, require the cooperation of personnel within the organization if they are to be introduced and used. A good example of a design change to internal systems is the introduction of *kanban* control of material flow. Another example might be the introduction of a CNC (computer numerically controlled) multi-axis machine cell to replace existing manufacturing equipment.

A top-level categorization of flexibility that distinguishes between people and internal systems is useful, therefore, inasmuch that it allows a distinction in turn to be made between altering how people behave (within the constraints of fixed systems) and altering the systems themselves to change the way that a business performs. This may be to achieve greater flexibility, or it may be for other identified purposes. By changing the systems that are employed, change is imposed upon the way that personnel behave to operate those systems. This is a theme that will be investigated in much greater detail in later chapters of this book when attention is concentrated specifically upon changeover issues.

2.3 Changeover and total productive maintenance (TPM)

Discussion in this chapter so far has highlighted the difficulty of isolating and defining manufacturing flexibility. As well as what occurs on the shopfloor, what happens beyond the immediate production environment, both within the organization and externally, may also influence the flexible manufacturing capability. Better change-over performance contributes to enhanced manufacturing flexibility, but it is not alone in doing so, and its specific contribution needs to be known. In particular, improved changeover performance enhances the response determinant of manufacturing flexibility, by making it easier to manufacture economically in small batches.

The impact of better changeovers need not be limited just to enhancing manufacturing flexibility. Previously in this chapter, for example, by analysing the cost relationships that allow an economic order quantity to be determined, it was shown that rapid changeovers can reduce inventory. Alternatively, better changeover performance might also be used to increase the number of products that are manufactured in a given time interval (to yield an increase in production volume). This potential option is investigated now in respect of a rigorously conducted Total Productive Maintenance programme.

TPM and rapid changeover as contrasting improvement initiatives?

TPM is aimed at maximizing productivity in both quantity and quality by maximizing overall equipment effectiveness.[80] In seeking to manufacture in greater quantity, TPM seeks to minimize all potential losses to productive capacity, and to operate equipment to its full design capability.[81] TPM also seeks to eliminate the production of defective products by targeting a zero product defect rate.[82] By contrast, as its title suggests, changeover improvement is highly specific and seeks to minimize only those losses (and typically only time losses) associated with changing over between the manufacture of different products.

TPM, like changeover improvement initiatives, is strongly advocated as a *kaizen*-orientated improvement programme.[83–86] The objectives of these initiatives, however, can be very different and a business should be able to assess in advance what each initiative potentially may offer, before it is embarked upon. Does the inclusion of changeover improvement only as an element of TPM mean that TPM potentially has a greater contribution to make? Alternatively, does the exploitation of rapid changeover within the confines of a rigidly adhered to TPM programme constrain its potential impact?

TPM to enhance manufacturing productivity

The use of TPM is such that it is sometimes interpreted as a programme of productive 'manufacturing'.[87] TPM aims to eliminate or minimize each of six identified major production and quality losses.[88] These losses are set out in Table 2.1, categorized into areas of Availability, Performance and Quality.

TPM inherently compromising improvement to manufacturing flexibility

Under TPM, changeover is considered as a potential loss to equipment availability – that is, the time that manufacturing equipment is available for use. TPM sets out to minimize this loss. Potentially this loss can be minimized both by conducting faster changeovers and by conducting fewer changeovers.

Questions are immediately raised in respect of frequent, responsive, small batch manufacture. If greater manufacturing flexibility is aspired to it would be sought, typically, to reduce changeover duration and at the same time conduct more changeovers. Whereas TPM promotes changeover time reduction it does not promote conducting changeovers more frequently. If all changeover losses are rigidly evaluated only in terms of lost equipment availability – where at the same time it is sought to minimize this loss – it is likely that manufacturing flexibility will be compromised.

This apparent conflict, upon first thought, can be addressed by differentiating between 'planned' and 'unplanned' changeovers. This mechanism can take into consideration the need to maintain or increase changeover frequency. Thus, for example, production planners might recognize the desirability of manufacturing, say, 10 batches on an identified line during a working day. Let us assume the mean changeover time is

Table 2.1 TPM's '6 Big Losses' affecting productivity[89]

Area	*Loss*
Availability	Breakdown Changeover (set-up)
Performance	Idling and minor stops Reduced speed
Quality	Start-up Quality defects, scrap and rework

known to be 30 minutes. If so, a time of 300 minutes could be set aside as 'planned changeover losses', and thus removed from further analysis. This however raises another potential problem: it is a mechanism that can largely relieve sustained direct pressure to improve changeovers.

There are further issues. First, as will be explained in detail in the next chapter, actual changeover times can be highly variable, because for different from-to changeovers often markedly different individual changeover tasks will need to be completed. Ideally any such time variability should be taken into account if 'planned changeovers' are to be separated from a TPM equipment performance analysis. This can be difficult to do. In addition, as also described elsewhere in this book, the run-up period (that typically exists as part of every changeover) can similarly make it difficult to determine a changeover's duration. Once again, therefore, it is difficult to determine a planned changeover time. Second, continuing with this theme of run-up, both production rates and product quality might be below acceptable standards during the run-up period (see Figure 1.2). How is this to be accounted for under TPM where, arguably, there is significant overlap with what are categorized as 'changeover losses' and 'start-up losses' (Table 2.1)? Finally, if 'planned changeovers' are taken account of, resistance to conduct 'unplanned changeovers', for whatever reason such changeovers are needed, might be greater than would otherwise be the case.

The exacerbating role of TPM's overall equipment effectiveness measure

A key element of TPM is its overall equipment effectiveness (OEE) performance measure. This measure is the product of performance rates in each of the three categories of loss identified in Table 2.1, that is:

$$\text{OEE} = \text{Availability} \times \text{performance efficiency} \times \text{quality rate}.$$

If a high OEE figure is to be achieved there is pressure to minimize the number of changeovers that are conducted, thus raising the availability figure. This pressure is intensified if OEE is used as an indicator of the company's 'world-class' status, or in circumstances where significant investment and other strategic decisions may be taken based on a site's OEE. Goldratt[90] has previously acknowledged the likely distorting influence – here on changeover behaviour – of any indicator that is accorded such a prominent status: 'Tell me how you measure me, and I will tell you how I will behave.'

TPM discouraging disposal of 'surplus' manufacturing equipment

A further possible effect of the OEE measure, where facilities are assessed only during periods of planned usage, is to encourage retention of poorly utilized parallel manufacturing facilities. In a situation where equivalent production lines are operated side by side, a rapid changeover capability might be used to dispose of superfluous equipment and concentrate manufacture on fewer, better utilized lines. TPM encourages retaining poorly used lines because, once again, by doing so a company can minimize the number of changeovers that occur – hence minimizing availability losses in the OEE equation. The OEE performance figure is thus elevated. Moreover, calculation of the OEE figure does not discriminate between an old line and a new line – on which the same products might be manufactured at, say, double the rate. On a line-by-line basis the OEE figure only indicates how closely a line is being operated to its capability – even though this capability may be extremely poor.

Reconciling flexibility and productivity

We argue above that within its primary objective to enhance productivity, and in sometimes subtle ways, TPM can apply pressure to reduce the number of changeovers that occur. In doing so manufacturing flexibility can be compromised. Certainly there is no incentive, within a TPM programme, to increase the number of changeovers that take place.

The exploitation of improved changeover performance either for increased productivity or for increased manufacturing flexibility is largely determined by the frequency with which changeovers occur. While changeover times remain significant (though perhaps still improving) the business must choose where it wishes to accrue benefit – a task that is made much easier if benefit in different areas can be quantified (see Chapter 4). As changeover times are steadily reduced, the inherent conflict between achieving either increased productivity or increased flexibility is steadily removed.

2.4 Options to achieve flexible small-batch manufacture

It has mostly been implied in the text so far that incremental improvement to existing equipment should be undertaken to achieve better changeover performance. This improvement route need not always prevail. Improved changeover performance may be bought as a feature of new, replacement manufacturing equipment. Additionally, responsive, small-batch manufacture may be realized by using product-dedicated 'surplus' equipment as and when required.

Two opposed, extreme options to achieve a small-batch manufacturing capability

Multi-product manufacture from a defined product range may be contemplated from two extreme positions:[91]

Position 1: To have an ability instantaneously to change manufacture between different products on a minimum possible number of manufacturing lines (where changeover performance is highly significant).

Position 2: To have separate manufacturing lines permanently set up for each individual product, and then to manufacture on these lines as required (where changeovers do not occur).

Each of these extreme positions faces operational problems.[92] In the first case a limited total manufacturing capacity is available, where it may be difficult to accommodate occasions of high demand (volume flexibility), and where machine breakdowns could significantly affect delivery performance. In the second case there is liable to be surplus, duplicate equipment in the factory that is rarely used. This equipment will take up space, require maintenance and is likely to impinge upon work-in-progress inventory and material flow (see Exhibit 2.2). Although the second option potentially will allow a business to respond more flexibly to changes to production volumes, volume flexibility by this means is typically highly dependent on changed numbers of operators being available to run additional equipment as required. Assuming that the company will not want to have otherwise unnecessary labour permanently on its books, it is quite likely that difficulty in securing this labour at short notice will limit the level of flexibility that this option can actually provide. Even so, even in instances where a rapid changeover capability exists, careful consideration should always be given to what degree of 'surplus' equipment is held,[93] striking a balance between these two extreme positions.[94] Constraints that may inhibit the way that changeovers are performed may also significantly influence this decision. These constraints will be investigated in Chapter 4, although Exhibit 2.2, which is an extreme case, alludes to some of the difficulties that may arise.

Exhibit 2.2 Difficulties in achieving product and volume flexibility – a case study of how things can get out of hand[95]

A long-established UK press factory manufactures decorative products. The majority of the presses that it uses are relatively small capacity OBI (open-backed inclined) presses. Two examples of its products are 'tin' trays and metal sweet boxes. The factory operates from a nineteenth-century multi-storey building that was originally used to mill seed. Other old, brick-built ancillary buildings on-site are also used. Many of the buildings are in poor condition. Space restrictions mean that almost all manufacturing lines are cramped and that part storage and transportation is difficult. Pre-printed (decorated) metal sheet is supplied by one of a number of metal printers in the same group. The factory prides itself on the range of products that it offers and in its ability to develop new products.

A particular feature of this business is its seasonal nature, where peak seasonal demand is more than twice the residual demand. This volume peak is sharp and occurs over a period of less than two months, predominantly stemming from the demands of the 'Christmas market'.

The factory have kept many of their old presses and tools. This stock of presses has been increased as former factories in the group have been closed down, where equipment from these factories has been picked up at very low cost. During 'quiet' months the number of presses on the shopfloor is less than half the total number of presses present on-site.

Only a few technicians know what tools can be used in what presses and have the ability to set them. Many tools and presses are in poor condition and require some level of refurbishment during setting if they are to produce goods of acceptable quality. Specialized knowledge in respect of the presses extends even to knowing where they are physically located.

The factory has evolved a 'flexible' manufacturing strategy whereby lines are built up in production areas as required from its stock of equipment. Hence

Exhibit 2.2 *(Continued)*

changeover frequently involves constructing lines from scratch (including dismantling existing lines and re-storing equipment, to create production space). Changeable product volume requirements are accommodated with a heavy reliance on short-term seasonal labour, although this does not include additional skilled technicians. During peak periods the existing technicians work at full stretch (and earn good overtime money) trying to keep this old equipment running. Breakdowns are common despite their efforts. Skilled staff are thus much in demand in the factory – not only to conduct changeovers, but also simply to maintain on-going production on lines that have already been set up. This latter task often competes with changeover duties, where sometimes, simply by the non-availability of a technician, completion of a changeover is delayed. Changeover performance, when investigated, was found to be extremely poor. The company's records showed that its ability to satisfy customer delivery and quality requirements was similarly extremely poor, such that the company's long-term viability, at the time of the study, was being called into question.

The replacement of existing manufacturing equipment

This book focuses primarily upon changeover improvement to an existing manufacturing system; that is, it sets out to move a manufacturing operation towards position 1 as outlined above (aspiring to instantaneous changeover on a small number of manufacturing lines). Such improvement of an existing manufacturing operation additionally can be contrasted with an option of buying and installing replacement manufacturing equipment. This replacement equipment may be technically more advanced than the equipment that it replaces, and its changeover proficiency, along with other determinants of its performance, can be specified and verified before purchase. Introduction of new or replacement equipment is a perfectly viable option to improve changeover performance. Examples of companies that have pursued this option are available,[96–100] but possible associated difficulties need to be fully assessed before the option is pursued. This can apply particularly to flexible manufacturing systems (FMSs), where complex computer controlled machinery may be used to replace more 'traditional' equipment.[101]

Deciding which approach to adopt

With different options available to seek better changeovers, a business is faced with the problem of deciding which option it should pursue. This is a decision that should be made by senior management. Some issues that should be taken into account are:

- **Appreciate the improvement to changeover time that is possible for the existing manufacturing system (an improvement 'blind spot')** It is not easy to predict in advance what level of improvement may be possible to the changeover performance of the existing manufacturing system. Early changeover consultants frequently claimed that improvement in excess of 50 per cent is readily achievable,[102–104] and these claims have been backed up in the literature.[105–108] Not least among published claims of improvement are those that have been made by Shingo.[109] We will describe later how the contribution made by organizational improvement alone (hence at low cost) may be estimated. We will also investigate that ever more substantive design changes may be required as ever greater improvement is sought.[110,111] In these circumstances the retrospective improvement option might start to lose its attractiveness, in comparison to installing changeover-proficient replacement equipment.[112–114]

- **Relate manufacturing system changes to business strategy and market needs** Ensure that the expected improvements are appropriate to the needs of the market and the development of the business.[115] If replacement equipment is being considered, is a better changeover capability let down by reduced equipment performance in other respects? If it is decided to improve the changeover performance of existing equipment, is the capability of the equipment in other respects adequate to compete successfully in the foreseeable future?
- **Assess the economic viability of each option** What is the cost/benefit assessment for each option?
- **Evaluate if the structure of the organization needs to be adapted** The structure of the existing organization and the skills within it may not be appropriate to support alternative, technologically advanced, production hardware.[116,117] The skills and structure of the organization also need to be taken into consideration even if retrospective improvement is being considered. Training, recruitment and the introduction of alternative flexible support systems may each be required, both to achieve and to exploit a new manufacturing capability.
- **Verify the dynamic capabilities of replacement equipment** Under some circumstances the dynamic capabilities of replacement equipment can be significantly below the level of performance anticipated from a static assessment of individual machine capabilities. For example, if a number of flexible, high-speed individual machines are grouped together for production, the performance of the new system as a whole may fall well below what is expected.[118] On an individual machine basis, the changeover claims that are made of any replacement equipment should always be rigorously tested. Performance testing should be done for changeovers that the customer decides, rather than simply observing a changeover from one product to another as chosen by the machine manufacturer.
- **Estimate the time required to achieve the desired level of improvement** The organization should review how long each option will take to achieve the results that are expected of it.
- **Question the ease of sustaining better changeover performance** There may be a tendency for changeover performance to lapse back towards its previous, pre-improvement, level.[119] This tendency is likely to be more pronounced where retrospective improvement is sought by emphasizing changes to work practice or, in other words, emphasizing organizational change in preference to using design to alter the existing manufacturing system.

This list is not necessarily exhaustive. For example, changeover performance may be identified, when undertaken as a team-based retrospective improvement exercise, as a possible vehicle to change workplace culture. Other site-specific circumstances may also influence the final decision, perhaps including some of the issues raised in Exhibit 2.2.

This checklist should be taken, at this stage in the book, as an introduction to issues that management should consider. These topics will be revisited. A precursor to making any such decisions is the undertaking of performance benchmarking, which will have indicated in what way, by how much and how quickly a company's performance needs to change for it to remain competitive.

Assessing changes to manufacturing hardware and changes to work practice

Physical changes to the manufacturing system represent one approach to improve aspects of manufacturing capability including, as we will demonstrate, changeover performance. Physical changes may involve either layout alterations or detail changes to individual machines. As noted above, updating of machines can occur to the extent of full machine replacement. In recent years manufacturing industry has experienced an increase in applications of both group technology (GT), or cellular manufacture, and flexible manufacturing systems (FMSs).[120]

Similarly, changes may be made in the deployment of labour, away from the 'Taylorism' of early mass production towards a more flexible, participative model. As with physical changes to the manufacturing system, changes in the use of labour may yield significant benefits, in addition to achieving an increased flexible manufacturing capability.[121,122]

If it is determined that better changeover performance should be achieved by retrospective improvement, a business is still faced with important questions regarding how retrospective improvement should be approached. Should retrospective improvement occur, for example, by internal *kaizen* activity by existing members of the workforce, or should a solution be imposed by external experts? Or should an expert facilitator be charged with leading an improvement programme? To what degree and under what circumstances should design changes be undertaken in preference to organizational changes? It has been asserted that in recent years that the pursuit of flexibility through hard automation has diminished, and that attention has turned to 'softer' managerial techniques, concentrating on changes to work practice.[123] Should this trend also apply to changeover improvement? If so, have all the factors that contribute to 'softer' organizational aspects been fully taken into consideration, and have difficulties that may arise from taking this approach been assessed? These issues will be fully addressed in later chapters of this book. Initially, in Chapter 3, we will describe the complexity of a changeover improvement initiative and the significant pitfalls that it may face.

2.5 Conclusions

For almost every business, manufacturing flexibility should be regarded as a multi-faceted and highly prized commodity. But flexibility that is not exploited remains just that: a commodity. An improved flexible manufacturing capability does not in itself make a business more competitive. Rather, it is how this capability is used – to satisfy customer demands, to accommodate internally or externally generated uncertainty or to manufacture more efficiently – that will give a business a competitive advantage.

Likewise, an improved changeover capability will not in itself benefit a business. As with enhanced flexibility, this capability needs to be exploited before it can be of service. In other words, it may be possible physically to change over the production facility more quickly, but there has to be an ability and a desire elsewhere within the business to use this capability competitively. If the business wishes to use better changeover performance to enhance manufacturing flexibility it needs to ensure there is also active parallel flexibility of resource elsewhere in the organization (most notably of planning and marketing). Flexibility of resource means both labour flexibility and flexibility inherent in the business systems that are employed. The decision how to exploit improved changeover performance will be dealt with in Chapter 4, where the application of better changeover performance to enhance manufacturing flexibility is just one among a number of competing options for its use.

Although retrospective improvement to the changeover performance of existing equipment is one way to achieve a more flexible manufacturing capability, this is not the only option available to manufacture more responsively in smaller batch quantities.

References

1 Schonberger, R. J. (1990). *Building a Chain of Customers*. Free Press, New York. That rapid changeover can be regarded, mistakenly, as a panacea to address flexibility problems has been recognized by Schonberger, who writes: 'Quick set-up enthusiasm has grown in industry to the point where people have begun to associate it with flexibility itself.'

2 Elmaghraby, S. (1978). The economic lot scheduling problem (ELSP): review and extensions. *Management Science*, **24**, No. 6, 587–598.

3 Abernathy, W. J. (1978). *The Productivity Dilemma: Roadblock to innovation in the automobile industry*. Johns Hopkins University Press, Baltimore, MD.

4 Naddor, E. (1966). *Inventory Systems*. Wiley, New York.

5 Naddor, E. (1966). *Op. cit.* ref. 4.

6 Naddor, E. (1966). *Op. cit.* ref. 4.

7 Hall, R. W. (1983). *Zero Inventories*. Dow Jones-Irwin, Homewood, IL. Hall writes 'The economics of lot sizing are not permanent economics'.

8 Voros, J. (1999). Lot sizing with quality improvement and set-up time reduction. *European J. of Operational Res.*, **113**, No. 3, 568–575.

9 Hall, R. W. (1983). *Op. cit.* ref. 7. Hall's book investigates the relationship between rapid changeover performance and inventory in detail. This is a theme that will be picked up again in Chapter 4 of this book, where we will present results from a case study.

10 Williams, B. R. (1996). *Manufacturing for Survival*. Addison-Wesley, Reading, MA. Many different costs may be associated with carrying excess inventory, including obsolescence; pilfering; damage or spoilage; insurance; handling (equipment and labour); cost of capital tied up in inventory; cost of space and fixtures allocated to inventory (including tax); cost of inventory control systems. If the quantity of finished goods inventory is significantly reduced the total inventory cost will similarly be reduced (which, like change to order costs, will affect the EOQ). Indeed, steps may be taken to change the inventory holding costs even if the level of inventory remains unaltered.

11 Hall, R. W. (1983). *Op. cit.* ref. 7.

12 Slack, N., Chambers, S., Harland, C., Harrison, A. and Johnston, R. (1995). *Operations Management*. Pitman, London.

13 Hall, R. W. (1983). *Op. cit.* ref. 7.

14 Anon. (1998). United Distillers, *Management Today* – Guide to Britain's best factories (supplement), **5**, 7–13. Four examples are cited (see also reference 15, 16, and 17), where improvements in changeover time of 50 per cent or more have been claimed. Many other examples exist.

15 Tully, S. (1994). Raiding a company's hidden cash. *Fortune*, **130**, No. 4, 58–63.

16 Jones, J. and Griffin, D. (1993). Achieving quick changeover, I Mech E Total Productive Maintenance Workshop, 25–29 October 1993, Stratford-upon-Avon, UK.

17 Schonberger, R. J. (1986). *World Class Manufacturing: The lessons of simplicity applied*. The Free Press, New York. Although no details are given on how it has been achieved, Schonberger offers one of the most dramatic claims of improvement (for a cough sweet production line changeover at Richardson-Vicks Home Care Products, Torshalla, Sweden) – from 1 shift to 18 minutes.

18 Mather, H. (1992). Reducing your changeover times. IMechE seminar, Birmingham, UK.

19 Shingo, S. (1985). *A Revolution in Manufacturing: The SMED system*. Productivity Press Inc., Massachusetts, USA.

20 Matupi, N. and Kay, J. M. (1996). Defining manufacturing flexibility requirements. *Advances in Manufacturing Technology, Proc. 12th National Conference on Manufacturing Res.*, eds Bramley, A. N., Mileham, A. R. and Owen, G. W., Bath, UK. Matupi and Kay introduce flexibility as a desirable attribute of a manufacturing system. They point out also that the concept of 'flexibility' is not a simple one, where there is difficulty in 'defining it, predicting it and measuring it'.

21 Olhager, J. (1993). Manufacturing flexibility and profitability. *Int. J. Prod. Economics*, **30**, No. 1, 67–78.

22 Bahrami, H. (1992). The emerging flexible organisation: perspectives from silicon valley. *California Management Review*, **34**, No. 4, 33–52.

23 De Meyer, A., Nakane, J., Miller, J. G. and Ferdows, K. (1989). Flexibility: The next competitive battle. *Strategic Management Journal*, **10**, 135–144.

24 Sethi, A. K. and Sethi, S. P. (1990). Flexibility in manufacturing: A survey. *Int. J. of Flexible Manufacturing Systems*, **2**, 289–328.

25 De Toni, A. and Tonchia, S. (1998). Manufacturing flexibility: a literature review. *Int. J. Prod. Res.*, **36**, No. 6, 1587–1617. This paper presents a comprehensive review of academic research on manufacturing flexibility. It highlights that flexibility may be considered in many different ways, and that research into the topic is by no means complete. Our own brief assessment of flexibility seeks to view it from different perspectives. This approach has been taken to gain an insight into its potential complexity. In doing so we are reflecting the many different ways that flexibility has been assessed in the literature. We have not attempted to present a full, academic appraisal – those seeking one may well choose to take De Toni and Tonchia's paper as a start point – but instead have drawn largely on our own experience.

26 Franza, R. M. and Gaimon, C. (1998). Flexibility and pricing decisions for high volume products with short life cycles. *Int. J. of Flexible Manufacturing Systems*, **10**, No. 1, 43–71. Occasionally the term 'changeover flexibility' might be encountered: 'We examine the competitive implications of a firm's ability to change over its facility for the manufacture of successive generations of high-volume products with short life cycles. This ability is known as changeover flexibility.' Caution is still needed in understanding what 'changeover flexibility' means. Here Franza and Gaimon are describing the ability of the manufacturing operation to set up for the manufacture of different (not previously manufactured) products. As investigated later in this chapter – see particularly the sec-

tion 'range and response' – this is just one facet, in the terms of this book, of 'changeover flexibility'.

27 Gerwin, D. (1993). Manufacturing flexibility: a strategic perspective. *Management Science*, **39**, No. 4, 395–410. Gerwin similarly describes changeover flexibility as 'an ability to quickly substitute new products for those currently being offered'. The term 'set-up' is used by Gerwin to distinguish changes made to individual machines to change over between products from and existing product range. These terms do not agree with the definitions we apply throughout this book. Additionally, Gerwin discusses the difficulty of measuring aspects of flexibility.

28 Alvandi, M. and Burgess, T. F. (1996). Investigating plant flexibility: Some results from a survey and case studies in the UK. *Proc. 3rd. Int. Conf. European Operations Management Association*, London, pp. 21–26. Alvandi and Burgess associate flexibility to the manufacturing philosophies presented in Chapter 1, asserting that: 'Flexibility is identified as a key ingredient of the evolving post-industrial manufacturing paradigm'.

29 Slack, N. (1991). *The Manufacturing Advantage: Achieving competitive manufacturing operations*. Mercury, London.

30 Newman, W., Hanna, M. and Maffei, M. (1993). Dealing with the uncertainties of manufacturing: flexibility, buffers and integration. *Int. J. of Operations and Production Management*, **13**, No. 1, 19–34.

31 Hill, T. (1985). *Manufacturing Strategy – the strategic management of the manufacturing function*. Macmillan, London. Hill describes the general desirability of revised technology (hardware) to manage change/uncertainty alongside attitude and procedural changes. While flexibility to accommodate uncertainty is becoming in part a property of the technology that organizations possess, flexibility is not necessarily intrinsic in the way that organizations use this technology.

32 Gunneson, A. O. (1996). *Transitioning to Agility: Creating the 21st century enterprise*. Addison-Wesley, Reading, MA.

33 CarnaudMetalbox (European division of Crown Cork and Seal, Inc.) (1996). Promotional brochure (untitled), Paris, France. This market-focused brochure assures CarnaudMetalbox customers of the company's ability to change its beverage can production rate in the event of a hot summer. This is an example (see later in the chapter) of volume flexibility. Although this is a defensive capability – to accommodate changes in customer demands – it is being marketed *offensively*.

34 Bateman, N., Stockton, D. J. and Lawrence, P. (1999). Measuring the mix response flexibility of manufacturing systems. *Int. J. of Production Res.*, **37**, No. 4, 871–880. The mix component of customer demands is described (see also later this chapter), for which good changeover performance typically is important.

35 CarnaudMetalbox (1996). *Op. cit.* ref. 33. Conditions that favour the manufacture of beverage cans in either steel or aluminium can be finely balanced. To assuage customer concerns that it might be unable to respond to changing economic circumstances (changes in the raw material cost of steel or aluminium) the company boasts that it is able to change over a beverage can production line from steel to aluminium (or vice versa) in less than 48 hours. Although this is apparently an excessive time by the standards of 'single-minute' changeovers, it is evidently considered to be competitive and worthy of publication.

36 Graham, I. R. (1990). *Just-in-time Management of Manufacturing*. Elsevier, Letchworth.

37 Boston, O. P., Court, A. W., Culley, S. J. and McMahon, C. A. (1996). Design information issues in new product development. In Duffy, A. (ed.), *The Design Productivity Debate*, Springer-Verlag, London. This refers only to design information issues. It indicates the extent (and the probable impact) of deficient information acquisition, communication and use.

38 Pratsini, E. (1998). Learning complementarity and setup time reduction. *Computers and Operational Research*, **25**, No. 5, 397–405. Despite the effect altered changeover perform-

ance can have on scheduling complexity (see also Chapter 3), greater flexibility brought about by more frequent, smaller batch manufacture should lessen the requirement for accurate long-term forecasting. Pratsini, for example, describes: 'By decreasing set-up time, lot-size and inventory level are reduced which makes it easier to manage changing work priorities.'

39 Schonberger, R. J. (1986). *Op. cit.* ref. 17.

40 Schonberger, R. J. (1982). *Japanese Manufacturing Techniques – nine hidden lessons in simplicity*. Free Press, New York.

41 Gerwin, D. (1993). *Op. cit.* ref. 27.

42 Son, Y. K. and Park, C. S. (1987). Economic measure of productivity, quality and flexibility in advanced manufacturing systems. *Journal of Manufacturing Systems*, **6**, No. 3, 193–207. The difficulty of measuring changing flexibility is described also elsewhere in this chapter. The Son and Park paper is an example of how the problem of quantifying enhanced flexibility can be tackled, although by no means every aspect of flexibility has been addressed (particularly in assessing offensive opportunities). An integrated approach to quantifying benefit in productivity, quality and flexibility together has been adopted. This has been done based on the definition of flexibility: 'Flexibility measures the adaptability to various changes in manufacturing environments.'

43 Swamidass, P. M. (1988). *Manufacturing flexibility*. Operations Management Association Monograph No. 2. Swamidass, for example, has written: 'Manufacturing flexibility refers to the capacity of a manufacturing system to adapt successfully to changing environmental conditions and process requirements. It refers to the ability of the production system to cope with the instability induced by the environment.'

44 Upton, D. M. (1994). The management of manufacturing flexibility. *California Management Review*, Winter, 72–89.

45 Gustavsson, S.-O. (1984). Flexibility and productivity in complex production processes. *Int. J. of Production Research*, **22**, No. 5, 801–808. This paper highlights one particular trade-off that we will discuss later in this chapter, namely that between flexibility and productivity.

46 Schonberger, R. J. (1982). *Op. cit.* ref. 40. Schonberger asserts that by adopting 'Japanese' manufacturing philosophy many of the trade-offs inherent in traditional Western manufacturing practice, including those involving flexibility, may be largely eliminated.

47 Mapes, J. (1996). Performance trade-offs in manufacturing plants. *Proc. 3rd. Int. Conf. European Operations Management Association*, London, pp. 403–408. This paper seeks to resolve support for Schonberger's position (above) and opposition to it. Mapes concludes that generally Schonberger's position on largely eliminating trade-offs is sound.

48 Upton, D. M. (1995). Flexibility as process mobility: the management of plant capabilities for quick response manufacturing. *J. of Operations Management*, **12**, No. 3–4, 205–224. Upton describes that process changeover times can vary because of the differing nature of the work that is involved for different changeovers. We shall later investigate this variability of changeover time as a function of the type of changeover that occurs. Upton describes a range of possible factors that can influence changeover performance under headings of: Managerial emphasis; Plant infrastructure; Operating policies (range produced, frequency of changeover, type of changeover); Plant structure.

49 McIntosh, R. I., Gest, G. B., Culley, S. J., Mileham, A. R. and Owen, G. W. (1994). Achieving and optimising flexible manufacturing performance on existing manufacturing systems. *Proc. 4th Int. Conf. on Factory Automation – Factory 2000*, York, UK, pp. 491–495. Our original definition has now been extended by inclusion of the word 'economic' and by inclusion of the term 'by any chosen personnel'.

50 Slack, N. (1988). Manufacturing systems flexibility – an assessment procedure. *Computer-integrated Manuf. Systems*, **1**, No. 1, 25–31. Care should be taken that 'response' is com-

patible with the definition of changeover provided in Chapter 1 – in other words it should represent the point when the product has been tested and accepted for quality, and full production has commenced. In other publications Slack has described changeover time only as the time from 'good piece to good piece'. By doing so he has failed to take account of the run-up period. As discussed here by Slack, the terms 'range' and 'response' are widely adopted. Some other authors may use the term 'mobility' in place of 'response'.

51 Lamming, R. (1993). *Beyond Partnership, Strategies for Innovation and Lean Supply*. Prentice Hall, Hemel Hempstead.

52 Greek, D. (2000). Suppliers meet carmakers halfway. *Professional Engineering*, **13**, No. 4, 38–39.

53 Correa, H. L. (1994). *Linking flexibility, uncertainty and variability in manufacturing systems*. Avebury, London. Correa categorizes two change agents to which the organization should respond: (1) environmental uncertainty and (2) variability in customer requirements. In practice environmental change can influence market conditions. As an example, macroeconomic conditions might curb public spending, which in turn might lead high street retailers to change the value and profile of orders placed with their suppliers. Like most other categorizations of flexibility, these two categories are to an extent interwoven. Indeed, a review of the literature shows the plethora of categorizations that have been made, and indicates that significant difficulty exists in isolating individual categories. Consequently, it is difficult to measure and assess flexibility. Although, in addition, Correa's categorization (above) is highly defensive, it remains useful as it allows an important customer-focused assessment of flexibility to be made.

54 Schofield, J. (1993). The chips are down at Big Blue. *The Guardian*, Section 2: Guardian Education, 12 January 1993, London, 2–3. Financial problems experienced by IBM at the start of the 1990s – one of the United States' largest and previously most successfully companies – are succinctly described. It is argued that IBM's failure to embrace changing technology significantly contributed to these problems; in this case failure to respond to rapidly developing microprocessor capability. As the IBM experience again demonstrates, environmental change and market change can occur hand-in-hand. For many customers, mainframe computers became a less attractive choice. These customers (a substantial proportion of the overall computer market) over time placed new demands on computer manufacturers – for smaller, cheaper computers. IBM could not competitively respond to this market demand, particularly in relation to other computer companies that had much more responsively embraced changing processor technology. Recalling an earlier era, Schofield likens IBM's lack of responsiveness to technological change to an ocean liner cruising serenely at the emergence of the age of jumbo jet travel.

55 Azzone, G. and Bertelè, U. (1989). Measuring the economic effectiveness of flexible automation: a new approach. *Int. J. of Production Research*, **27**, No. 5, 735–746.

56 Bateman, N., Stockton, D. J. and Lawrence, P. (1999). *Op. cit.* ref. 34.

57 Bartezzaghi, E. and Turco, F. (1989). The impact of just-in-time on production system performance: an analytical framework. *Int. J. of Operations and Production Management*, **9**, No. 8, 40–62.

58 Gunneson, A. O. (1996). *Op. cit.* ref. 32. This is an extension of a theme presented by Gunneson. See also note 65 (Schonberger) and note 66 (Williams) below.

59 Bartezzaghi, E. and Turco, F. (1989). *Op. cit.* ref. 57. Many commentators and academics isolate volume flexibility as a component of manufacturing or overall business flexibility, as here, for example, by Bartezzaghi and Turco.

60 Mair, A. (1994). Honda's global flexifactory network. *Int. J. of Production Research*, **14**, No. 3, 6–23. Mair describes that Honda's production lines are each able to manufacture a number of different models (in batches), and that the demand volume per model, which varies over time, can be accommodated in part by changing the proportion of the different

models on each production line. High production volume for any one model means that corresponding production volumes for other models is reduced.

61 Mair, A. (1994). *Op. cit.* ref. 60. A description was given in the notes above that 'changeover flexibility' can refer to the introduction of new (previously unmade) products into manufacture (see references 26 and 27). Mair describes that there are two 'extreme types' of flexibility, citing these types of flexibility in conjunction with a time dimension: '[There are] two extreme types of flexible product change at Honda factories in Japan: first, short-term responses to shifting demand for different models in the range; second, long-term restructuring of the product type made at a factory.' An observation made by Mair is that Honda deliberately stage the introduction of new models to the model range to attempt to maintain a relatively even total production load.

62 Schonberger, R. J. (1986). *Op. cit.* ref. 17. 'World class' is for many businesses an aspiration: many will lack the resources genuinely to achieve this status. Use of the term to mean 'as highly competitive as possible' is appropriate in many cases – to at least a level that ensures continued survival and growth. Or, as Schonberger puts it, world-class manufacturing means 'continual and rapid improvement'.

63 Slack, N. (1991). *Op. cit.* ref. 29. Slack presents a useful overview of 'the doctrine of competitiveness' at the start of his book, including discussion of benchmarking world-class performance and evaluating the need for strategic change to the business.

64 Slack, N. (1991). *Op. cit.* ref. 29.

65 Schonberger, R. J. (1990). *Op. cit.* ref. 1. Schonberger describes the 'multiple dimensions of quality' that need to be fulfilled to ensure customer satisfaction (Conformance to specifications; Performance; Quick response; Quick-change expertise; Features; Reliability; Durability; Serviceability; Aesthetics; Perceived quality; Humanity; Value). Although it has been provided under the auspices of quality, this list is an alternative representation of the criteria that we have outlined previously. Schonberger's list is an extension of work originally conducted by Garvin. It is interesting to note the prominence that Schonberger assigns to rapid changeover as part of his categorization.

66 Williams, B. R. (1996). *Op. cit.* ref. 10. Williams provides a succinct overview of what is meant by 'quality' in terms of customer satisfaction, citing the work of Garvin, Crosby and Juran.

67 Slack, N. (1991). *Op. cit.* ref. 29.

68 Hayes, R. H. and Pisano, G. P. (1994). Beyond world class: the new manufacturing strategy. *Harvard Business Review*, January/February, 77–86.

69 Stalk, G., Evans, P. and Shulman, L. E. (1992). Competing on capabilities: the new rules of corporate strategy. *Harvard Business Review*, March/April, 57–69.

70 Forth, K. D. (1994). Quick die change helps auto stamper produce to order. *Modern Metals*, **50**, No. 9, 30–35. Forth describes a highly mechanized motor panel stamping facility in North America. The company cites a desire to use better changeover performance to serve a large variety of small/medium volume markets (automotive, truck and off-highway equipment for a variety of major US assemblers): 'Producing small batches on these presses is important in the low-to-medium volume market that we serve... As changeover times shrink, we can run more small lots.'

71 Anon. (1997). Small runs, big business. *Graphic Arts Monthly*, **69**, November, 79. A brief insight is given into the role of rapid changeover to help develop a small batch book binding business.

72 Field, K. A. (1997). New plant cuts cycle time by a third. *Modern Materials Handling*, **52**, No. 3, 31–33. Rapid changeover, along with other improvements, is identified as a way to improve lead time performance for a US gaming machine manufacturer.

73 Anon. (1997). Quick change key to just-in-time delivery at Nissan. *Manufacturing Engineering*, **118**, No. 2, 81–83.

74 Gerwin, D. (1993). *Op. cit.* ref. 27. Gerwin writes: 'Achieving the flexibility to deal with uncertainties in the product mix reduces to achieving small setup times.'

75 Sekine, Y., Koyama, S. and Imazu, H. (1991). Nissan's new production system: Intelligent body assembly system. *SAE Transactions*, **11**, Part 5, 790–800. Sekine *et al.*, describing the need for flexibility in Nissan's production system (in which sophisticated automated manufacturing equipment is present), argue that '. . . there must be flexibility in the production system in terms on being able to produce any volume of any product, anywhere, at any time and by anyone'.

76 Mair, A. (1994). *Op. cit.* ref. 60.

77 Atkinson, J. (1985). Flexibility: planning for an uncertain future. *Manpower Policy and Practice*, **1**, Summer, 26–29. Atkinson specifically identifies labour flexibility as an issue affecting manufacturing flexibility.

78 De Toni, A. and Tonchia, S. (1998). *Op. cit.* ref. 25. De Toni and Tonchia interpret a previous paper by Newman *et al.* to distinguish two categories of flexibility: (1) flexibility arising on technological grounds and (2) flexibility arising on managerial grounds. The separation of people from systems is not absolute in this distinction but, as the breadth of categorizations in this review paper shows, no single categorization is strong enough to enjoy common acceptance. Our categorization is similar to that identified here by De Toni and Tonchia. It is useful in that it provides a further insight into how flexible manufacturing performance might be improved. The reason that so many difficulties exist in pinning down flexibility is that the myriad factors which influence flexible performance are very highly interwoven. It is thus extremely difficult to isolate and categorize primary factors which are not themselves under the complex influence of other factors.

79 Slack, N., Chambers, S., Harland, C., Harrison, A. and Johnston, R. (1995). *Op. cit.* ref. 12. We have used the word 'design' here in the sense described by Slack *et al.*: 'Design activities are those which define the physical form of the operation and its products and services. They shape the "architecture" of the organisation – the parts of the operation of which it is composed, and the way in which they relate to each other.' This is a global description of design. Later in the book we will be much more focused on the use of design in a strictly mechanical engineering sense (which we will define) to alter products and manufacturing processes to improve changeover performance.

80 Kelly, A. and Harris, M. J. (1993). Uses and limits of total productive maintenance. *Professional Engineering*, **6**, No. 1, 9–11.

81 Nakajima, S. (1988). *Introduction to TPM*. Productivity Press, Cambridge, MA.

82 Nakajima, S. (1988). *Op. cit.* ref. 81.

83 Schonberger, R. J. (1990). *Op. cit.* ref. 1. Schonberger, like many other authors, argues very forcibly in favour of *kaizen*-orientated improvement activity. He specifically relates operator involvement as important when undertaking TPM programmes, asserting: 'The only really new concept – which is the all important foundation of TPM – is that the operator assumes responsibility, and ultimately comes to feel a sense of "ownership" of the equipment.'

84 Nakajima, S. (1986). TPM – Challenge to the improvement of productivity by small group activities. *Maintenance Management International*, **6**, No. 2, 73–83.

85 Sekine, K. and Arai, K. (1992). *Kaizen for Quick Changeover*. Productivity Press, Cambridge, MA. The topic of incremental *kaizen* improvement will be investigated in much greater detail later in the book. To indicate the importance of this approach at this stage, however, it is difficult to do better than cite the title of Sekine and Arai's book.

86 Mileham, A. R., Culley, S. J., McIntosh, R. I., Gest, G. B. and Owen, G. W. (1997). Set-up reduction (SUR) beyond total productive maintenance (TPM). *Proceedings* Institution of Mechanical Engineers, Part B *Journal of Engineering Manufacture*, **211**, 253–260.

87 Koelsch, J. R. (1993). A dose of TPM. *Manufacturing Engineering*, **110**, No. 4, 63–66.

88 Shirose, K. (1992). *TPM for Workshop Leaders*. Productivity Press, Cambridge, MA.
89 Shirose, K. (1992). *Op. cit.* ref. 88.
90 Goldratt, E. M. (1990). *The Haystack Syndrome*. North River Press, New York.
91 Boyer, K. K. and Leong, G. K. (1996). Manufacturing flexibility at the plant level. *Omega*, **24**, No. 5, 495–510. The two positions that we set out might be seen as a 'response' determinant of manufacturing flexibility, based on a similar, but more general, distinction made by Boyer and Leong: 'Managers have two basic alternatives for addressing the challenge posed by variable demand: (1) build manufacturing plants with excess capacity and/or stock excess goods in inventory to help smooth over fluctuations in demand, or (2) increase the flexibility of their manufacturing plants so that production can be varied more easily to meet demand.'
92 Owen, G. W., Gest, G. B., Culley, S. J., McIntosh, R. I. and Mileham, A. R. (1995). Manufacturing flexibility – a case for excess capacity. *Advances in Manufacturing Technology IX – Proc. 11th Nat. Conf. on Manuf. Res.* DeMontford University, pp. 366–400, Taylor and Francis, London.
93 Mapes, J. (1992). The effect of capacity limitations on safety stock. *Int. J. of Operations and Production Management*, **13**, No. 10, 26–33.
94 Boyer, K. K. and Leong, G. (1996). *Op. cit.* ref. 91. Manufacturing flexibility is discussed in the context of variable product volume demand.
95 Gest, G. B. (1993). Attachment progress review. University of Bath internal report, GG/VR/100.
96 Rover Group Ltd (1991). *Triaxis Press* (promotional publication), Swindon, UK. The Tri-axis press is used to manufacture vehicle body panels that were previously stamped on less sophisticated conventional press lines which were not equipped for automated tool changeover.
97 Anon. (1993). Orchid blossoms in Nottingham. *Sheet Metal Industries*, **70**, No. 9, 21.
98 Hartman, L. R. (1997). Plumbing parts tap bagging mettle. *Packaging Digest*, **34**, No. 11, 72–76. A non-press system example of more sophisticated replacement equipment being installed, among other benefits, to improve changeover time.
99 Anon. (1997). Quick change key to just-in-time delivery at Nissan. *Manufacturing Engineering*, **118**, No. 2, 81–83.
100 Remich, N. C. (1997). From 15 min. to 90-sec. die changeover. *Appliance Manufacturer*, **45**, No. 7, 44–46. Automated die changing equipment was installed on presses manufacturing air-conditioning components, replacing a previous procedure that involved the use of fork-lift trucks. Improvement is cited as 'part-to-part set-up' time and 'die changeover' time. Previous 'part-to-part set-up' time was between 20 to 30 minutes. Previous 'die changeover' time was 15 minutes. Improved times, after installation of the new equipment, respectively are 5 minutes and 90 seconds.
101 Luggen, W. W. (1991). *Flexible Manufacturing Cells and Systems*. Prentice Hall, Englewood Cliffs, NJ.
102 Hay, E. J. (1987). Any machine setup time can be reduced by 75%. *Industrial Engineering*, **19**, No. 8, 62–67.
103 Robins, R. M. (1989). Quick changeovers – fast paybacks. *Manufacturing Systems*, **53**, No. 3, 53–55.
104 Shingo, S. (1985). *Op. cit.* ref. 19.
105 Dunn, J. (1990). Stiff measures to save Scotch. *The Engineer*, **270**, No. 6992, 60–63.
106 Miller, W. H. (1992). Chesebrough-Ponds. *Industry Week*, **241**, October, 43–44.
107 Szatkowski, P. M. and Reasor, P. E. (1991). The SMED system for setup reduction – A case study. *Proc. Int. Industrial Engineering Conf.* Inst. Industrial Engineers, USA, pp. 123–129.
108 Vicens, I. (1989). SME/FMA Quick Die Change Clinic. 2–3 May, Detroit.

109 Shingo, S. (1985). *Op. cit.* ref. 19.

110 Nunn, R. E. and Eichhorn, K. H. (1985). Quick mold change – manufacturing flexibility for injection molding. *Proc. Injection Molding – the next five years*, Pittsburgh, USA, pp. 67–81. The relationship between both cost and design input, on the one hand, and improvement in changeover time, on the other, has rarely been investigated in the literature. Nunn and Eichhorn discuss cost and levels of automation to achieve faster changeover times in the injection moulding industry. A typical changeover on equipment that existed in 1985 was estimated as taking approximately 10 hours to complete. In comparison, a 'semi-automatic' changeover (for a 'modest capital investment') was assessed as taking 1 hour. A 'fully automatic' changeover machine (for a 'higher capital investment') was estimated to reduce this time to 6 minutes.

111 Hughes, D. and Hobbs, S. (1994). SMED – Assessing the cost effectiveness of set-up reduction. *Proc. 27th Int. Symp. on Automotive Tech. and Automation.* Aachen, 31 Oct. – 4 Nov., pp. 601–607. Hughes and Hobbs' paper concludes that for changeover improvement 'you get what you pay for': if greatly improved changeover performance is sought, expensive design changes are likely to be required.

112 Quinlan, J. P. (1987). Shigeo Shingo explains 'single-minute exchange of die'. *Tooling and Production*, Feb., 67–71. The best approach to adopt, when also in mind of the level of improvement that is being sought, is not generally agreed. There are those who strongly favour an incremental, team-based approach, in which design change is often accorded a very low priority. Quinlan, for example, argues that ingenuity, not automation, is the best route to changeover improvement. Reporting that Shingo sees mechanization as being adopted only as a last resort, Quinlan writes: 'Mechanising and inefficient set-up operation will achieve time reductions, but will do little to remedy the basic faults of a poorly designed set-up process.' These and other issues (see also the next two references) will be analysed in detail in the current book.

113 Mather, H. (1993). Product design: simplification before automation. *Manufacturing Systems*, June 58–62. Despite knowledge that design change can bring about improvement, some writers urge caution in respect of employing new, more technologically advanced equipment where process improvements (process simplification) is not undertaken first.

114 Johansen, P. and McGuire, K. J. (1986). A Lesson in SMED with Shigeo Shingo. *Industrial Engineering*, **18**, No. 10, 26–33. Johansen and McGuire are similarly cautious in adopting technology-led change for changeover purposes, before first fully evaluating what can be achieved by retrospective improvement of existing hardware: 'In fact, where companies approach reduction in set-up through costly investments, this usually fails to bring the type of dramatic results we had just observed. It is brain power that is important in "SMED", not fancy machinery.'

115 Hill, T. J. and Chambers, S. (1991). Flexibility – a manufacturing conundrum. *Int. J. of Operations and Production Management*, **11**, No. 2, 5–13.

116 Hayes, R. H. and Jaikumar, R. (1988). Manufacturing's crisis: New technologies, obsolete organisations. *Harvard Business Review*, **66**, No. 5, 77–85.

117 Jaikumar, R. (1986). Postindustrial manufacturing. *Harvard Business Review*, Nov.–Dec., 69–76. It is described that a mismatched management/business system structure can inhibit manufacturing performance when new, advanced equipment is installed: 'With few exceptions, the flexible manufacturing systems installed in the United States show an astonishing lack of flexibility. In many cases they perform worse than the conventional technology they replace. The technology itself is not to blame; it is management that makes the difference.'

118 Boyle, A. and Kaldos, A. (1994). Analysis of the operation and performance of flexible manufacturing systems. *Proc. 10th NCMR*, Loughborough, UK, pp. 366–370.

119 McIntosh, R. I., Culley, S. J., Gest, G., Mileham, A. R. and Owen, G. W. (1996). An assessment of the role of design in the improvement of changeover performance. *Int. J. of Operations and Production Management*, **16**, No. 9, 5–22. We argue that sustainability can be a problem. Despite numerous reports of significant changeover performance gains, no known follow-on study has been undertaken to determine how well these improvements have been sustained.

120 Shimokawa, K. (1994). *The Japanese Automobile Industry: a business history*, Athlone Press, London. Shimokawa, discussing the evolution of the Japanese automobile industry, describes a recent trend towards 'multi-kind' and small-lot production (to meet the demands of JIT manufacturing). The significant role of FMSs in this production regime is strongly identified: 'The so-called FMS (flexible manufacturing system) was also a factor in the acceleration of this trend. This freed the automobile industry from conventional and rigid production systems and made it more adaptable to changes in the market environment and ups and downs of both quantity and kind.'

121 Melcher, A., Acar, W., Dumont, P. and Khauja, M. (1990). Standard-maintaining and continuous-improvement systems – experiences and comparisons. *Interfaces*, **20**, No. 3, 24–40.

122 McIntosh, R. (1993). Attendance at a factory changeover presentation. 15 February 1993, Bath University changeover project, File note. Our own experience of changing workplace culture – specifically, more participatory, less antagonistic, staff/management relationships – was never more forcefully conveyed than at an internal factory changeover presentation we attended: 'The environment of the forum was refreshingly open. Everyone made a contribution at one stage or another and notable deficiencies in the organisation were tackled without rancour. This, in the opinion of both members of the management team and line supervisors, is a considerable achievement. [The factory general manager], at the end of the session, commented to the author that some of the things that were being said would previously have landed the staff into trouble, but here the intention was to bring difficulties out into the open for discussion and solution, not retribution.'

123 Alvandi, M. and Burgess, T. F. (1996). *Op. cit.* ref. 28.

3

Addressing diverse changeover issues: An overall methodology for changeover improvement

This chapter is presented in two parts. Part 1 describes the extent of the improvement task that the business faces. Part 2, based on this earlier discussion, sets out an overall methodology for changeover improvement.

First, in Part 1, the chapter appraises the scope and complexity of what lies ahead. Significant issues are outlined that will need to be considered as part of any comprehensive, competently undertaken initiative. These are issues that any changeover improvement practitioner should be aware of, but they are selected in this chapter for particular attention by business managers (later chapters will address other issues that are of particular significance to those who actively participate in seeking improvement on the shopfloor).

In Part 2 our overall methodology for changeover improvement is described. By adhering to the overall methodology a business will have a better chance of achieving and sustaining significant improvement to changeover performance. In addition, the overall methodology seeks to ensure that better changeovers are exploited to the maximum benefit of the business.

The overall methodology deliberately seeks to provide a global framework in which to conduct improvement, embracing every aspect of an initiative.* Use of the overall methodology obliges consideration of issues that are described throughout this book. By appraising globally what is likely to have to be done, when, and by whom, the chapter explains why a three-phase methodology structure is adopted. The second part of the chapter describes in detail the individual steps of the overall methodology in each of its three constituent phases – *strategic*, *preparatory* and *implementation*.

---

* The overall methodology may be contrasted with other tools or methodologies which are more specific in their nature, concentrating upon particular aspects of an improvement programme. The 'SMED' methodology, which is discussed later in this chapter, is a notable example.

PART 1 Addressing diverse changeover issues

3.1 Initiative overview

Much of what has been written in the past has failed to take adequate account of the scope of a successful changeover improvement initiative. Certainly, little or no reference has previously been given in the literature to failing changeover initiatives, or an assessment of why these failures have occurred. In this chapter we seek to redress these omissions, where we will discuss that failure can result for many different reasons. Case studies will be presented to support our claims. By use of references we will also briefly draw parallels with the experience gained in TQM (Total Quality Management) initiatives, where failures have previously been reported.

Achieving lasting improvement to changeover performance is a wide-ranging and complex task. In this chapter we argue that it is an overall task that extends significantly beyond the confines of the immediate shopfloor environment, and hence beyond a reliance on shopfloor improvement teams alone, no matter how well trained and how empowered these teams are.

The scope of a changeover improvement initiative

Figure 3.1 presents an overview of components of a successful improvement initiative. An initiative is liable to falter or struggle to be successful if one or more of the four components shown in Figure 3.1 is missing. It is likely that the success of an initiative will also be compromised if constituent elements of these four components are either missing or inadequately represented. Figure 3.1 indicates at the outset the scope of the task that a business confronts.

The complexity of a changeover improvement initiative

In the past it has been written that changeover improvement initiatives should be strongly concentrated on the shopfloor.[1,2] Authors have stressed that senior managers should not participate in shopfloor improvement activity, rightly arguing that it is shopfloor personnel who understand in fine detail the day-to-day operational intricacies of their equipment. Direct management involvement in improvement activity is argued to be unwise if a full contribution of those who operate the equipment is to be encouraged.

With the proviso that management support for the team's efforts must be seen to occur and, perhaps, that engineering or planning staff might also be included, we generally support this stance on team composition. However, any perception that an improvement initiative should only take place on the shopfloor is to be avoided. This book will describe that a far wider perspective needs to be adopted. Where it is appropriate to do so, improvement should also involve other personnel, including the possible use of external agencies. As part of a wider overall initiative, minimal direct senior management involvement, as encouraged in respect of shopfloor activity, would not be sensible.

- A workplace culture that is receptive to change and on-going improvement – both within senior management and on the shopfloor

- An awareness of the contribution of improved changeover performance to alter the company's competitive position
- An awareness of factors/issues that need to be addressed that might otherwise render an initiative less than wholly successful
- An awareness that improvement in changeover performance may be achieved by different approaches, where each separate approach needs to be carefully assessed

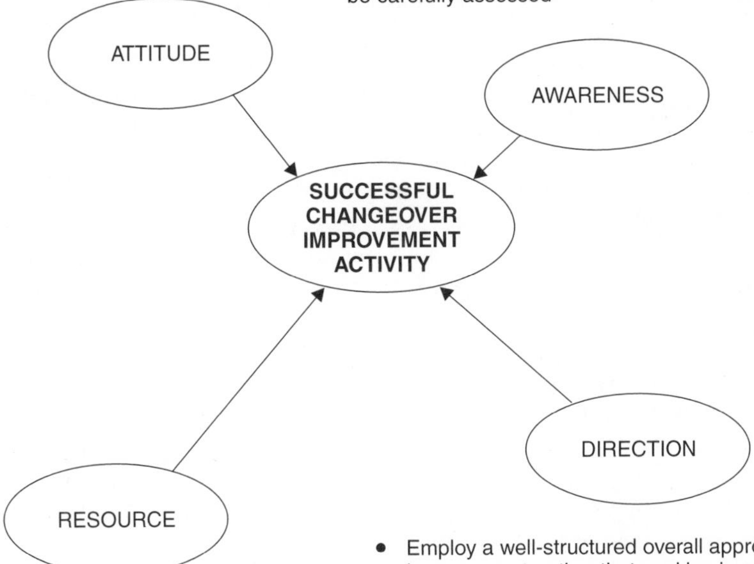

- Time
- Money
- Personnel
- Training, including training in the use of possible tools to assist an improvement initiative (overall methodology for changeover improvement; design rules for changeover; 'Reduction-In' strategy; matrix methodology; reference changeover models; 'SMED' methodology)
- Additional equipment/hardware

- Employ a well-structured overall approach to improvement, rather that seeking improvement in an *ad hoc*, haphazard fashion
- Senior management leadership/commitment
- Improvement team leadership/commitment
- An ability to identify problems with existing practice
- Select which machinery needs to be worked upon
- An ability to identify and rank improvement opportunities
- Setting (and working to achieve) appropriate targets

Figure 3.1 Components of a successful retrospective improvement initiative

In particular, senior managers have a crucial role in defining the overall purpose and direction of an initiative, and ensuring that momentum for improvement is maintained.

These observations provide little help in formulating how an initiative is to be tackled. For example, if it is accepted that participation from shopfloor personnel will probably be necessary for retrospective improvement initiatives, how should this participation be structured and undertaken to make the most of every individual's potential contribution – and to make sure that those involved, in turn, get the most back from it? This is just one issue that the business needs to confront. The business should ask a number of searching questions of itself at the outset, to assess the overall task ahead. These questions should embrace strategic issues. Once the initiative has progressed through to the manufacturing environment, issues of who is to participate and how improvement is to be sought will also need to be addressed.

Some questions that the business might consider are set out below. This series of questions should not be regarded as exhaustive:

- Is the business clear why (for what purpose) improved changeover performance is being sought? What level of improvement is required, and in what timescale? Is the business able clearly to identify and prioritize the manufacturing processes to which improvement activity should apply? Should the business attempt to improve just one line (or machine) and then 'roll out' the improvements, or seek improvement in an alternative way?
- How should a changeover improvement initiative be ranked in importance against alternative initiatives that the company may wish to undertake? Is it possible or desirable to undertake more than one initiative together at the same time – in which case which ones?
- Should the company opt for retrospective improvement of existing equipment, buy alternative equipment or, possibly, seek to use a degree of surplus equipment to limit the number of changeovers that occur? Can (and should) scheduling be used to some extent to limit changeover losses? If it is not appropriate to use additional machines, are there elements of machines that should be replicated to benefit changeover performance?
- If retrospective improvement is pursued, should the company emphasize design improvement or low-cost organizational change to changeover tasks and procedures? If design and organizational change are to be employed in unison, on what basis should this occur? Should incremental improvement by team activity be undertaken? Who should lead team improvement activity? Is a workplace culture present that is conducive to improvement by this means?
- Who is to be involved in changeover improvement activity, within the organization itself (from the shopfloor and elsewhere), and including the possible use of external specialists (notably management and design consultants)? How are members of different shifts to be involved and kept up to date with progress that is being made? How is progress to be communicated to all other factory personnel?
- What training should be conducted, including training to change the workplace culture? To what extent should these issues influence the overall approach that is adopted?
- Has an overall cost been assigned to achieving improvement targets (level of improvement to changeover performance and the date by which this improvement is to have been completed)? What resource provision, including time allocated to personnel engaged in an initiative, should be made available by the company to achieve the level of improvement that it seeks?
- How should an improved changeover capability on the shopfloor be supported by commensurate changes in working practice (notably in work scheduling and

marketing) in other areas of the business so as to maximize the benefit from it – in other words, are possible skill or system deficiencies elsewhere within the organization likely to inhibit changeover performance, or the benefit that is derived from it?

- What changeover performance records are there? How accurate are these records? How should existing changeover performance be assessed? By whom? How should on-going improvement be monitored?
- What tools are available to identify problems with existing practice? What tools are available to match possible solutions to these identified problems?
- How can a business access the work that others have conducted previously (to save 'reinventing the wheel')? What solutions are available 'off the shelf' as sub-assemblies or devices that could be incorporated into the improvement programme?

Phases of a retrospective improvement initiative

At one level this series of questions might be used as an initiative 'checklist'. By asking itself these questions the business would be led to consider important issues. The business would be served better still, however, if the issues that these questions represent are addressed in sequence in a structured manner. It would be unwise, for example, to buy a new flanging machine to attain faster changeover performance without first having justified this expenditure, without having validated the manufacturer's changeover claims for this new equipment and without first having evaluated all possible alternative improvement options. Similarly, it would be foolish to direct a local team to improve the changeover performance of, for example, a welding machine if it had not first been identified that this machine warranted the team's attention – either as a matter of priority or as a training vehicle (with the intention of later applying the newly learnt improvement expertise elsewhere).

To give structure to the issues raised in the questions above it is useful to categorize an overall improvement initiative into distinct phases. The most prevalent improvement route (at least as purveyed by changeover consultants) is to seek retrospective improvement to changeover practice by use of internal improvement teams.[3] Figure 3.2 applies to improvement by this means and presents the three-phase structure of our overall methodology (as yet individual steps within each phase are not described). As shown, these three phases are: strategic; preparatory; implementation. Figure 3.2 also indicates who may participate in these individual phases and what the global purpose of each phase is. The three-phase structure is used to separate activities, to identify the people that are typically concerned with those activities, and to allow progress to occur in an ordered sequence. It indicates that much needs to be done before physical improvement to process machinery and/or improvement to changeover procedures is actually undertaken.

The overall methodology applied to non-retrospective changeover improvement options

As noted, Figure 3.2 applies to retrospective improvement that is conducted largely by internal factory personnel. It was indicated in the previous chapter that this is not

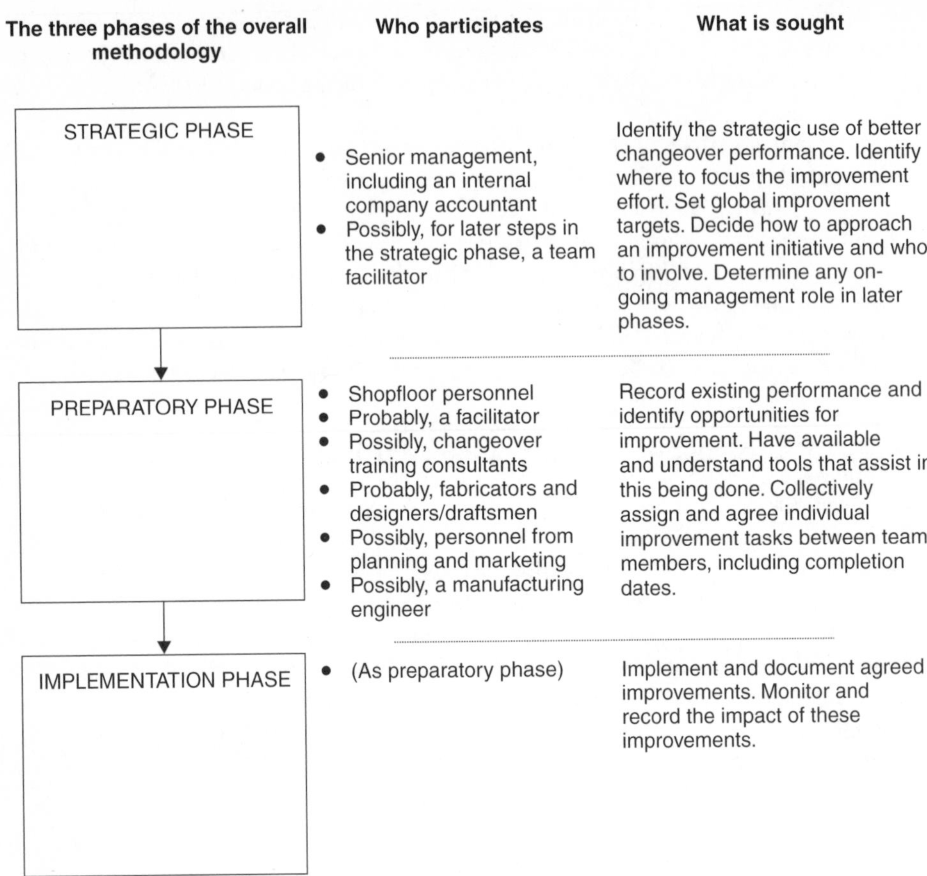

The three phases of the overall methodology
Who participates
What is sought
STRATEGIC PHASE
• Senior management, including an internal company accountant
• Possibly, for later steps in the strategic phase, a team facilitator
Identify the strategic use of better changeover performance. Identify where to focus the improvement effort. Set global improvement targets. Decide how to approach an improvement initiative and who to involve. Determine any on-going management role in later phases.
PREPARATORY PHASE
• Shopfloor personnel
• Probably, a facilitator
• Possibly, changeover training consultants
• Probably, fabricators and designers/draftsmen
• Possibly, personnel from planning and marketing
• Possibly, a manufacturing engineer
Record existing performance and identify opportunities for improvement. Have available and understand tools that assist in this being done. Collectively assign and agree individual improvement tasks between team members, including completion dates.
IMPLEMENTATION PHASE
• (As preparatory phase)
Implement and document agreed improvements. Monitor and record the impact of these improvements.

Figure 3.2 Retrospective changeover improvement: overview of activities in each phase of the overall methodology

the only way that improvement may be achieved. First, a 'big bang' solution may be sought, where external expertise and labour is brought in and given full responsibility to improve changeover performance – be this by overseeing better discipline to changeover procedures or by undertaking design changes to the existing equipment (see Chapter 5). Alternatively, equipment replacement may be undertaken. It is also possible to utilize a degree of 'surplus' parallel equipment to benefit responsive, small-batch manufacture. Each of these options does not involve the existing workforce, save to gain acceptance of the changes that have been made and to ensure operational competence.

In each of these situations the overall methodology would not progress through to the 'preparatory' and 'implementation' phases. While a strategic review would occur as before, a conclusion of this review would be that one of these alternative improvement routes should be adopted. This is shown in Figure 3.3, where each of the possible options that have been described are presented.

In practice, more than one option effectively may be employed simultaneously. For example, external changeover consultants may be brought in to conduct training or to

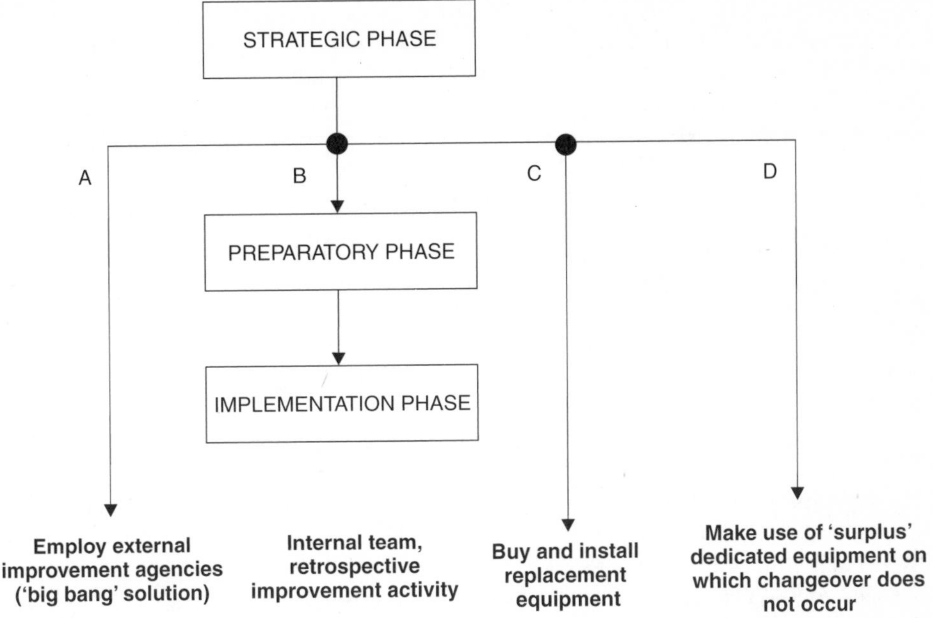

Figure 3.3 Improvement strategies (options 'A', 'C' and 'D') that do not progress through to the latter preparatory and implementation phases

lead factory personnel in what is otherwise wholly internal improvement effort. Also, while wholesale equipment duplication to allow the use of dedicated lines may not occur, replication of identified equipment components or sub-assemblies to assist changeover performance, by allowing conversion of tasks to external time, might be desirable. In other cases responsibility for updating the mechanical capability of equipment might be sub-contracted to external design agencies, while responsibility for defining 'best practice' changeover procedures might be assigned to a local site team (comprising principally machine operators).

In any situation where different approaches are combined, the business must be very clear how it is to manage the overall improvement effort. The way that external changeover consultants are employed needs to be given particular attention. This is one topic that we will consider now as attention is turned to some specific difficulties that a typical improvement initiative can face. Case study and other research evidence that is presented below will be used to define in greater detail the component steps of each of the three phases of the overall methodology. The complete overall methodology for changeover improvement (showing all its component steps) will subsequently be presented in Part 2 of this chapter.

3.2 Specific issues for management consideration

The current authors have researched changeover improvement in field studies throughout Europe. Far from the success that has been heralded in the academic and trade literature, many of the improvement programmes that have been studied have struggled both to achieve and then to sustain significant levels of improvement. In some cases improvement initiatives have either faltered or petered out entirely.[4] These initiatives can mimic what has also been found to have occurred when attempting to implement TQM[5–8] or, sometimes, what has been reported when implementing JIT.[9]

Issues that warrant consideration include:

1. Involving internal factory personnel: training, skill and application
2. Changeover knowledge dissemination
3. Loss of momentum
4. Sustaining gains in changeover performance
5. Measuring changeover performance
6. Safety
7. Integrating changeover improvement throughout the organization
8. Changeover performance specification for new equipment
9. The type of changeover that is conducted: sequence dependency/independency
10. Scheduling considerations
11. The dominant use of the 'SMED' methodology
12. The role of changeover improvement consultants
13. Communication
14. Financial benefit analysis (introduction)
15. Changeover item quality
16. The ability of shopfloor teams to bring about change

In particular these are issues that are of immediate concern to senior management. They are issues that should be considered as an initiative commences, not once the initiative is well under way. This list is not fully representative of all the potential difficulties that an initiative faces. Difficulties that are more directly concerned with shopfloor activity will be assessed in later chapters.

1 Involving internal factory personnel: training, skill and application

The way that the workforce is to take part in an initiative warrants particular attention. If this issue is not dealt with correctly the business is likely to face poor levels of active participation and contribution from those personnel that it chooses to involve. Changeover improvement correspondingly will be poor. Worse than this, if staff involvement is mismanaged the business faces engendering apathy or resentment for any future collaborative activity. If a changeover improvement initiative is poorly undertaken first time out it is likely to be much harder to reinstate it, if attempted, in the future.[10]

All staff within the business will have different skills which they can bring to an initiative. The business has options of trying to match different individuals' skills to the different requirements of the initiative and/or conducting training to heighten skills where these are believed to be deficient.[11] This will be a decision that only the business itself can make, but it is a decision that should always be made with an awareness of what training can contribute.

Typically, training is necessary. Where little or no prior training has been conducted, in our experience, teams engaged on changeover improvement will struggle. In these circumstances personnel will often fall back on a 'brainstorming' exercise which, unless focus is concentrated on potentially fruitful areas of improvement,[12,13] is liable to be, at best, inefficient. If training is to be undertaken, therefore, it should be on the basis of a coherent analysis/improvement methodology. As noted previously, Shingo's ostensibly simple 'SMED' methodology[14] (see below) is frequently used for this purpose.

Even when training occurs and when tools such as the 'SMED' methodology are employed, significantly variable practitioner performance, initiative take-up and results are still probable. Exhibits 3.1 and 3.2 indicate the extent to which this can occur. These exhibits draw attention to possible issues of attitude and competence of the personnel that are involved, and to the corresponding content/approach of the training programme that is employed. Other factors might also influence the success that an initiative enjoys including, for example, the nature of the equipment on which improvement is sought.[15–17] All factors that may impinge upon an initiative's success ideally should be considered in the context of the training that is conducted and, more generally, in the decision whether or not even to train and employ site changeover teams in the first place in preference to using external expertise. Training, if conducted solely on the basis of teaching the 'SMED' methodology, particularly if this is done under the auspices of low-cost organizational improvement, does not guarantee a successful outcome.

Exhibit 3.1 Different responses in parallel initiatives

A large UK manufacturing plant sought to improve changeover performance within different departments on the same factory site.* A common training programme based on the 'SMED' methodology was employed, on which personnel from these different departments jointly participated. The initiative as a whole was championed by an influential production manager, but did not have prominent support from some other colleagues, nor from more senior business managers. A supervisor was involved in the overall programme, who was charged with overseeing the developments in each department.

There was a marked difference in the success that each department enjoyed. In the best case the changeover on one line was brought down by more that 80 per cent (from 8 hours to 90 minutes). Equally significantly, this performance was sustained: 2 years later there had been no decline in this changeover time. In an entirely different department, involving different personnel, and where entirely different equipment was employed, improvement took longer to achieve and was less dramatic (a line changeover time improvement of just 30 per cent was recorded).

* It was noted that one department was deliberately not chosen to be involved in this initiative because the attitude of the workforce in that department was not considered suitable to allow meaningful improvement to occur.

Exhibit 3.2 Different responses by personnel improving the same equipment

An American multinational company sought to improve changeover performance across its UK sites. A forum was set up, with the intention that delegates from up to nine different sites should meet once every month to pool their experiences. Attendance on the forum was voluntary. There was an emphasis on involving shopfloor personnel directly in the forum and excluding both line management and senior management personnel. Attendance needed to be sanctioned by the different sites, who thereby effectively allocated time and finance to their shopfloor staff for this activity to occur. Venues rotated from month to month. The forum was chaired by a member of staff from a non-manufacturing divisional office. There was a ground rule that all attendees had to make a contribution during a meeting. At its inception the forum was used to describe the 'SMED' methodology collectively to all participating delegates, which was done by internal company training staff.

Five sites participating on the forum had examples of near-identical equipment. One site team gave a presentation on work that they had conducted on this equipment, highlighting the changes that had been made and the significant improvement that they had achieved. Despite the attention given to this equipment by the forum including, on one occasion, the lead site physically demonstrating to other site teams how this changeover was conducted, similar results were not achieved on the near-identical equipment at the other four sites. One site experienced no improvement at all to its own changeover performance on this equipment.

Both Exhibits 3.1 and 3.2 indicate that there may be a raft of complex issues involved in what was, for each exhibit, an apparently straightforward situation. Widely divergent results were experienced, even though joint training programmes had taken place in each case. This indicates that the training that occurred did not necessarily adequately address all the complex issues that can prevail when changeover improvement is sought. It also demonstrates it can be dangerous to expect that widely claimed levels of improvement of 75 per cent or more are always easily achieved.[18–20]

Attributes of attitude and competence of those who are to be involved in training programmes may be difficult to predict in advance. We have found it is sometimes those who are least expected to do so who contribute most to a team's achievements. A positive attitude separately towards both changeover and change itself equally is important in respect of senior managers.[21] If those who are involved in an improvement programme, at whatever level of seniority, do not believe that a changeover initiative is in their interests they are likely to give less than full support to it. For all relevant production personnel, therefore, it is often wise to describe why changeover improvement is being conducted and what benefit both the business and the operators themselves will derive from it. More senior personnel should determine for themselves how the business will accrue benefit from better changeover performance (see the next chapter). For example, there was evidence that management at the participating sites in Exhibit 3.2 attached different importance to the changeover performance on their equipment, influenced in part by their own differing operational circumstances. Conversely, if a positive attitude to seek improvement is prevalent, those who participate on a training programme are likely to be more receptive to the training material that is presented.[22]

As well as being either receptive or non-receptive to training that is conducted, it may also be, of course, that some training programme participants will be more competent in understanding and applying the material that is presented. As cultures will vary from site to site, it may be necessary to tailor a training programme (in terms of both style and content) to those who participate on it; in other words 'off-the-shelf' core training material may need to be enhanced individually for each business that uses it.[23–25]

2 Changeover knowledge dissemination, evaluation and application

In general, the forum described in Exhibit 3.2 was not particularly successful. It wound down and was closed after 18 months despite strenuous efforts by its chairman to keep it running. The purpose of the forum was principally to share changeover experiences. Our research in respect of this forum presented some evidence that improvement practitioners can find it difficult either to assess or apply changeover knowledge with which they have been provided.

For the most part the ideas that were presented on the forum tended to be fairly site-specific. The forum delegates demonstrated difficulty in using these ideas as inspiration for improvement in their own particular circumstances. For example, a delegate suggested using a power winder in one application. There was no subsequent known uptake of a power winder on different machines, either at the originating site or for machines at alternative sites. This, of course, might have been because applying this and other basic ideas elsewhere was inappropriate. Nevertheless, the very low rate at which ideas were adopted in different circumstances was found to occur for the majority of the work that the forum did. Even where ideas could be translated directly to other sites, poor uptake was identified at those sites where the improvement did not originate (see also Exhibit 3.2). This varied in its degree from partial uptake through to the expression of negative opinions or even hostility – where a 'not invented here' attitude could be identified.[26]

The pervading improvement theme expressed by the forum delegates was the desirability of translating tasks into external time, echoing Shingo's principal thrust for improvement,[27] and stressing arguably the key message of the training that they had been given. The same delegates, though, as described above, did not always demonstrate an ability to understand and share the often subtle techniques by which this objective could be effected. Neither did the forum show any particular inclination to embrace Shingo's improvement concept of 'streamlining' changeover operations.[28] It might be argued that these observations reflect inadequate training in respect of the 'SMED' methodology, hence yielding these poor results. Nevertheless, the whole of one day had originally been devoted to this topic, including the issuing of copious reference notes. An alternative judgement is also possible: that the 'SMED' methodology itself does not necessarily adequately equip the improvement practitioner to be receptive to, or to be able to identify, all possible improvement opportunities. Or, considered another way, the 'SMED' methodology, as understood by recently trained changeover practitioners, may not always allow improvement ideas to be assessed in context, and hence easily to be related to alternative changeover situations.

Other observations were made of this forum's work. While the delegates generally expressed their ideas well verbally, in many cases using supporting diagrams, documented procedures and photographs were often poor or non-existent. Detailed engineering drawings were rarely observed being made available to other participating sites. Where such drawings were available, unambiguous parallel improvement to identical machines could be made with greater reliability and greater ease. The same applies in documenting in detail improvements to changeover procedures, and thereby being able to make these changes exactly known elsewhere. This was despite the importance that had been stressed on documenting new procedures in this particular company's own internal changeover training material.[29]

These findings prompted the authors to undertake a trial introduction of a 'browse-through' catalogue of changeover improvement techniques and examples, including, in addition to procedural changes, the presentation of potentially useful mechanical devices and their suppliers. A new and more expansive classification of improvement practice was a feature of this work, and in general the catalogue was well received by the forum delegates.[30] Both in respect of the classification that it provided and in terms of being an accessible means of communicating others' ideas, the catalogue was regarded as a potentially important tool in the quest to drive down changeover times.[31–34] We intend to update this work (see Preface).

3 Loss of momentum

Changeover improvement does not simply happen: it has to be worked for. Effort is required from personnel throughout the organization. As has been discussed, this effort needs to be applied in a structured manner. Equally important, effort needs to be maintained if progress is to continue to be made. If effort falters, then so too the initiative's momentum will falter.

An analogy may be made of a ball ascending an incline. As presented in Figure 3.4, the higher the ball rises, the greater the improvement in changeover performance that has been achieved.

The ability of the ball to ascend the slope is enhanced if it is already underway as the slope is first encountered – signifying that work has been undertaken before actual changes to changeover activity are made. If at any point the effort is withdrawn, the ball will quickly slow and then start to reverse. To then get it to regain its original direction will require greater effort than was first required. The slope of the incline is likely to change as the initiative progresses, which is analogous to it becoming ever more difficult to gain incremental improvement to changeover performance. Once the target level of performance has been achieved the business can start to relax its efforts for improvement. It should be cautious of withdrawing attention entirely, however, especially if organizational improvements have been concentrated upon. There will be a natural tendency to revert to the state of relative disorder that previously characterized changeover activity. Withdrawing effort will result in the ball once again starting to descend the slope – at a rate determined by the extent to which effort has been

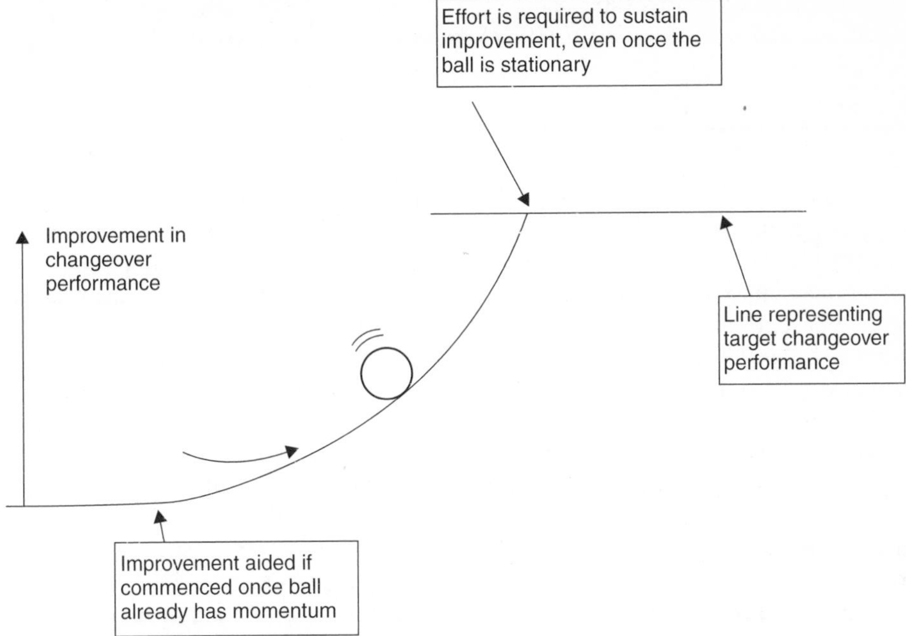

Figure 3.4 The 'rolling ball' analogy

relaxed. Conversely, if better changeover procedures are practised over and over again (for which effort is required) progress along the changeover 'learning curve'[35,36] will continue to be made – and changeover performance should continue to improve.

A modified version of Figure 3.4 is shown as Figure 3.5. This revised analogy incorporates 'design plateaus' – that is, it represents physical changes made to the equipment that modify and 'freeze' the changeover procedure in such a way that lapses back to the original performance level are prevented (see later).

Many different situations can contribute to faltering or inhibited momentum. The most usual are probably to do with declining interest by initiative participants. This can occur for many different reasons. One possible reason is initiative fatigue – a situation

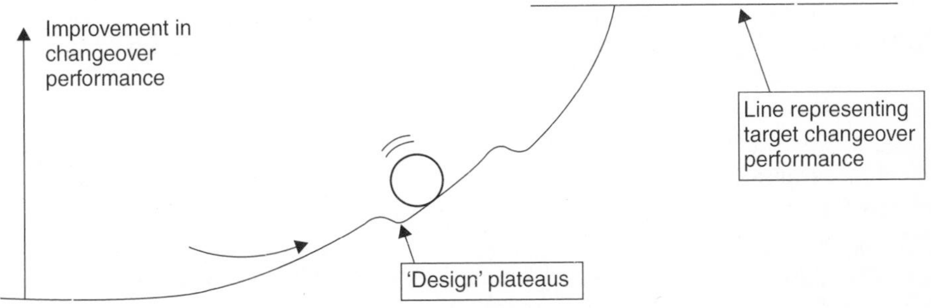

Figure 3.5 The 'rolling ball' analogy, incorporating 'design plateaus'

where too many initiatives are being (or have recently been) attempted. Exhibit 3.3 presents an example of initiative fatigue that we have encountered.

Exhibit 3.3 Initiative fatigue

Management at an East Anglian (UK) factory were prone enthusiastically to embrace many of the business-improvement programmes of which they were successively becoming aware. They were also very keen to involve shopfloor staff in business improvement as fully as possible. Focus upon changeover performance was apparently selected in these circumstances more because it was in fashion than as a carefully considered initiative in response to identified business needs. A junior production manager was assigned to select and lead an internal changeover team.

This improvement programme was limited in its success. A particular problem was that it was very difficult, once nominated, to get team members to participate. This difficulty, of course, might have arisen for a host of different reasons, but it was acknowledged that at that time most operators were involved in at least one site-improvement initiative to which they were often expected to contribute in their own time. A few operators were simultaneously involved in three or four separate programmes.

This poses many problems, not least of which is in distinguishing between important and relatively unimportant initiatives. In this climate initiative fatigue is likely, where all the pending improvement programmes, as a result, can suffer.[37–39]

4 Sustainability

Original research by the authors has indicated the extent to which difficulty can be experienced in sustaining improvement to changeover performance. The example described in Exhibit 3.4 has been published previously,[40] and shows performance reverting back almost completely to its original level. A feature of this particular initiative is that it had, apparently, every chance of success according to changeover improvement wisdom that then prevailed.

Exhibit 3.4 The difficulty of sustaining changeover performance

In 1989 a UK factory seriously engaged itself in improving changeover performance. A well-known consultant organization was employed to lead the improvement programme. Significant numbers of shopfloor personnel were engaged in training activity that featured, among other components, Shingo's 'SMED' methodology. Specific equipment was targeted for improvement, and it was ensured that all operators of this equipment were involved in the training programme. A study mission to Japan was also undertaken, on which a number of members of staff participated.

By interviewing factory personnel and corroborating factory records, poor sustainability of changeover gains was determined. This is shown in Figure 3.6.

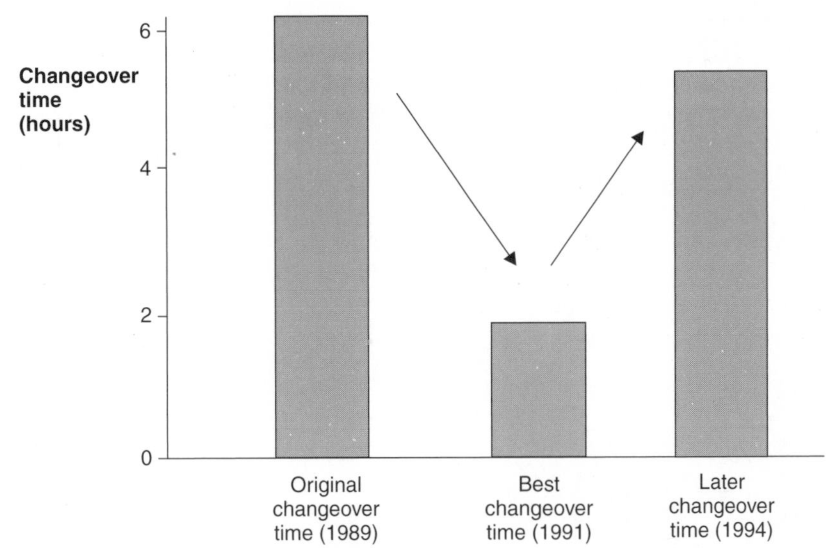

Figure 3.6 Changeover performance through time at a factory that had previously rigorously undertaken to improve its changeover performance

This is not the only example that has been observed. Exhibit 3.5 demonstrates considerable improvement for a similar consultant-led initiative at a Spanish factory. Subsequent candid discussions with the factory manager elicited that performance, as with the previous UK example, had rapidly lapsed back to be close to its previous level. This difficulty of sustaining improvement has also been recorded by changeover consultants themselves.[41] Such (rare) instances of failure reporting support our own belief, based on our industrial experience, that difficulty in holding on to improvements is common.[42] Other evidence is also available to us. We have a copy of one consultant's presentation overheads, wherein a changeover improvement methodology is described. On a sheet entitled 'Create a Climate for Change' it is described (without further qualification) that a surprisingly high rate of initiative failures can be expected. Elsewhere, a notable admission of the difficulty of sustaining changeover gains was located in documents exchanged internally by yet another group of consultants.[43] Suggested steps by these latter consultants to try to arrest declining performance included writing and enforcing standard changeover procedures, generating changeover information displays and forcing planners to maintain pressure on the new changeover times. Despite these internal reservations, these same consultants were still keen only to promote the very best level of performance that they had achieved – in other words, before problems of sustainability became an issue. It was, of course, in their interests to do so. In the wider academic literature to date very little has been published on if and how gains in changeover performance have been sustained through time.

Exhibit 3.5 Further evidence of deteriorating changeover performance

A Spanish plastic moulding factory engaged consultants at the outset of its own changeover improvement initiative. A difference between this initiative and that described in Exhibit 3.4 was the deliberate decision to achieve rapid improvement in a concentrated burst of improvement activity. As with Exhibit 3.4, the full involvement of machine operators was sought, although in this case only for those personnel directly associated with the equipment (rather than also including some of the more peripheral shopfloor personnel).

A spectacular improvement was reported over the short duration of the consultant's direct involvement, as presented in Figure 3.7.* It did not last. The factory manager was quizzed some 6 months later, and admitted that performance had returned almost to pre-improvement levels.

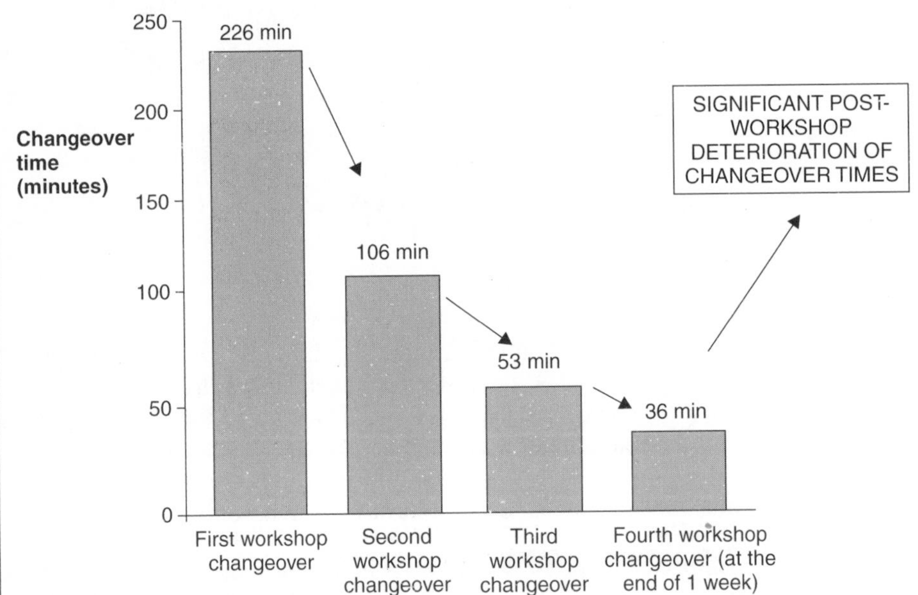

Figure 3.7 Changeover performance during and after a 1-week duration concentrated initiative

* Questions may be asked of the validity of these data (see elsewhere in this chapter).

Despite the lack of published research, it is possible to speculate on why sustaining improved changeover performance may be a problem. Among many potentially influential factors, the importance of 'cultural issues', including an attitude to pursue and accept change, should not be underestimated. Alternatively, taking a different perspective, sustaining changeover performance might be assessed in terms of chaos theory, which dictates that, unless prevented from doing so, systems will revert to an ever more disordered state. If improvement is largely generated by ordering the relative chaos of an existing changeover it is reasonable to argue that there will be a natural inclination for good changeover practice to decline, or lapse back, unless deliberate action is taken to ensure that this does not happen. Included in 'ordering relative chaos' will be steps to move and convert tasks to external time. By contrast, one particular benefit of employing design solutions is that change is imposed on the way that a changeover is conducted that cannot be undone unless the physical changes that were made to the manufacturing system are also undone. In this case conscious steps have to be taken to revert to how the changeover was previously conducted – rather than by allowing this to occur by failing to inhibit natural pressures for incremental degradation in working practice through time (see Figures 3.4 and 3.5). By this token, as we will argue in detail later, particularly in Chapter 5, any tendency for changeover gains to lapse is largely subdued by the application of design changes.

5 Measurement of changeover performance

It is important to measure changeover performance. Measurement will inform business managers (and those engaged in improvement activity on the shopfloor) of the progress that is being made. Measurement also provides data that should be fed to both marketing and planning functions within the organization.

Different aspects of a changeover can be measured. Principally the business is likely to want to measure the total elapsed changeover time (see Chapter 1). It is also useful separately to record both the duration of the set-up period and of the run-up period. These latter data allow a more detailed assessment of what is occurring to be made. A business will be able to correlate changeover performance (for both the set-up and run-up periods) with line performance parameters such as subsequent line speed, product quality, scrap rates and line breakdown. If the run-up period is extensive, as we shall later explain, particular attention to issues of adjustment might be warranted.

Measuring changeover performance can be difficult. For any serious initiative though it must take place. If nothing else, measurement should occur and be seen to occur to keep changeover in focus – applying pressure either to maintain or to continue to improve changeover performance.[44]

A number of potential problems are associated with measuring changeover performance. Specific difficulties may arise in areas of:

- Having reliable historic data against which to assess improvement
- Determining when the changeover has been completed
- Overcoming any inclination to fraudulently record changeover performance
- Measuring 'repeatable' performance, not 'forced' performance (see also Section 6: 'Safety', below)

Having reliable historic data against which to assess improvement

Many companies might think that they have reliable, accurate historic changeover data. Often this is a mistaken belief. Historic changeover data need to be treated with considerable caution if they have not been gathered in the same way – and to the same high standard – that current data are being gathered.

Determining when the changeover has been completed

Figures 3.8 to 3.10 present real line output graphs for changeovers in three separate industries. For each industry different requirements will have applied as part of the changeover, hence influencing the run-up curve that is experienced. Each one of these three figures shows the difficulty of determining when the changeover has been completed. Typically it is straightforward to determine when the changeover commences (point A) and when the first piece of the new product has been manufactured (point B). It is much less easy to determine when production simultaneously satisfies both quality and line volume requirements (point C). This applies even to changeovers that are nominally complete once the line has been synchronized with other production equipment.[45] For example, Figure 3.8 shows synchronization first occurring where marked, but poor line performance immediately subsequent to this time was identified as changeover-related problems (and hence, strictly, the changeover is not complete).

The determination of when the changeover is truly complete is a problem on which it is very difficult to give guidance. Each business should derive its own system for determining when this occurs. Consistency in making this judgement from one changeover to the next is important.

As noted elsewhere in this book, what occurs in the set-up period, particularly in terms of the way that components are accurately located and settings are quickly and accurately achieved, affects the run-up period.[46] By recording the set-up period (A to B) and the run-up period (B to C) separately, data will be available to correlate set-up performance and run-up performance. How changeover performance varies when identical changeovers are repeated may also be studied. In the long term the business may use these data to justify spending what might otherwise be considered an overlong period conducting the set-up, knowing that the benefits of doing so are likely to be a reduced run-up period and a reduced total elapsed changeover time (and, perhaps, better subsequent overall line performance).

Overcoming any inclination to fraudulently record changeover performance

Changeover can sometimes be seen by management as a task whose duration is solely determined by the physical performance of the operator(s) that conduct it. This view is almost without exception flawed: although operator performance is undoubtedly important, factors contribute to the duration of almost every changeover that are beyond the operator's control.

In circumstances where management hold the view that operator performance alone determines the time that a changeover takes to complete, and especially if perceived or real sanctions are feared by the operator(s), false recording of changeover duration by those who conduct it is a possibility. Exhibits 3.6 and 3.7 present two instances where this has been observed. The second of these exhibits demonstrates

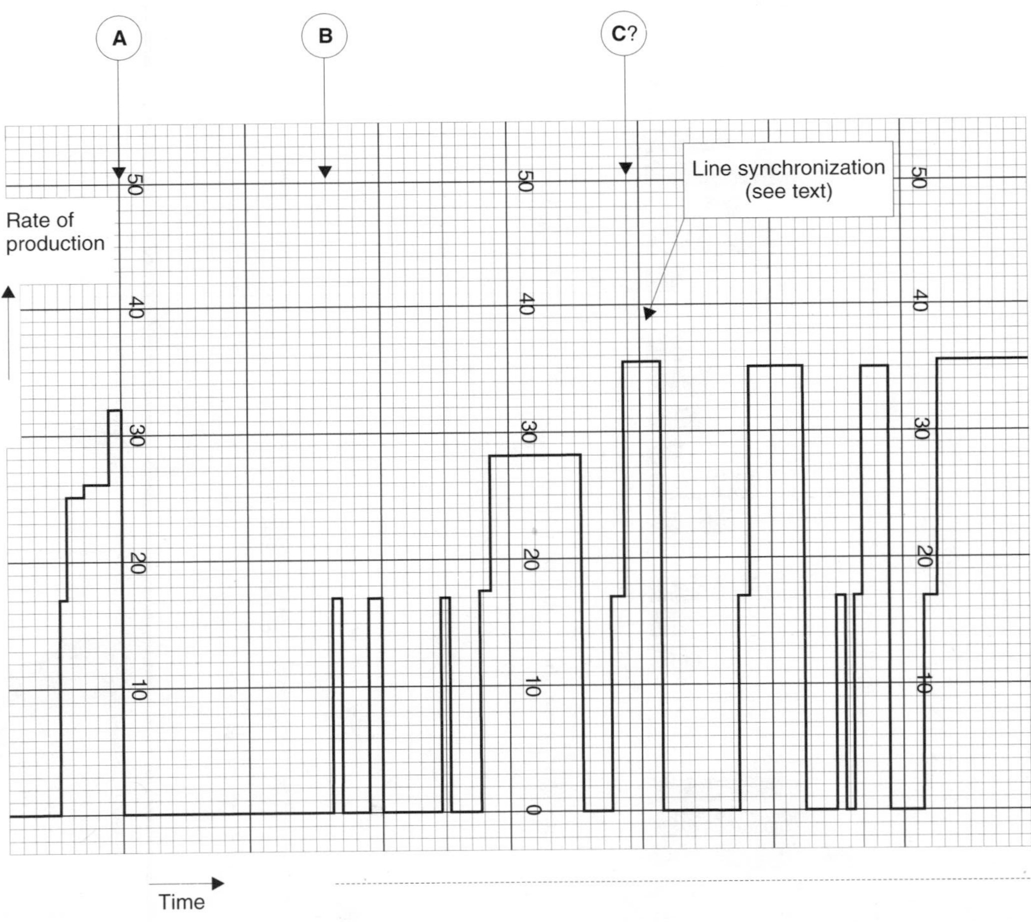

Figure 3.8 Line changeover, demonstrating the difficulty of assessing run-up (1)

that false recording of data can extend further up the organizational hierarchy. For this reason alone line output that is automatically recorded during the changeover period is desirable, from which true changeover time may be independently, consistently and accurately assessed.

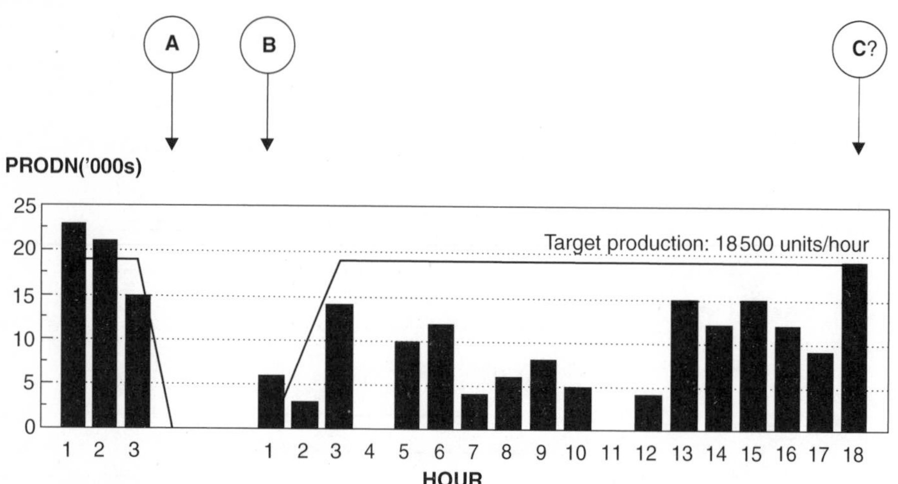

Figure 3.9 Line changeover, demonstrating the difficulty of assessing run-up (2)

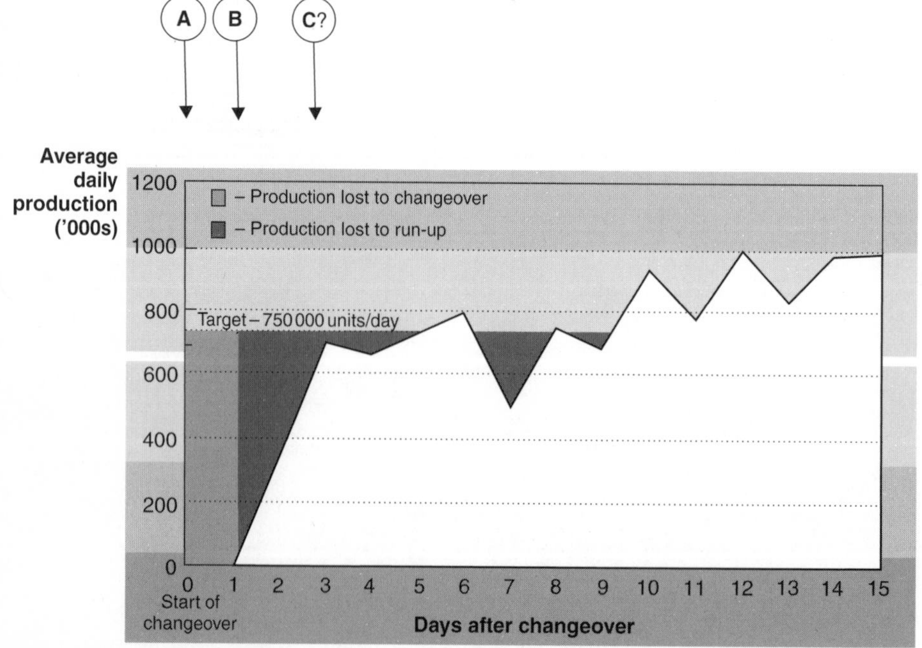

Figure 3.10 Line changeover, demonstrating the difficulty of assessing run-up (3)

Exhibit 3.6 Fraudulent recording by operators

A UK company had recently focused attention on the changeover performance of its manufacturing lines as part of a TPM initiative. There were few historic changeover performance data. The company accepted that what data it did have should be put to one side and that new records should be taken. The company was also keen to record other line losses under TPM. It was decided to issue a form to the line operators on which all losses, including changeover, could be manually recorded.

The authors investigated changeover at this site by participating for one month as improvement team members on the shopfloor. It soon became clear – and was acknowledged by the operators – that it was in their interests to emphasize 'breakdown' losses in preference to 'changeover' losses on these forms, because breakdown performance was believed to be less directly attributable to their own performance. Although difficult to put a figure on, the changeover data were probably distorted by at least 50 per cent.

Exhibit 3.7 Fraudulent recording by more senior personnel

A Dutch company had had expensive new manufacturing equipment installed. This equipment's claimed changeover capability (by its manufacturers) had been accepted by senior company executives. For financial reasons there were substantial pressures to achieve this expected changeover performance once the equipment was in use.

Despite the need to monitor carefully true changeover times, a manual data logging system was employed. Changeover events, including the problems that were experienced, were classified on the data logging sheets, where entries were to be made in 5-minute increments. In practice – understandably – these sheets were filled in long after the changeover had been completed. In terms of duration and of the categories in which the operators chose to record changeover events, the completed sheets were often highly unrepresentative of what had actually occurred.

This, though, was incompetent, not fraudulent. Pressure to record fraudulently was applied by the fact that the actual changeover performance was well in excess of the time (the manufacturer's claimed time) that had been set into company manufacturing schedules and budgets. As before in Exhibit 3.6, there was a temptation (which was succumbed to) to enter losses into categories that were subsequently argued to be because of problems with the machine – not problems with the changeover. This distinction was forcibly argued.* Moreover, it was line management, not the operators, who were distorting what was presented. It was not only the use of problem categories that was being used to do this. In one case a changeover that was witnessed as having lasted slightly over 2 hours was recorded on the data sheets by a senior line supervisor as having taken just 15 minutes to complete.

* Staff continued to argue this position even though it was at odds with the company's own definition of a changeover (which corresponds to that given in this book in Chapter 1) which clearly stated that all events that occur within the total elapsed time must be considered as changeover events – no matter what their cause.

Management can sometimes believe that all changeovers that take place, irrespective of which products are being involved, should be of equal duration (this topic is discussed under the heading 'Type of changeover', below). If operators are expected always to achieve consistent changeover times, this too can provide pressure to falsely record changeover data.

Irrespective of the pressures that can lead to changeover times being falsely documented, it is a practice that will serve nobody well in the long term. The business should take care to ensure that it does not occur, particularly by encouraging openness and honesty in a non-threatening participative environment.

Measuring 'repeatable' changeover performance

Different personnel conducting the same changeover are likely to perform differently, even when it is attempted to introduce and maintain standard changeover procedures. In addition, the same personnel repeatedly conducting the same changeover might not perform consistently from one changeover to another. The business should take care to assess and integrate into its planning schedules 'repeatable' changeover performance times that take this variability into account.[47] The business should not reasonably expect the best ever recorded changeover time to be repeated on every occasion.

A particular problem can occur when management personnel or their representatives come on to the shopfloor armed with stopwatches and study sheets.[48] Under these circumstances the operators responsible for physically conducting the changeover are likely to perform atypically, particularly if encouraged to set a record. Short cuts may be taken that compromise safety. These same short cuts might also compromise subsequent product quality or line performance. If the management team are focused only on the set-up period, failing to appreciate that run-up constitutes part of the changeover (as we have witnessed), steps may be taken by the operators to reduce this part of the overall changeover – at the likely expense of an extended run-up period and subsequent line performance. What is witnessed (and recorded) can be significantly different from what occurs during unscrutinized activity. In these circumstances management can take away a highly distorted impression of what constitutes a changeover and how long it takes to complete. It is a false impression which may later be very hard to dislodge. On all occasions, changeover performance should only be recorded using well-founded work study practices.

6 Safety

Safety has been given very little overt attention in the discussion of changeover improvement in the literature.[49] Our own research has provided evidence that safety can sometimes be neglected to a greater or lesser extent in the drive to reduce changeover times. This can occur in one-off changeover situations when particular pressure is applied. Occasionally it may also apply systematically when, in seeking ever better performance, unsafe new procedures (either formal or informal) are adopted. Certainly we have been present in situations when teams of operators, especially when under pressure of peer groups or of production schedules, have worked in a less than totally safe manner – and certainly in a way that they would not normally expect (or be expected) to conduct their tasks.

Management has a responsibility to ensure any changeover procedure that is used is inherently safe, and that dangerous short-cuts are not permitted. Also, design may often be used to enhance safety[50] (which should, of course, always be to the fore as any design or other changes are sought). Ultimately, in instances where fully automated changeovers occur, there should be no direct operator exposure to safety risks.

7 An integrated business approach to use and sustain better changeovers

Chapter 2 addressed the topic of exploiting improved changeover performance. It is a point that is worth repeating: when it remains unused, any improved capability on the shopfloor serves no purpose. The role of marketing in this respect should not be underestimated.[51,52] The way that the planning function views and exploits an improved changeover capability similarly is crucial (particularly in modelling when changeovers are to occur and in applying pressure to sustain improved levels of performance).* In other words, an integrated business approach to rapid, high-quality changeovers is a necessity if the business is both to maintain pressure on its changeover capability and derive the maximum benefit from it. It may be that new internal systems need to be devised and implemented to enable this to happen.[53,54]

8 Changeover performance specification for new equipment

Where a new machine is sought – in preference to improving changeover performance of existing factory equipment – the changeover specification of this new equipment needs to be rigorously tested. Our experience is that changeover performance claims made by manufacturers often represent only the best case situations – where a full complement of skilled and experienced changeover operators are present and, particularly, where an optimum (easiest/quickest) type of changeover is being conducted. Sometimes the manufacturer's claims can be more optimistic still, being wholly unrepresentative and never being likely to be achieved with the equipment that is provided – unless subsequent retrospective design changes are made to modify it. Chapter 10 will present a case study where this occurred, where the company that bought a machine failed to test fully the manufacturer's changeover claims and hence, once the machine was in use, imposed upon itself very significant operational and financial difficulties.

Although in general terms flexibility, including greatly improved changeover performance, is becoming more and more an intrinsic feature of new equipment (as opposed to the organization which uses that equipment), it is wise to have any

---

* Operations research models that might be used to schedule when changeovers should take place can be extremely complex, and for many factory personnel will be difficult to understand. The accurate data that many models require may also be difficult to source, particularly as some of these data can be constantly changing.

equipment's changeover capability demonstrated by equipment suppliers as exhaustively as possible before a purchase is made.

9 Type of changeover: sequence dependency/independency

A changeover from product A to product B need not be the same as a changeover on the same line from product C to product B, nor indeed the same as a changeover from product B back to product A. If different tasks are involved, these all represent different types of changeover. By introducing some additional tasks, or by eliminating others different types of changeover will be of differing complexity.

A major reason for verifying manufacturer's claimed changeover performance (above) is that different types of changeovers on the same equipment will usually take different times to complete. In other words, changeovers are sequence-dependent.[55,56] Sequence dependency can occur in a high proportion of changeover situations.[57] It can allow, as will be investigated in the next section, opportunities to sequence changeovers to minimize overall changeover losses.

The times for different types of changeovers can be markedly different. Evidence of the extent to which the type of changeover that is conducted affects its duration is provided in Exhibit 3.8. Similarly, Figure 3.12 presents a further example of variable changeover time, once again probably largely resulting from the differing types of changeover that were known to be occurring. Both the intrinsic variability of different types of changeover and the possible extent of this variability are not always fully appreciated.

Exhibit 3.8 The impact of different types of changeover

On one production line a company manufactured products for which four major parameters could be varied. Although each of these parameters only varied between a few values, this alone gave rise to the possibility of manufacturing at least a hundred distinctly different products.

Variation of each of these parameters impacted significantly on the work that was conducted during a changeover. For example, a consistent size dimension from one scheduled product to the next eliminated a significant body of changeover work that would otherwise have applied (if this size was to change). This same observation applied to each of the other three parameters.

Selected changeovers were modelled in conjunction with line staff. It was determined that, provided they were conducted free of other problems, changeovers on this line should take between 11 minutes and 47 minutes to complete, dependent on the type of changeover that was undertaken.

Actual changeover performance for this line is shown in Figure 3.11. These times were considerably greater than the times determined above as significant problems that were yet to be overcome still affected changeover performance. For this reason the variation changeover performance shown in Figure 3.11 cannot be attributed directly to the different types of changeover that were taking place. Nevertheless, the variation that is inherent in changeover times on this line is clearly apparent.

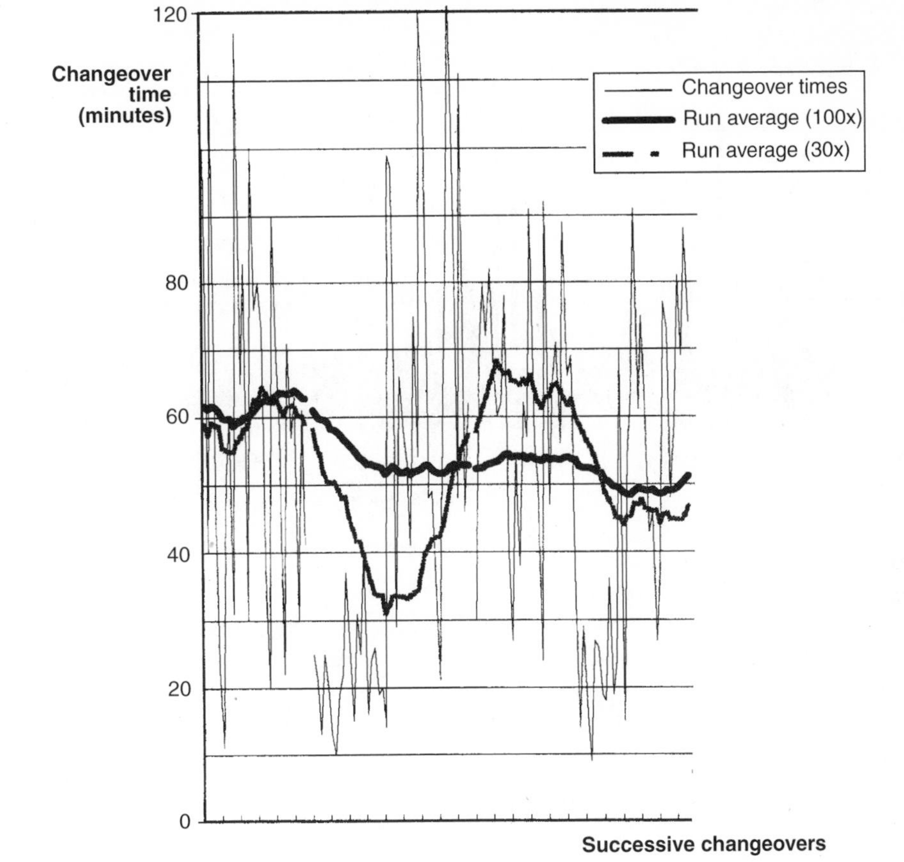

Figure 3.11 Variation in changeover time, which is strongly influenced by the type of changeover that occurs (approx. 100 changeovers shown). (*Note*: Run average 30x/100x describes the plot of the mean of the previous 30/100 changeover times)

Changeover type variability gives rise to two main problems. First, it is difficult to integrate variable changeover times into production schedules (including deciding what batch size to manufacture). Second, it can be difficult to define and use standard changeover procedures: in some cases procedures for each type of changeover can vary considerably and/or the number of changeover procedures can be too numerous realistically to expect that each and every one of them will always be fully documented and adhered to.

10 Scheduling considerations: constraining changeover frequency and constraining the type of changeover that occurs

A traditional 'head-in-the-sand' approach to reducing changeover losses has often been to minimize the frequency with which changeovers occur – rather than to seek

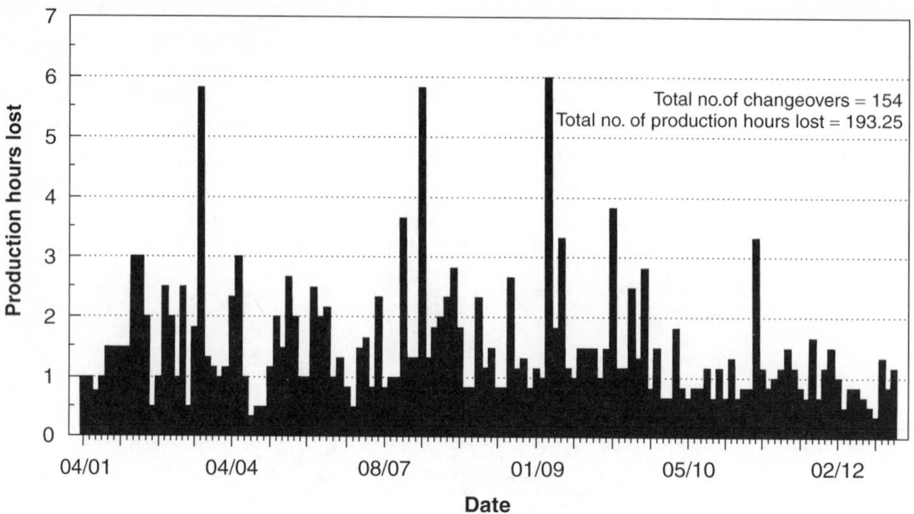

Figure 3.12 A further indication of variation in changeover time

changeover improvement. This is not to say that some restriction in the frequency at which changeovers take place should not be imposed because for the great majority of businesses, without near-instantaneous changeovers being possible, unit batch sizes will not be viable. Considerable research continues to be conducted on deciding when changeovers should take place.

A business that wants to increase the number of changeovers that it conducts is not, by definition, seeking to reduce losses by limiting when changeovers occur. Instead changeover losses need to be tackled by another means. To this point we have described that changeover performance may be improved by emphasizing organizational or design improvements. In either case faster, higher-quality changeovers should result. Another option is also available to the business, although this does not in itself improve individual changeovers. It is an option that allows for changeover losses to be reduced, but at the probable expense of limiting manufacturing flexibility to a greater or lesser extent. This option is to constrain the type of changeover that occurs.

This option has two elements. On the one hand, the business can constrain the type of changeover that occurs by limiting the range of products that it manufactures.* In addition the business can take care to schedule its changeovers in a sequence that minimizes their impact in terms of time losses on the shopfloor.

---

* Constraining the type of changeover that occurs is considered, perhaps informally, whenever any business decides to embark upon multi-product manufacture on limited production facilities. In any decision to limit the variety of products that it manufactures the business has already placed constraints upon the types of changeover that are to take place. Limiting the variety of products manufactured on specified equipment is formalized when a decision is taken to pursue cellular manufacture, in which families of products are manufactured in assigned cells.[58] Changeover time savings by adopting cellular manufacture can be considerable,[59,60] but are likely to be offset in part by the detrimental impact of reduced routing flexibility.[61]

Table 3.1 Characteristics of five products manufactured on a common production line

Product	*Length (mm)*	*Fascia*	*Heat shield*
A	500	Plain	Partial
B	500	Stencilled	Partial
C	650	Plain	Partial
D	650	Stencilled	Full size
E	720	Stencilled	Full size

Table 3.2 Possible changeover tasks and their duration for the products of Table 3.1

Changeover task	*Duration (min)*
Change product length (any)	9
Change from plain fascia (any) to stencilled fascia (any)	17
Change from stencilled fascia to stencilled fascia (different length)	23
Change from stencilled fascia (any) to plain fascia (any)	3
Change heat shield (partial to full size)	18
Change heat shield (full size to partial)	15

Examples may be used to illustrate these points. The first situation is straightforward. Consider a company that manufactures electrical transformers. This company might decide to concentrate one line's manufacturing capability exclusively on a limited range of 10VA transformers. If it had decided to allocate the same manufacturing line in addition to larger and more complex transformers, the types of changeover that this line would experience would almost certainly increase (because additional parameters associated with the manufacture of these alternative transformers would also come into consideration).

The next option of using scheduling to limit changeover time losses for the range of products that is manufactured likewise is a simple one to understand but, in all probability, a difficult one to optimize. Different changeovers, as we have explained, comprise different tasks. Careful scheduling from one product to the next can lead to some product manufacturing sequences requiring less changeover work to be expended overall. Consider the range of products A, B, C, D and E in Table 3.1. Each product is manufactured on the same line. The products have the characteristics as shown and, if appropriate, require the changeover tasks shown in Table 3.2 to occur.

Let us say, for the sake of this example, that successively one batch of each of these five different products is to be manufactured. This will require four changeovers to occur. Irrespective of the size of the batches or the rate at which production takes place, different changeover losses will be incurred dependent on the sequence in which these batches are manufactured. A–D–B–E–C and C–E–B–D–A are each worst-case sequences where 168 minutes in total would be lost to changeover. The best case sequence would be E–D–C–A–B, where just 76 minutes in total would be lost – comfortably less than 50 per cent of the worst-case time.

An appreciation of the impact that enlarging batch sizes (reducing changeover frequency) can have can be gained as well from this example. While reducing batch sizes can be highly desirable, for all the reasons that are explained within this book, larger batch sizes do lead to changeover time savings, as well as likely simplification of both logistics and scheduling.[62] Consider again the above example, with a requirement now to manufacture two identical batches of each of the five different products ($2 \times$ identical batches of product A, B, etc.). To minimize changeover losses it is clearly better to manufacture in the sequence E–E–D–D–C–C–A–A–B–B, rather than, say, A–D–B–E–C–A–D–B–E–C.[63] It needs to be resolved *how* small the batch sizes are to be. Bearing in mind also scheduling complexity and other factors, there needs to be a carefully determined batch size target, alongside determination of changeover improvement targets.

Any business that wishes to use scheduling to reduce changeover time losses is faced with the problems of recording the differences between all possible different types of changeover and then, using these data, optimizing production sequences. Neither might be particularly easy to do. Inherent difficulties in this approach also include accommodating 'emergency' schedule changes and continually having to update task data as the changeover is otherwise continually improved (assuming that an on-going improvement programme is also taking place).

Despite these potential difficulties, limiting the type of changeover that occurs can be important in many different industries, for example in the print industry.[64,65] The two examples presented in Exhibits 3.9 and 3.10, respectively, show how both controlling the job sequence and limiting the range of products that can be manufactured might be employed to apparent advantage. While in each case the print line that is described experienced reduced changeover time losses, there was also a trade-off in terms of slightly reduced flexibility. It was a price, evidently, that the company believed was worth paying.

Exhibit 3.9 Product scheduling to limit the impact of the type of changeover that occurs

A six-colour print line was operated as a small-batch, rapid changeover facility. The line typically would employ two 'solid' colours that varied from one job to the next, alongside four more consistent 'process colours' – cyan, yellow, magenta and black – to reproduce photograph-like images. Not every one of the six print stations was used for every job that was printed. Different changeovers comprised markedly different tasks dependent on how many print stations were put into use and what colour changes took place (for example, dark blue to yellow in contrast to white to crimson). The sheets that were being printed also varied in their size, for which further differing changeover tasks were involved.

The average changeover time for the line was approximately 72 minutes. This time applied to job sequencing that paid little or no attention to these issues. Indeed, the line was intended to be operated initially with a very high degree of flexibility, and placing constraints on what job was to follow what previous job was believed to compromise the quick response time that the line offered to its customers.

The impact of sequencing on the changeover times for this line was discovered almost by accident. Unusually, a run of print jobs needed to be completed for one customer that were each of an identical size and used identical colours (both the colours that were employed and the number of print stations that were used). Each batch was particularly small and the sense of scheduling these jobs to follow one another was identified. The average changeover time between these particular print jobs fell to just 19 minutes.

Exhibit 3.10 Limiting the range of products to limit the impact of the type of changeover that occurs

The same print company as in Exhibit 3.9 identified an opportunity to reduce its changeover losses by employing a 'hexachrome' ink system.[66,67] 'Hexachrome' uses six colours which combine to simulate solid colours and allows images to be printed simply by adding a special green and orange to the four 'process' colours. The effect is to reduce or eliminate the need to blend solid colours. This in turn can reduce or eliminate ink changes. Were 'hexachrome' inks to be used exclusively it is likely that there would be a trade-off between faster changeovers and the quality of the finished job, which would be slightly below that achieved with existing inks – thereby compromising the company's ability to accept and process work from its most discerning customers.

The topic of scheduling should always include consideration of how to deal with 'emergency' production requirements. In a truly flexible manufacturing environment the disruption caused by an unexpected priority batch will be negligible. Most manufacturing environments are not able to operate in this way. If a major customer, for example, demands immediate attention, what should the company's response be? We have conducted research in a European factory where 'emergency' situations were so frequent (and the scheduling thus significantly disrupted) that a high proportion of all changeovers had to be conducted twice – once to set the job up in the first place and then once again to complete the job after it had been interrupted. Some batches that were being affected in this way originally required a changeover of approximately 2 hours' duration to be followed by just 90 minutes of production. Interruption of these batches meant that the changeover had to be repeated, now taking 4 hours in total. This meant that the ratio of production time to changeover time could fall as low as 25:75 per cent – in other words, that three times as much time was allocated to changeover as to production.[68] In effect the company concerned was striving to provide a level of responsiveness that was beyond what it could deliver. In attempting to accommodate fully the demands of different customers the company was incurring significant difficulties and, in all probability, at the same time compromising its service to its other customers. Moreover, scheduling control was so poor within the company that responsibility for changing schedules was frequently assumed by different personnel – from the CEO through to the night shift foreman. These problems could have been alleviated by imposing better scheduling discipline and by not being as accommodating – as flexible – to all conflicting demands placed on the business.

In general terms the topic of changeover and scheduling remains a complex one.[69] Notwithstanding the considerable literature that assesses changeover as a component of the overall scheduling problem, better changeover performance might be argued to make scheduling either more or less difficult to conduct. In itself, if approximately the same changeover frequency is maintained, faster changeover times should allow for easier scheduling.[70] Conversely, for the situation of greater changeover frequency, smaller lot sizes and frequent production runs may increase the scheduling problem.[71] In addition, large changeover time variability, sequence dependency or, under certain circumstances, changeovers that represent a high percentage of overall job production time, might also add to scheduling complexity.[72–74] Finally, scheduling will become more difficult still if the full implications of the run-up period, as part of the changeover, are considered.

11 Dominant use of the 'SMED' methodology

Galbraith,[75] arguing that monetarism is flawed because it places absolute faith in a classical free market, asserts 'If one loves something enough, or is sufficiently so conditioned by education, one can attribute to it virtues that others cannot see or which do not exist'. The same danger exists beyond economics. For changeover improvement there is a considerable thrust to improve changeover performance by use of Shingo's 'SMED' methodology.[76] This needs to be put into perspective. The 'SMED' methodology is a possible tool to improve changeover performance – no more.[77–79] Although it is widely understood and widely advocated, its prominence should not imply that it does not have limitations.[80,81] It remains an approach to changeover improvement that should be considered alongside possible alternative approaches.[82–85]

The 'SMED' methodology benefits from being very simple. Its primary focus – both as apparently intended by its originator[86] and as interpreted by those who apply it in industry[87–89] – is to separate (move) and convert tasks to external time. Similar emphasis on achieving improvement by distinguishing between internal and external time is advocated by academic writers.[90–93]

While the 'SMED' methodology can no doubt yield significant results[94,95] it is wrong, in our opinion, to bestow upon it a status where 'SMED' and 'changeover improvement' are perceived as one and the same thing. In Chapter 7, once the potential of applying design changes has been investigated in much greater detail, we will discuss certain limitations of Shingo's methodology. Prominent among these limitations is the way that it does not readily allow solutions to be identified that reduce the duration of existing changeover tasks or eliminate them altogether. These and other limitations have prompted us to develop and test an alternative improvement methodology – one that allows a changeover practitioner to evaluate a full range of possible improvement opportunities on merit, at all stages of an improvement initiative. This approach will be described in detail in Chapter 7.

12 The role of changeover improvement consultants

Businesses frequently choose to employ consultants to assist in alleviating changeover problems, particularly in training internal personnel how to go about achieving improvement. The business may select which training package/consultant it wishes to use from the considerable number that are available.[96]

A primary business requirement will almost invariably be the reduction of changeover times.* The degree of apparent improvement will be a major criterion on which the consultant's performance will be judged. A consultant will be aware, however, that there can be subtle ways of enhancing customer approval further still, without apparently compromising this objective. Specifically, higher approval is likely if the consultant's work matches the customer's expectations of how improvement is to be achieved. This and a number of other issues are discussed now in relation to changeover consultant activity.

Dominant use of the 'SMED' methodology

The dominant use of the 'SMED' methodology has been highlighted previously. Its prominence is such that its use is invariably expected by the customer.[97] For this reason alone consultants are likely to continue using it, or at least demonstrate that its key concepts of separating and converting tasks to external time are being employed.[98] In this way the methodology's use becomes almost self-fulfilling. The 'SMED' methodology may or may not be the best tool to use in terms of its effectiveness, but its use, because of its prominence, is still likely to be anticipated in advance by the customer. No matter that specific improvement targets may not have been agreed when the consultant was engaged nor that, once the consultant has departed, there may be difficulties in sustaining the improvement that has been achieved.

That use of the 'SMED' methodology alone might not always represent the best route to changeover improvement is demonstrated by the number of experienced consultants who have choosen to augment it in their own programmes in industry. Particular issues that are given added attention include adjustment, communication, line layout, clamping techniques and mechanization.[99–105] Similarly, despite some authors having previously argued that improvement should be a low-cost, organization-based exercise, these same authors sometimes acknowledge that additionally there can be a place for system improvement by design. Design might be used, for example, to eliminate existing tasks, or to simplify existing tasks. Attention can be required to issues of clamping, adjustment and mechanization (among others) more prominently than the 'SMED' methodology encourages.[106–108]

The 'SMED' methodology does not set out to address numerous wider issues pertaining to an overall improvement programme, such as machine or process identification for changeover, conducting a changeover analysis, improvement team selection and deciding improvement team responsibilities. Experienced, professional consultants will typically be able to provide significant assistance in these and other areas.

The culture of organizational improvement

Much of our book describes the potential role of design to improve changeover performance. This is a topic that, to date, has received relatively little overt attention in the academic literature or in trade publications. The current pervasive culture of participative, incremental improvement in industry[109–114] might be used partly to explain

---

* Alternatively, occasionally, the business may seek primarily to improve the quality of the changeover or perhaps use a changeover improvement programme as a means to change the workplace culture.

the reason why, if design truly influences changeover performance to a significant extent, design solutions have not previously been more vigorously promoted. What we have termed previously as a 'big-bang' solution (deploying and imposing externally generated design or organizational changes) would seem to be at odds with such an approach. Further, substantive internally generated design solutions (not necessarily by shopfloor personnel) are also apparently incompatible with it.

In this wider context any decision to seek improvement only on the basis of continuous, incremental, internal, team-inspired activity represents a restriction of how improvement may be undertaken. In many instances a *kaizen* approach[115] may indeed represent the best way forward, but this need not always be so, where it can be argued that it is not always appropriate.[116–119] A business needs to understand that alternative improvement approaches may also be adopted.[120–123]

Our research has shown that changeover consultants (with the notable exception of those whose focus is directly on design solutions) strongly favour employing a team-based continuous improvement approach.[124] A similar stance is frequently adopted by academic writers.[125] In respect of *kaizen* improvement, it should be noted that the 'SMED' methodology can easily be interpreted and emphasized by consultants and others as a tool that is ideally suited to this approach.

Low-cost improvement

It may be speculated that in terms of 'selling' their programmes to industry, many changeover consultants are aided by pointing out that theirs is a low-cost/high-reward package.[126–128] The low-cost aspect of consultant-led programmes could not be claimed so readily if substantive design changes,[129] substantive equipment duplication[130] or new equipment purchase[131] were advocated.

Sometimes a company can be preoccupied with cost to the extent that it will actually dictate that only low-cost improvement should occur[132] even though, beyond a company's own internal policies, this may not necessarily be considered as wise.[133] Any significant pressure on cost will inevitably tend to exclude seeking design changes (for which some expenditure is almost always required). Through their own positions on cost, therefore, both industrial changeover consultants and the companies that they serve might each separately be actively diminishing the role of design, effectively excluding substantive design improvements (or other high-cost improvement options) from taking place.

Design consultants

Notwithstanding the previous comments, what can best be described as design consultants for changeover do exist. These consultants seek to improve changeover performance not by instigating better procedures but, instead, by altering existing equipment.[134] In addition, numerous devices are available on the market which, to a greater or lesser extent, can be employed to allow faster/better quality changeovers to occur. In some cases devices are designed specifically with this purpose in mind.[135–137]

Such a design-led approach to changeover improvement is apparently at odds with the approach described elsewhere in this section. Not once in our experience (although we are not saying that it never happens) have we been aware of both design consultants and 'organizational' changeover consultants being employed simultaneously. In practice these two apparently conflicting approaches are far better viewed instead as different positions as part of a unified approach. By doing so business managers and the individual shopfloor practitioner (at strategic and implementation levels respectively) will both be in a better position to evaluate all possible solutions on merit, rather than being conditioned to favour any one particular improvement route.

13 Communication

There are different facets of communication as part of a changeover initiative. At one level training is largely to do with communication, as are tools such as the 'SMED' methodology. Both aim to communicate to others how best to seek improvement. The better the training and the better the improvement tools that are employed, the better the results that are likely to ensue.

Disclosure between personnel of what is occurring is a further aspect of communication. It should be encouraged.[138] This communication might be in terms of: who is involved; what ideas have been put forward and used; how improvement is being sought; what production lines/machines are being selected for the initiative; why these lines are being selected; what benefit both the practitioner and the business stand to gain; what improvements are pending. The use of prominent progress display boards, for example, will allow what is taking place to be communicated, encourage acceptance of this work, and quite possibly elicit new improvement ideas from different sources. Demonstration of support by senior management is also to be encouraged.[139] Positive communication is important in other ways. Agreement and communication of targets can be used as a mechanism to maintain momentum. The generation of improvement ideas similarly will be assisted by good communication of work that others have done elsewhere in the past (see Exhibit 3.2).

The business should also be aware that in other respects, when specific information is being handled, excessive or poorly conducted communication can inhibit how a changeover occurs.[140,141] Communication can be error-prone or imprecise and, as such, can cause delay to a changeover.[142] Changeover can be greatly assisted if as few as possible communication steps occur and if information is conveyed in a clear, accessible and unambiguous manner.[143] A case study example of how greatly improved communication can assist a changeover will be presented in Chapter 10. The handling of specific information for changeover purposes is often related to aspects of changeover logistics (ensuring the right components for a changeover are in the right place at the right time). Communication in respect of operating procedures and issues of adjustment are also likely to be important.

Components that need replacing due to wear or damage can hinder changeover performance. Maintenance issues may be identified. Communication avenues also need to be in place to ensure that such defective components are attended to.[144]

14 Financial benefit analysis (introduction)

JIT, 'agile' manufacture, responsiveness, small batch sizes and flexibility are all strongly promoted expressions of modern manufacturing practice. A business, though, should not seek to improve changeover performance (which can contribute to achieving these objectives) simply because it is perceived to be 'a good thing'. Instead, any business should investigate what exactly better changeover performance can contribute to its overall competitiveness. It can only do this non-judgementally if it financially quantifies the benefit that better changeover performance will bring. This can be a complex task if every facet of potential benefit is to be investigated, and this topic of conducting a financial benefit analysis will be investigated in detail in the next chapter. Not least of the reasons for conducting a comprehensive benefit analysis is that it allows a business to assess how much it can spend to achieve the improvement that it seeks. Substantive design changes or new equipment purchase may not necessarily feature in a changeover improvement programme, but once sound financial data are available the business will be in a position to sanction expenditure for these purposes if they are considered appropriate.

15 Changeover item quality

Another important issue that business management should be aware of is what we term changeover item quality. It is highly desirable that all items supplied and used during a changeover should be of an appropriate quality[145] (fit for the purpose for which they are to be used). If this is not done changeover performance can be severely affected.

Figure 3.13 presents four categories where quality problems that can affect changeover performance may arise. Each category is illustrated by means of an exhibit. Item quality applies both to consumables (such as printing inks) and to assemblies that are exchanged (such as die sets). Neither are permanent features of the machine. Item quality can also be an issue for machine components that remain in place from one changeover to the next. Equally, the part-completed incoming product may need to be within specification if the changeover is to proceed unhindered.

Exhibits 3.11 to 3.14 illustrate that poor item quality may either add delay to a changeover or lead to a changeover being abandoned completely. The potential impact of components that are not apparently directly associated with the changeover should not be underestimated. In particular, wear of machine components can be a problem, where worn parts may affect the degree of adjustment that has to take place although, as demonstrated in Exhibit 3.11, this is not the only way that a changeover might be hindered. The desirability of having an 'in specification' machine represents one way in which maintenance programmes and changeover might impinge upon one another.

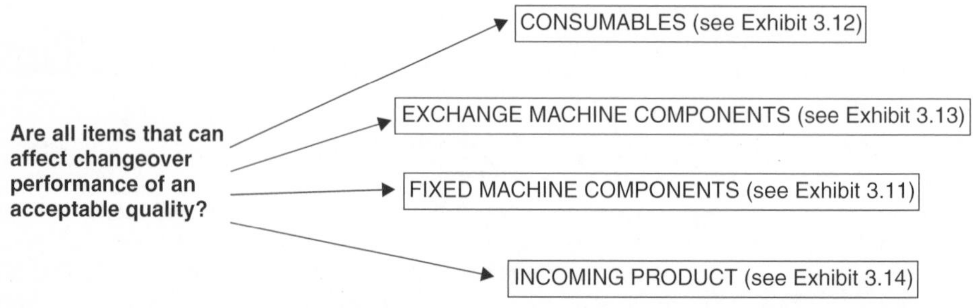

Figure 3.13 Categories in which items of poor quality might affect changeover performance

Exhibit 3.11 Poor changeover by virtue of poorly maintained equipment

A print line was operated as a small-batch, rapid changeover facility. The line was claimed by its manufacturers to have been designed with an excellent changeover capability. The company operating the equipment had incorporated the manufacturer's claimed performance into its scheduling programmes. This claimed changeover performance had been a major consideration in the equipment's original purchase.

Many of the tasks associated with changeover on this machine were fully automated. One such task was roller washing. The machine was supplied whereby the operator could select cleaning programmes of differing duration (on the basis of the from–to colour changeover that was being made on successive print stations).

There was a problem that this automatic cycle was not cleaning the rollers as it should. Thus the longest wash cycle was always being selected (4½ minutes compared to 1½ minutes for the shortest cycle). More than this, it was almost always found necessary to repeat this longest cycle. This was a further 4½ minutes lost per changeover, plus additional time for a task that should not have been required: checking that the rollers were sufficiently clean. It took time for the company to understand the causes of this problem. Eventually two factors were identified. First, while the wiper blades associated with each roller (which are fixed, non-changeover components) were correctly mounted, they were found not to be in optimum condition. When original specification replacements were used there was an immediate improvement, but the longest cycle was still usually being selected and, sometimes, having to be repeated. Different blades to an alternative design were then fitted, which were found to be more durable, but these too did not solve the problem. Eventually it was found that the cleansing fluid used in the wash cycles was not to specification and therefore, understandably, was not working as it should. Unfortunately there were health and safety questions regarding the manufacturer's recommended fluid – this being the reason why it was not being used in the first place. The whole episode showed once again that considerable vigilance is needed in validating manufacturer's changeover performance claims. Similarly, given the difficulty in isolating this problem, this change to machine specification (the cleaning fluid) had not been adequately communicated.

Exhibit 3.12 Poor changeover by virtue of poor quality consumable items

A five-station UV-cure print machine in a German factory was being operated specifically as a rapid changeover facility (as are many of the machines that are highlighted in this book). Its manufacturers specified that it could be changed over in 2 hours or less. In the majority of cases the actual changeover time significantly exceeded 2 hours.

In many print equipment installations a notable aspect of a changeover is the need to remove all the unused old ink, clean the surfaces with which it had been in contact, place new inks into the ink ducts and finally distribute the ink over the application surfaces. This particular print machine was appreciably different in

Exhibit 3.12 *(Continued)*

that silk-screen printing occurred – hence eliminating the need for most of these operations (wiper blades and the silk screens themselves were almost always replaced with new components during each changeover). Like most other print installations, though, there was a requirement to match the colour of the ink that was actually being applied to that of a 'proof' colour master. No parallel off-line facility was available to do this on-site and, although considerable effort was made to try to ensure that ink formulations were consistent from one batch of ink to the next, exact colour matching still remained a problem. In particular, any colour mismatch only became apparent during the changeover once it was nearing completion. Sometimes, when colour mismatching occurred, attempts were made to change the colour formulation (to bring it in line with that demanded by the 'proof') on a trial-and-error basis. This inevitably took quite a long time to do. Often, however, this failed to achieve the desired results and it then became necessary to resort to remixing the ink from scratch. When this occurred, approximately 4 hours was added to the changeover.

Exhibit 3.13 Poor changeover by virtue of poor quality exchange items

A lithographic print line was operated with a policy of manufacturing new printing plates for every new job that was commenced (even when repeat orders were made the old plates were never re-used).* High levels of precision are required for successive colour image registration (from print station to print station) to ensure that the final multi-colour image is of an acceptable quality.[146] This precision has a number of components. It is important first that the successive plates must all be installed square to one another. The machine is designed to move square plates up and down and from side to side in relation to one another to get each colour of the final multi-colour image to overlay correctly. What the machine cannot compensate for is plates that incur an angular misalignment, or plates whose images are distorted relative to other plate images in the set (for example, images that are skewed). Angular misalignment might occur in the installation of the plates into their holders or might be a function of the way that the plates are manufactured.

Many of the changeovers that were recorded for this machine showed considerable time was often required to get the images from each of the print stations to line up. Often this led to a 'best-fit' compromise for the final multi-colour image, that was below the quality that would normally be expected. Often in these circumstances a factory supervisor had to be called over to give approval for the best final image that the printers were able to achieve. Sometimes, if the supervisor was engaged in other tasks, this could add up to 30 minutes to the already extended changeover. On many occasions, either on the supervisor's instruction or on the printer's own initiative, it was decided that the image could not be adjusted to an acceptable quality. The whole job was then delayed while new plates were made. Sometimes the job was broken down from the press, to be reinstated later. Either event added at least 2 hours to the changeover – on top of delays that the original poor quality plates had already caused.

---

* The situation of using new plates for each new print job is not universal. One company in particular that we have researched was experiencing considerable difficulty during changeover when re-using plates that had become damaged. It was sometimes found – at the time of the changeover – that plates had deteriorated beyond use.

This same company was further encumbered in its changeover operations by the fact that some of its print customers insisted on first-off approval (in place of ceding authority for proof matching). In such circumstances the changeover time was extended by the time that the customer took to arrive on-site and give approval for the job to commence. This was highly costly in terms of lost production. Also, significant problems were imposed on the company's scheduling operation.

Exhibit 3.14 Poor changeover by virtue of poor incoming product

Metal food/drink can-end manufacturers have long used primary and secondary scroll operations as described in Figure 3.14 to enable circular blanks to be manufactured using as high a percentage as possible of coil stock (to approach as closely as possible the 90.69 per cent utilization that is the maximum possible attainable from an unbounded (infinite) sheet).* Different-sized blanks are required for different products and so, typically, both primary and secondary scroll shear lines require changeover operations to occur.

The changeover time for the secondary shear operation is dependent in part on the dimensional tolerances that are achieved for the incoming primary scrolled sheet. Incomplete final round blanks – blanks with their edge removed, even by the slightest amount – are totally unacceptable because when used to manufacture can-ends these would lead to insecure food packaging with significant health risk implications. If the coil cut length (a function of the primary scroll operation) or the coil width (as supplied by the coil manufacturer) deviate from their specified values the adjustment that is required to set the secondary scroll cut edge becomes either more or less critical – and can take more or less time to complete.

---

* Primary and secondary scrolling are relatively complex additional operations in the manufacturing process for food can-ends. These operations are justified by raising material utilization in comparison to that which would be achieved were square cut sheet to be used (which would yield 78.54 per cent utilization for a perfect, edge-to-edge, shearing operation). Even with primary and secondary scroll operations, partly because of the material overlap that is needed, typical material utilization still falls appreciably short of the 90.69 per cent that is mathematically possible.

Exhibit 3.14 (*Continued*)

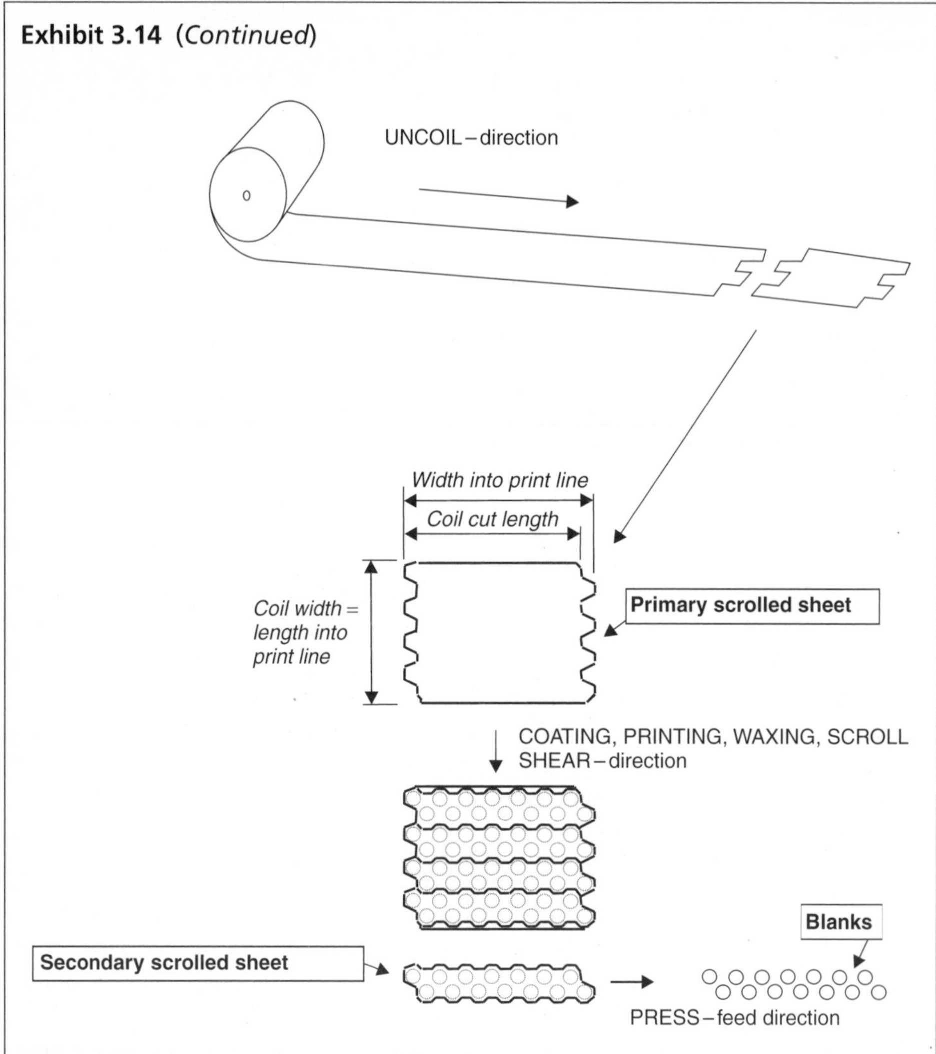

Figure 3.14 Schematic of primary scroll and secondary scroll operations to manufacture circular metal blanks with high raw material utilization

As well as Exhibit 3.13, Exhibit 2.2 earlier in this book similarly provided an example of an exchange machine component (a die set) that equally should have been fully prepared in readiness for the coming production run. Instead of this, die sets in this poorly organized 'flexible' press shop were refurbished during the changeover (thereby contributing to its duration). If the die sets had been of an adequate quality – in an 'in specification' state of repair – when they were selected for use, the changeover would have been significantly quicker. One way that die sets are likely to fall out of specification is because of wear, but wear is not the only reason that the desired specification might not be achieved. Exhibit 3.14 describes the need for metal sheet sizes to be within specification. This time, for brand-new items, the quality problems that are described are because of manufacturing problems. A third potential problem is damage, which can sometimes also lead to significant difficulty in establishing machine settings during a changeover.

As these exhibits indicate, a high proportion of changeover problems can be traced back to problems of item quality. Procedures may be implemented during a changeover to overcome these problems, but this is usually a second-best option to ensuring that all applicable items are within specification in the first place. In one situation, for example, we observed a procedure that called for two people to be present to push against a large die set to man-handle it out of a press on an air bearing (a carpet of compressed air that lifted the die set to reduce friction). Originally, when the equipment was new, this operation could be conducted by just one operator. Over time the bearing surfaces had become scored. Bringing the equipment back to 'as-new' condition would have eliminated the need for this two-operator procedure, instead allowing a simpler procedure to be used (getting two people to work together is inherently more complex).

As indicated there are distinct categories into which problems with item quality may arise. By classifying the type of item quality problem that is present, the business is likely to be better able to identify a solution for it. Figure 3.15 presents an overview of possible problems and some generic approaches that may be applicable to overcome them (and hence improve changeover performance).

Figure 3.15 describes item quality problems in terms of items being out of specification. While ever tighter item specification may well mean that faster, higher-quality

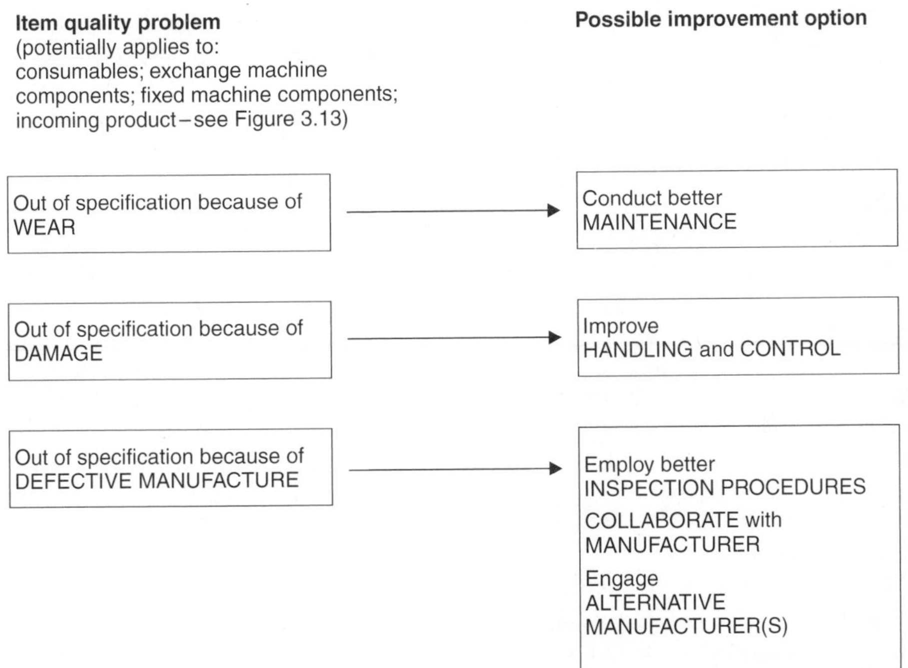

Figure 3.15 Categorizing item quality problems and possible options to overcome them

changeovers result, better specification typically will have to be paid for. If it wants to pursue this option, the business needs to evaluate at what point tightening specification of changeover items no longer becomes cost effective overall.

The problems that can arise from the poor quality of changeover items should not be underestimated. In one study the authors ranked issues affecting changeover performance to put the impact of poor item quality into perspective. Poor item quality was found to rank equally in its impact on line changeover in this study alongside poor logistics (delivering items required for changeover on time, in the right location, in the right quantity and correctly identified) and poor operator performance (conducting changeover tasks in a haphazard, inefficient manner, including not separating internal and external tasks).

16 The ability of shopfloor teams to bring about change

A significant feature of poor-quality items is that a need for improved quality might not be immediately apparent and, even if this need is identified, it may only be personnel away from the shopfloor who can cause improvement to come about.[147]

A potentially limited capability of shopfloor teams to actually bring about change does not end with issues arising from poor-quality items. For the significant logistics problems that were noted in the study cited immediately above, for example, the team themselves were unable (not least for reasons of experience, influence and time) to bring about what others considered the only realistic long-term solution – implementing an entirely new, wide-ranging *kanban* system. Other examples of a limited ability of shopfloor teams to bring about change, particularly in respect of design change, are noted elsewhere in this book.

The problem of progressing changeover improvement thus should not be assessed only in terms of team motivation, aptitude and influence. Or, in other words, a team's results should not necessarily be attributed simply to a lacklustre performance from a group of individuals (remembering too that the decision of who is in the team is largely a management responsibility): other factors are also likely to contribute to a team's ability to 'get things done'. Notable among these other factors will be the changeover improvement tools that are employed. The 'SMED' methodology has been used in every changeover initiative that has been described in this chapter. This methodology could (and did) direct the shopfloor team in the previously described study to suggest improvements to the logistics problems, within the bounds of the existing system that was in use. In this case, though, as more generally with physical changes to manufacturing equipment, the 'SMED' methodology did not show itself as a tool to assist substantive redesigning of the existing system. This should not be an unexpected conclusion: it is not intrinsically what the methodology sets out to do.

In terms of typical narrow changeover team selection and of typical training and use of the 'SMED' methodology, a company's internal programme is not always strongly placed to address all potential improvement opportunities. Hence the programme's results might fall significantly below what could otherwise be possible.

PART 2 Undertaking improvement

3.3 An overall methodology for changeover improvement

The overall methodology for changeover improvement sets out to give structure to the global improvement task.

Figure 3.16 clearly shows the three-phase approach of the overall methodology. The initial strategic phase should be undertaken irrespective of the subsequent route (see Figure 3.3) that is chosen for the improvement initiative: retrospective improvement

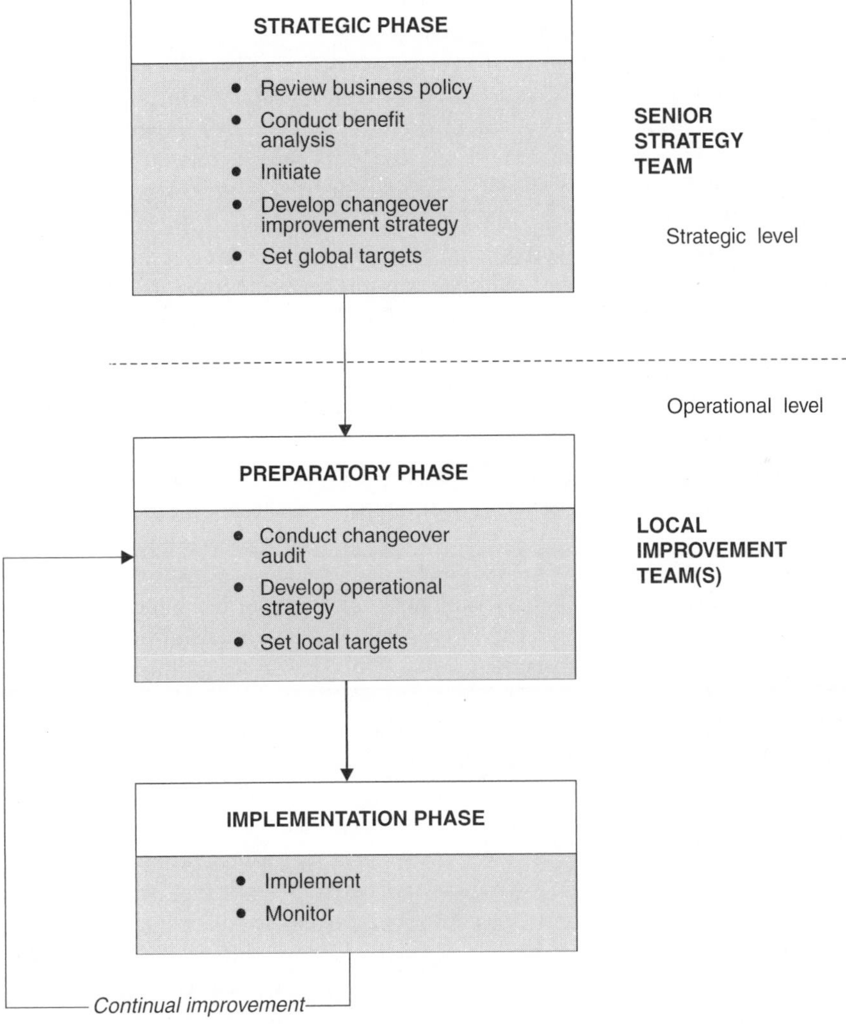

Figure 3.16 The overall methodology for changeover improvement

of existing equipment; employing dedicated parallel equipment; buying in and imposing externally generated solutions; purchasing more changeover-capable replacement equipment. The overall methodology applies in its entirety only to the retrospective improvement route.

The overall methodology is described in this section in summary only. Text elsewhere in this book (notably in this chapter, Chapter 4 and Chapter 7) provides more detail of how to use the methodology. Some improvement tools are also described in detail later in this book that are applicable to individual steps of the methodology.

The strategic phase of the overall methodology

The activities of the strategic phase are primarily undertaken by senior business managers. A facilitator might also be involved in the final steps of 'develop changeover improvement strategy' and 'set global targets'.

A **review of business policy** is the first step, where it will be assessed how enhanced changeover performance might assist the business to accomplish its strategic objectives. The business should ask itself where and how it wishes to compete. It will need to appraise its performance in the market relative to its competitors and decide on what basis it wishes to develop its business. In doing this it should seek to understand the potential contribution of better changeover performance towards, for example, achieving a JIT manufacturing environment, or otherwise to assess the role of changeover to enhance manufacturing flexibility and responsiveness to customer needs.

Next, the business should quantify the benefit that better changeover performance will bring, by conducting a thorough **benefit analysis**, perhaps using the procedure described in Chapter 4.

The information generated in the financial benefit analysis will be used in the decision of whether or not it is viable to proceed with a changeover improvement initiative. Once the decision to go ahead has been made, it is desirable formally to **initiate** a changeover improvement programme, which is highlighted as a separate step to allow the business to demonstrate that it is an issue that is being taken seriously. A high-profile launch is a statement of intent and as such a powerful way to convey management commitment. At this point additional personnel might also be introduced to help in undertaking the final two steps of the strategic phase.

A **changeover improvement strategy** should be developed without undue delay. In doing this senior management will need to assess all possible routes to achieving rapid, high quality changeovers. Is, for example, retrospective improvement of what now occurs the best way to proceed? Alternatively, should a combination of approaches be pursued (see Figure 3.3)? If so, how is the overall improvement effort to be managed (and by whom)? The business will need a full grasp of the diverse factors that typically influence all these decisions. Or, expressed in another way, the business must take full account of its own particular situation. Attention should be given to the issues of sustaining improved performance and integrating the programme across a wide range of business functions. Thought should be given to who should be involved, including personnel beyond the immediate shopfloor environment, and how they should work together. Lines/processes/machines on which improvement is to occur need to be identified. Possible training programmes will also need careful consideration, including who should participate (both the trainers and the trainees) and what format the training programme is to take (deciding also what improvement tools to incorporate as part of the training programme). Training can serve the function, additionally, of reinforcing top management commitment to a changeover initiative.[148]

Based on these decisions (and also making use of data from the financial benefit analysis), the senior managers will then **set global targets** to define what is required of the improvement programme. Targets will include: the time reduction that is sought; the resources to be allocated to reach this level of reduction; the time scale in which this level of reduction is to be reached.

The preparatory phase of the overall methodology

As stated, the preparatory and implementation phases only apply to retrospective changeover improvement activity on existing manufacturing installations. Preparatory phase activities are undertaken principally by personnel in a shopfloor improvement team. Design or manufacturing engineering staff may also be involved. Senior or middle managers play no active role, but should be available and prepared to afford any support requested of them. They should also demonstrate an on-going commitment to the improvement programme.

It will be necessary to record how changeovers currently occur, and to do this the improvement team will need to **conduct a changeover**. This will include an assessment of: what changeover tasks are conducted; what times are taken for these tasks; in what sequence the tasks occur; what skills are needed to conduct these tasks; what interdependencies apply; what problems can occur; what hand tools or devices are used; which personnel are involved; where adjustments are made and why; how change parts are handled; what product standards apply. Use can be made of the tools that we have developed to assist with this work. A comprehensive audit might also include an understanding of operator aptitudes, to assess who is best able to conduct which changeover tasks, and whether specialist skills are required.*

Having completed the audit, the team should then **develop an operational strategy**. Aspects of the changeover that may be improved need to be identified. A number of competing improvement options should then be set out. The operational strategy seeks to consider all potential improvement options, and select, based on a number of criteria, which options to proceed with. Tools are again described later in the book (Chapter 7) that can assist in conducting this work.

Once it is known how improvement is going to be undertaken (during the current iteration) it will be necessary to **set local targets**. These are used formally to set out who is doing what, and when completion is expected.

---

* For example, in one German factory we observed a specific task that required considerable physical strength on the part of the changeover operative. With this information the improvement team could then decide whether to continue to assign stronger operatives to this task in the future, or whether the requirement for physical strength could be alleviated by changing the tasks, tools or procedures that were employed.

The implementation phase of the overall methodology

The implementation phase has been separated deliberately from the preparatory phase to highlight who is doing what (and why), on what aspect of the changeover, before improvements are commenced. This phase comprises two self-explanatory steps. The first of these is to **implement** the improvement decided from the preparatory phase; by changing working practice or by changing, by design, existing equipment. To ensure that the initiative is maintaining satisfactory progress, the successful completion of each improvement should be **monitored**. This will almost certainly involve the use of some documentation. For design improvement, documentation typically should include the updating or preparation of engineering drawings.

Continuous improvement

When considered in its full complement of three phases, the overall methodology advocates a continuous improvement (CI) cycle, as indicated by Figure 3.16.[149,150] Upon completion of tasks within the current iteration the improvement team would look to update the 'changeover audit'.

We envisage that continuous improvement should not be deliberately limited by cost (by an upper cost ceiling) or by any other specific constraint – in other words, if a particular solution is warranted and can be justified then it should be permitted.[151] This rationale therefore does not exclude design modifications at any stage in the overall improvement process. What solution is adopted is likely to vary as the initiative proceeds (as successive improvement iterations are completed). In the earlier stages it is probable that small 'no-cost' procedural changes will be made. It may be only at later stages in the initiative that design comes more to the fore (although not necessarily so). Cost will be a major consideration in selecting what solution to adopt from possible options that are identified during each iteration, but cost alone should not determine how to proceed (see Chapter 7). CI, conceptually, should continue until the desired improvement in changeover performance has been achieved. The business should be aware, however, that it can be very difficult to predict a CI initiative's progress. As is noted elsewhere in this chapter (particularly in the References section), undertaking team-based CI need not be straightforward.[152,153]

Also, because changeover times can be difficult to sustain, a business will need to ensure that focus on changeover performance does not end once the desired level of improvement has been achieved.

The overall methodology and the 'SMED' methodology

It should be noted that there is no correlation of stages 0 to 3 of the 'SMED' methodology with individual steps of the overall methodology.

3.4 Conclusions

This chapter has investigated many diverse issues, each of which can inhibit what a changeover improvement initiative is able to achieve. By subscribing to an overall methodology in which these and other issues may in turn be addressed, a business is more likely to achieve significant and lasting improvement. Nevertheless, adoption of the overall methodology in itself will not ensure that an initiative will enjoy success.

The chapter has deliberately concentrated upon issues that are of particular relevance to senior managers. For the most part these issues need to be at least considered before the initiative is passed to the shopfloor, and retrospective improvement proper commences. The chapter argues that a higher degree of 'front-end loading' of the initiative should occur than is now common.

A comprehensive initiative will encompass more than a shopfloor improvement team working alone. There should be recognition that team members are probably limited in what they will be able to contribute (and this limit might be greater than is anticipated). Just as it is unreasonable to expect unsupported team activity always to achieve significant and lasting improvement, so too it is not sufficient for the blame for existing poor changeovers to be laid exclusively at the feet of those who currently conduct them. There is much more to changeover performance than merely how well the operators conduct their tasks.

Changeover consultants are quite likely to feature in any serious initiative. Their role needs to be carefully considered, as does the training (including the tools that the training is based on) that they may conduct. In particular, care needs to be taken that no artificial restriction is imposed on where and how improvement is sought, or upon the way an initiative is to progress.

References

1 Hay, E. J. (1987). Any machine setup time can be reduced by 75%. *Industrial Engineering*, **19**, No. 8, 62–67.

2 Johansen, P. and McGuire, K. J. (1986). A Lesson in SMED with Shigeo Shingo. *Industrial Engineering*, **18**, No. 10, 26–33.

3 McIntosh, R. I. (1998). *The Impact of Innovative Design on Fast Tool Change Methodologies*. PhD thesis, University of Bath, UK. How changeover improvement is currently undertaken in industry was studied.

4 McIntosh, R. I. (1998). *Op. cit.* ref. 3.

5 Boyer, K. K. (1996). An assessment of managerial commitment to lean production. *Int. J. of Operations and Production Management*, **16**, No. 9, 48–59. It is instructive to reflect on research that has been conducted in respect of TQM implementation in industry. What TQM may be able to teach a business contemplating changeover improvement has to be very carefully qualified. Nevertheless, it is possible to identify likely overlap between TQM and changeover improvement initiatives, from which TQM experience may serve to warn of potential changeover implementation pitfalls. This overlap – which Boyer briefly touches upon – can occur particularly in terms of:

- Manufacturing infrastructure (including workforce attitude; skill; training)
- Managerial leadership and commitment
- Team-based work organization, under a continuous improvement regime, to solve problems and develop new production methods.

Some points are highlighted in the small cross-section of references provided immediately below. Other references in this chapter also describe TQM experience.

6 Longenecker, C. O. and Scazzero, J. A. (1993). Total quality management from theory to practice: a case study. *Int. J. of Quality and Reliability Management*, **10**, No. 5, 24–31. Lack of a suitable environment in which a TQM programme can develop has been cited by Longenecker and Scazzero as a potential reason for its failure.

7 Taylor, W. A. (1997). Leadership challenges for smaller organisations: self-perceptions of TQM implementation. *Omega*, **25**, No. 2, 567–579. Taylor writes: 'Few would argue about the important role which senior managers play in TQM. Their grasp of its purpose and intent is indicative of the amount of time they have given to its consideration, and by implication therefore, their level of involvement.'

8 Lascalles, D. M. and Dale, B. G. (1990). The key issues of the quality improvement process. *Int. J. of Production Res.*, **28**, No. 1, 131–143. The impact of an unrealistic expectation of time scale and cost (arising because of inadequately determined targets) on the outcome of a TQM improvement programme has been described by Lascalles and Dale. We describe in this chapter that changeover improvement targets need to be set in terms of:

- The level of reduction that is sought
- The time allocated to reach this level of reduction
- The resource allocated to reach this level of reduction

These three separate targets impinge upon one another and need to be set together.

9 Koufteros, X. A. and Vonderembse, M. A. (1998). The impact of organizational structure on the level of JIT attainment: towards theory development. *Int. J. of Production Research*, **36**, No. 10, 2863–2878.

10 Gest, G. B. (1995). *The Modelling of Changeovers and the Classification of Changeover Time Reduction Techniques*. PhD thesis, University of Bath, UK.

11 Choi, T. Y. and Liker, J. K. (1995). Bringing Japanese continuous improvement approaches to US manufacturing: the role of process orientation and communications. *Decision Sciences*, **26**, No. 5, 589–620. The training of staff for continuous improvement (CI) activity is compared with selecting specific personnel for this activity. The decision on whether to train or whether to select personnel for a changeover initiative might be heavily influenced by the likely need to engage those personnel who work directly on the equipment upon which the initiative is to focus.

12 Lee, D. (1986). Set-up reduction: making JIT work. *Management Services*, May, 8–13. Lee describes that internal time/external time considerations are important. In terms of achieving improvement Lee describes how a set-up can be characterized as: Clamping elements; Adjustment; Other; Unclamp; Cleandown; Problem. He describes that, based on this classification, the improvement team should 'brainstorm' appropriate solutions, from which the best ideas should be implemented.

13 Hall, R. W. (1983). *Zero Inventories*. Dow Jones-Irwin, New York. Hall exhorts: do the easy things first, then the things that contribute most to changeover time. Hall's selection of what areas of the changeover to work on closely mirrors the approach based on a Pareto analysis of a changeover as advocated by Lee (above).

14 McIntosh, R. I. (1998). *Op. cit.* ref. 3. The use of Shingo's 'SMED' methodology has been found to be extremely widespread.

15 Hall, R. W. (1983). *Op. cit.* ref. 13. Hall observes that typical site improvement teams are sometimes unable to identify possible improvement options for more complicated items of equipment.

16 McIntosh, R. I. (1998). *Op. cit.* ref. 3. An improvement team was found to be unable to identify possible improvements for one piece of complex, compact equipment. Subsequent major improvement in changeover performance (from more than 30 minutes to just 10 seconds) was achieved by design changes conceived by a site production engineer.

17 Brown, C. R. (1992). Easing the impact of just-in-time inventory management on machine set-up with quick change tools. *American Machinist*, February, pp. 54–57. Brown presents a review of tool design (for use with CNC equipment). The difficulty of achieving low-cost, team-based improvement to such equipment where fully automated changeover tasks prevail will be discussed in Chapter 5 of this book.

18 Shingo, S. (1985). *A Revolution in Manufacturing: The SMED system*. Productivity Press Inc., Massachusetts. Shingo writes of the 'SMED' methodology (described by Shingo here as the 'SMED system'): 'Although not every setup can literally be completed in single-digit minutes, this is the goal for the system described here, and it can be met in a surprisingly high number of cases.'

19 Hay, E. J. (1987). *Op. cit.* ref. 1. A 75 per cent reduction in changeover time – as described in the title of Hay's article – represents a frequently set target. In other cases the defined target is a 'single-minute' changeover (although this is normally referred to as a 'single-minute' *set-up*, possibly, in some cases, not taking account of the run-up – see Chapter 1). Hay is also forthright about the way that this level of reduction should be achieved and about its universal validity. With reference to his own adaptation of Shingo's methodology, Hay states: '. . . regardless of whether current equipment changeover takes 24 hours or 12 minutes, this methodology can reduce setup time of any piece of equipment by 75% without the need to buy a solution'.

20 Hay, E. J. (1989). Driving down downtime. *Manufacturing Engineering*, **103**, No. 3, 41–44. Hay revisits the topic of targets (see also above) a little later. Rather than making unqualified claims of improvement, Hay is now a little more circumspect, writing: 'Realistic targets, milestones and checkpoints are set; the goal is success within a controlled environment'.

21 Johansen, P. and McGuire, K. J. (1986). *Op. cit.* ref. 2. Johansen and McGuire describe attitude in terms of challenging the root cause of problems and accepting that improvement can be achieved. The attitude of management in recognizing the importance of changeover and acting accordingly is reflected upon. Attitude is also described in terms of the spirit that is engendered among changeover team members.

22 Mather, H. (1992). Reducing your changeover times. IMechE seminar, Birmingham, UK. Some changeover consultants express reservations about working with companies who cannot demonstrate a sufficiently strong attitude either to work for change, or to accept change throughout the organization.

23 Rzevski, G. Meeting at the Open University, 25 February 1993. Rzevski pointed out the extent to which changeover improvement is typically a highly complex 'socio-technical' activity. For an active improvement programme, highly interactive groups typically work upon a major task (changeover improvement) for which many possible different approaches and solutions may be adopted. In addition, different skills will be called upon from those who are engaged in improvement activity, where sociological/psychological factors may also impinge upon what occurs. One site (or even different departments within the same site) might be very different in terms of the abilities and attitudes of those who work there. A training programme is likely to have more impact if it can be tailored to take the situation of those that are engaged on it into account. Good changeover improvement consultants will be able to contribute in this regard.

24 Elgemal, M. A. (1998). An examination of organization and suborganization readiness for total quality management. *Int. J. Technology Management*, **16**, No. 4–6, 556–569. (From a perspective of TQM implementation.)

25 Sohal, A. S., Samson, D. and Ramsay, L. (1998). Requirements for successful implementation of total quality management. *Int. J. of Technology Management*, **16**, No. 4–6, 505–519. (Again, from a perspective of TQM implementation.)

26 Whalen, C. E. (1991). A cost effective approach to reduction in set-up times. *Proc. Conf. of Plastics Engineers – ANTEC 91*, Montreal, Canada, Ch. 693, pp. 2656–2658. Whalen observes that a 'not invented here' attitude can detrimentally affect changeover improvement when it is attempted to introduce externally derived solutions. In particular, from observations made within the injection moulding industry, Whalen describes that prejudices can exist in introducing 'packaged Quick Mould Change (QMC) systems' – representing a 'buy-in' technological solution to the changeover problem (see also Figure 3.3). More generally in our own experience a 'not invented here' attitude can apply to solutions originating even from other parts of the same company's workforce.

27 McIntosh, R. I., Culley, S. J., Mileham, A. R. and Owen, G. W. (2000). A critical evaluation of Shingo's 'SMED' methodology. *Int. J. Production Research*, **38**, No. 11, 2377–2395.

28 Shingo, S. (1985). *Op. cit.* ref. 18. 'Streamlining' represents the final 'conceptual stage' of the 'SMED' methodology. It is difficult to determine from Shingo's book exactly what 'streamlining' aims to achieve – where no explicit definition of 'streamlining' is presented. It seems, particularly with reference to Shingo's Figure 5–12 (page 92), that Shingo intended for 'streamlining' to be the opportunity to reduce the duration of all existing changeover tasks, once their internal time/external time status had been altered. A fuller discussion is presented in McIntosh *et al.* (reference 27, above). Shingo's 'SMED' methodology is also discussed further within the current chapter.

29 McIntosh, R. I. (1998). *Op. cit.* ref. 3.

30 Gest, G. B. (1995). *Op. cit.* ref. 10.

31 Mileham, A. R., Culley, S. J. and Shirvani, B. (1996). *Set-up Reduction Techniques: A practical approach with 200 ideas for implementation.* University of Central England. This catalogue was based very largely on Graham Gest's original work. It is relatively unrefined and, certainly, there are innumerable other possible examples that could be included to take the total of 200 ideas ever upwards. Although this work can be refined, it still serves to prompt corresponding ideas from changeover improvement practitioners that are applicable to their own particular situation.

32 Sekine, K. and Arai, K. (1992). *Kaizen for quick changeover.* Productivity Press, Cambridge, Massachusetts. Sekine and Arai's book represents an example of a 'techniques-led' approach to changeover improvement (although a higher-level conceptual 'framework' under which improvement might be conducted is also described). Their book relates directly to press systems and possibly has restricted use beyond this industry.

33 Shingo, S. (1985). *Op. cit.* ref. 18. Shingo assigns improvement techniques to the different stages of the 'SMED' methodology. Techniques and examples are described throughout Shingo's book. The second part of the book is dedicated to presenting examples of changeover improvement. The catalogue noted above (Mileham *et al.*, reference 31) likewise concentrates on presenting changeover improvement examples, but makes use of alternative classifications to those used by Shingo.

34 Womack, J. P. and Jones, D. T. (1996). *Lean Thinking*. Simon and Schuster, New York. Womack and Jones describe that rapid changeover is a key component of lean manufacturing, acknowledging the prominence of the 'SMED' methodology as a means to improve changeover performance. The importance of the techniques that Shingo sets out throughout his book are highlighted by Womack and Jones' description of 'SMED' as '. . . a series of techniques pioneered by Shigeo Shingo for changeovers of production machinery in less than 10 minutes'.

35 Wright, T. P. (1936). Factors affecting the cost of airplanes. *J. of Aeronautical Sciences*, **3**, No. 4, 122–128.

36 Yelle, L. E. (1979). The learning curve: historical review and comprehensive survey. *Decision Sciences*, **10**, No. 2, 302–328. Although many different learning-curve functions

are possible, Yelle shows that the classical log-linear function has been the most commonly used.

37 Steudel, H. J. and Desruelle, P. (1992). *Manufacturing in the Nineties – how to become a lean, mean world-class competitor*. Van Nostrand Reinhold, New York. The participation of shopfloor personnel in total on *kaizen* initiatives comfortably exceeded what Steudel and Desruelle describe as the benchmark level of a 'world-class' business of 40–60 per cent.

38 Beer, M., Eisenstat, R. A. and Spector, B. (1990). Why change programs don't produce change. *Harvard Business Review*, **68**, No. 6, 158–166. Beer *et al.* give the same message as Steudel and Desruelle (above), asserting that the credibility of any one programme should never be diluted by the sheer volume of other concurrent (but less important) programmes.

39 Sheridan, J. (1997). Lessons from the best. *Proc. Annual Int. Conf. American Production and Inventory Control Soc.*, Washington, pp. 126–131. Disagreeing with Steudel and Desruelle, Sheridan maintains that 100 per cent participation in empowered work teams should be a benchmark. Even so, there is no mention of employees participating simultaneously in more than one team.

40 McIntosh, R. I., Culley, S. J., Gest, G., Mileham, A. R. and Owen G. W. (1996). An assessment of the role of design in the improvement of changeover performance. *Int. J. of Operations and Production Management*, **16**, No. 9, 5–22.

41 Swoyer, S. (1998) (web site accessed 3 Nov. 1998). Set-up reduction. http://www.tbmcg.com/tech_setup.html. Swoyer, a changeover improvement consultant, is unusually forthright, writing: 'I have come to the conclusion that the greatest challenge is not generating the dramatic setup time reductions, it is sustaining them.'

42 McIntosh, R. I. (1998). *Op. cit.* ref. 3.

43 McIntosh, R. I. (1998). *Op. cit.* ref. 3.

44 Lord Simon of Highbury (former chairman of BP) (1998). Acceptance speech during conferment of Degree of Doctor of Laws *honoris causa*, 9 July 1998, University of Bath, UK. The importance of performance measurement in industry was noted.

45 Lingle, R. (1996). Bottling changeovers keep Humko humming. *Packaging Digest*, **33**, No. 4, 56–58.

46 Smith, D. A. (1991). *Quick Die Change*. SME, Dearborn, USA.

47 Freund, J. and Simon, G. (1992). *Modern Elementary Statistics*. Prentice Hall, Englewood Cliffs, NJ. Good work study practice will recognize that completion of task times will be variable. This variability should be taken account of, with specific focus on the mean time that a task takes to complete. Assuming that times obey a normal distribution, their standard deviation might also be determined.

48 Lee, D. (1986). *Op. cit.* ref. 12.

49 Smith, D. A. (1991). *Op. cit.* ref. 46. Smith is one of the few authors who directly discuss safety in the context of better changeover performance.

50 Anon. (1988). Nicht nur, um zeit zu sparen. *Plastverarbeiter*, **39**, No. 5, 148–149. Highly automated mould changeover is discussed. It is shown that automated mould changeover on injection moulding machines has been viewed primarily from a time-saving angle to date. The greater comfort and, in particular, safety of the fitter have been of secondary importance. The author draws attention to potentially major safety issues. Examples include releasing the mould clamping screws that secure the mould against a vertical platen, open-circuit operation of the hydraulic hoses and accidental mixing up of heating/cooling lines. It is described how design can substantially overcome these safety risks.

51 De Groote, X. (1994). Flexibility and marketing/manufacturing coordination. *Int. J. Production Economics*, **36**, No. 2, 153–167.

52 Kingsman, B., Worden, L., Hendry, L., Mercer, A. and Wilson, E. (1993). Integrating marketing and production planning in make-to-order companies. *Int. J. Production*

Economics, **30**, No. 2, 53–66. Kingsman *et al.* discuss in depth the need for integration of sales/marketing and planning functions, particularly in relation to 'make to order' companies.

53 Robins, R. M. (1989). Quick changeovers – fast paybacks. *Manufacturing Systems*, **53**, No. 3, 53–55. Robins writes: 'An operation that has successfully reduced its set-up time has dramatically changed its fundamental operating capability. That additional productivity potential isn't completely viable unless management systems for master scheduling or inventory control are returned to take full advantage of the changes. In other words, quick changeover practices must be woven into an overall re-assessment of the plant's planning and control framework.'

54 Gerwin, D. (1993). Manufacturing flexibility: a strategic perspective. *Management Science*, **39**, No. 4, 395–410. Gerwin describes that enhanced flexibility requires change to: production equipment; product design; work organization; planning and control procedures, material management; information technology. Some of these categorizations can be related directly to process equipment on the shopfloor and the staff who operate this equipment. The scope for change, if anything, is wider than that described by Gerwin. A design emphasis and an organizational emphasis for these separate categories might also be argued.

55 White, C. W. and Wilson, R. C. (1977). Sequence dependent setup times and job sequencing. *Int. J. Production Research*, **15**, No. 2, 191–202. White and Wilson identify that different types of changeover take different times to complete. The term 'sequence dependent set-up times' in the literature, usually in the context of complex mathematical scheduling models, describes an awareness that variable changeover times exist. White and Wilson write in respect of changeovers on machine tools, but their observations can be generalized: 'On many machine tools, the time to set-up for a part depends upon how the machine is currently set-up. Thus the time devoted to set-up is often dependent upon the sequence of the part types being processed.' They continue: 'When set-up time estimates are available for production planning, they are usually expected values over all part types rather than estimates which predict sequence-dependent set-up times.'

56 Allahverdi, A., Gupta, J. N. D. and Aldowaisan, T. (1999). A review of scheduling research involving set-up considerations. *Omega*, **27**, No. 2, 219–240. Allahverdi *et al.* similarly describe the difference between sequence-dependent and sequence-independent set-ups (changeovers): 'For a separable set-up, two problem types exist. In the first type, set-up depends only on the job to be processed, hence it is called sequence-independent. In the second, set-up depends on both the job to be processed and the immediately preceding job, hence it is called sequence-dependent.'

57 Panwalker, S. S., Dudek, R. A. and Smith, M. L. (1973). Sequencing research and the industrial scheduling problem. In Elmaghraby S. E. (ed.), *Symposium on the Theory of Scheduling and its Applications*. In this paper, sequence-dependent changeovers are determined to occur frequently. Approximately three quarters of managers surveyed reported that at least some operations require sequence-dependent set-up times, while approximately 15 per cent of managers surveyed reported that all operations require sequence-dependent set-up times. Although old, these findings are still noteworthy. With current trends to ever greater product diversity, the occurrence of sequence-dependent set-ups (changeovers) possibly has increased.

58 Wemmerlöv, U. and Johnson, D. J. (1997). Cellular manufacturing at 46 user plants: implementation experiences and performance improvements. *Int. J. Production Research*, **35**, No. 1, 29–50. A description of a manufacturing cell is given: 'A manufacturing cell is a production unit for which a group of functionally dissimilar equipment are located in close proximity and dedicated to the manufacture of a family of parts and/or products with similar characteristics.'

59 Wemmerlöv, U. and Hyer, N. L. (1989). Cellular manufacturing in the US industry: a survey of users. *Int. J. Production Research*, **27**, No. 9, 1511–1530. As in the previous reference, it is shown that, based on survey results, adoption of cellular manufacturing is undertaken partly to reduce set-up (changeover) times. Resulting reduction in changeover times by moving to cellular manufacture was found in this earlier survey to be 41 per cent.

60 Burbridge, J. L. (1975). *The introduction of group technology*. Wiley, New York.

61 Shafer, S. M. and Charnes, J. M. (1997). Offsetting lower routeing flexibility in cellular manufacturing due to machine dedication. *Int. J. Production Research*, **35**, No. 2, 551–568.

62 Mair, A. (1994). Honda's global flexifactory network. *Int. J. Production Research*, **14**, No. 3, 6–23. Mair describes manufacture of Honda cars in batches of between 30 and 60 identical vehicles (colour, engine specification, right-hand drive, etc.). He further describes that manufacture in this way: '... simplifies logistical planning in parts supply compared with production in lots of one and still offers high market flexibility'.

63 Bateman, N., Stockton, D. J. and Lawrence, P. (1999). Measuring the mix response flexibility of manufacturing systems. *Int. J. Production Research*, **37**, No. 4, 871–880.

64 Rizzo, K. (1993). Trimming press makeready time. *Gatfworld*, **5**, Issue 4, 1–8. Rizzo describes print industry changeovers, where scheduling, in his view, is significant enough to interpose as a separate stage in Shingo's 'SMED' methodology (described elsewhere in the current chapter). More generally, in respect of colour scheduling, cleansing is far less critical (and thus more quickly achieved to an acceptable standard) if changing over from a 'light' colour to a 'dark' colour. We have conducted major research at three widely different print operations. For each situation a loose 'rule-of-thumb' scheduling policy that exploited this factor was in use.

65 Cook, F. (1990). Dyeing fundamentals, controls highlight AATCC. *Textile World*, **1**, No. 140, 85–88. Similarly to Rizzo (above), Cook describes scheduling in terms of colour sequencing.

66 Charlesworth, K. (1996). Six appeal. *Printweek*, August, 22–25.

67 Brown, M. (1996). A six-sided argument. *The Canmaker*, May, 23–24.

68 Escott, R., Davis, C., Bazlinton, P., Lambkin, T. and Ko, T. (2000). Coldform equipment changeover – interim presentation, Bath University changeover project, 22 February 2000. A ratio as low as 25:75 per cent production time to changeover time need not always arise because of scheduling difficulties. A coldform equipment changeover is being researched at which, currently, a production time to changeover time ratio of 20:80 per cent can be experienced: specifically, an 8-hour tool changeover for each 2-hour production run.

69 Graves, S. C. (1981). A review of production scheduling. *Operations Research*, **29**, 646–675. Graves describes the difficulty of integrating batch size (lot size) and sequencing decisions. Much has been published in the operations research literature, where different components of this overall problem have been tackled in analyses of varying complexity.

70 Kim, S. C. and Bobrowski, P. M. (1994). Impact of sequence dependent setup time on job shop scheduling performance. *Int. J. Production Research*, **32**, No. 7, 1503–1520. Kim and Bobrowski report that 'Reduction of set-up time would reduce the scheduling complexity'. This assertion, though, needs to be considered also in the context of increasing the frequency with which changeovers occur.

71 Sohal, A. S. and Al-Hakim, L. A. R. (1994). Implementation of JIT technology. In Zaremba, M. B. and Prasad, B. (eds), *Modern Manufacturing: Information control and technology*. Springer-Verlag, New York. Conversely (see also the preceding reference), Sohal and Al-Hakim describe a contrasting effect on scheduling complexity: 'Smaller lot sizes mean frequent production runs. This in turn causes a major scheduling problem.'

72 Aghezzaf, E. H. and Artiba, A. (1998). Aggregate planning in hybrid flowshops. *Int. J. Production Research*, **36**, No. 9, 2463–2478.

73 Laguna, M. (1999). A heuristic for production scheduling and inventory control in the presence of sequence-dependent setup times. *IIE Transaction*, **31**, No. 2, 125–134. Laguna states: 'The consideration of sequence-dependent setup times is one of the most difficult aspects of production scheduling problems.'

74 Jensen, J. B., Malhotra, M. K. and Philipoom, P. R. (1998). Family based scheduling of shops with functional layouts. *Int. J. Production Research*, **36**, No. 10, 2687–2701.

75 Galbraith, J. K. (1981). Free market mystics. *New Statesman*, 9 October, 14–15.

76 Shingo, S. (1985). *Op. cit.* ref. 18.

77 David, I. (1997). The quick-change artists. *Professional Engineering*, **10**, No. 21, 31–32. David provides a brief overview of some of the approaches that are employed to tackle changeover problems. While the 'SMED' methodology features strongly, it is not the only approach that is used. As we describe elsewhere, the 'SMED' methodology does not encompass all that is involved in an overall initiative. Nor, in the opinion of some of those sampled in David's paper, are aspects such as mechanization and adjustment accorded an appropriate status as part of the methodology (which we shall similarly investigate further).

78 Steudel, H. J. and Desruelle, P. (1992). *Op. cit.* ref. 37. Steudel and Desruelle present a comprehensive description of changeover improvement. It can be seen that the 'SMED' methodology is not slavishishly adopted: that some issues are given greater prominence, and a wider perspective of improvement proposed.

79 Sekine, K. and Arai, K. (1992). *Op. cit.* ref. 32. Sekine and Arai provide an example of an approach to changeover improvement that does not incorporate the 'SMED' methodology at all. This approach is likely to be limited to press systems. Although Sekine and Arai do not include the 'SMED' methodology, the word 'SMED' is used to represent a target improvement time.

80 McIntosh, R. I., Culley, S. J., Mileham, A. R. and Owen, G. W. (2000). *Op. cit.* ref. 27. This paper critically analyses the 'SMED' methodology. A number of potential deficiencies are investigated. The paper argues that there is scope for an improved methodology to be developed to supersede it. Another point can be made, that the paper misses. Traditional 'SMED' improvement practice places strong emphasis on separating and converting changeover tasks into external time. The ratio of production time to changeover time needs to be assessed in this context (two examples of very poor changeover time/production time ratios are given in this chapter). In some situations, where short manufacturing batches are being scheduled, 'externalizing' all possible changeover tasks simply cannot represent a wholly viable improvement option – there will come a point, assuming that line operators are involved in the changeover of their own equipment, when it becomes impossible, in the periods when the line is running, to conduct all necessary external changeover tasks alongside all necessary tasks to maintain line production.

81 McIntosh, R. I., Culley, S. J., Mileham, A. R. and Owen, G. W. Reinterpreting Shingo's 'SMED' methodology (paper in preparation). We are also writing a second paper assessing the 'SMED' methodology. The focus of this second paper is different, where it is argued that two distinct changeover improvement mechanisms exist: first, altering the time at which a task is conducted (essentially without otherwise altering that task), and second, altering (improving) existing tasks. It is further argued that the 'SMED' methodology is not adequately structured to exploit these two mechanisms which, prima facie, it sets out to do (changing when tasks are conducted in stage 1 and stage 2; altering tasks in stage 3). Both this paper and the preceding paper represent our own criticism of the 'SMED' methodology and its interpretation. Other work that we refer to within the current chapter additionally demonstrates that Shingo's work can be adapted or revised. In some instances substantially different approaches are adopted.

82 Soros, G. (1998). Interviewed on the *Newsnight* television programme, BBC2, 7 December. The topic of changeover is not alone, of course, in having strong adherents to one particular stance or model. Many years after Galbraith's argument of the flaws of modelling macroeconomics on the concept of a perfect free market, Soros, concurring with Galbraith's position, describes that 'market fundamentalists' still exist. Unwavering support of any particular stance in the presence of alternative positions that others favour is only valid if the trouble has been taken to assess and then reject these alternative positions.

83 Smith, D. A. (1991). *Op. cit.* ref. 46. Smith draws attention to many options to improve changeover performance for press systems, some of which the 'SMED' methodology does not directly highlight, or to which it gives low prominence.

84 Jenkins, R. (1995). Quick die changing. Conf. on leading edge strategies for the metalforming industry, Metalform '95, Chicago, in Smith, D. A. (1991). *Op. cit.* ref. 46. This is one example of a design-led approach to changeover improvement that does not feature moving and converting tasks to external time.

85 Johnson, R. I. and Cousins, G. (1989). Presentation at SME Quick Die Change Clinic, 2–3 May, Detroit, in Smith, D. A. (1991). *Op. cit.* ref. 46. Although Shingo himself was present on this GMC initiative, the 'SMED' methodology was not exclusively concentrated upon. This paper describes how a much wider approach was adopted that also incorporated *Ishikawa* analyses of existing problems.

86 Shingo, S. (1985). *Op. cit.* ref. 18. For example, Shingo asserts that stage 1 of the 'SMED' methodology alone ('separating internal and external set-up') usually accounts for a 30–50 per cent reduction in changeover time. Shingo argues strongly, although he does not put a figure on it, that a further significant contribution will be made as the methodology's stage 2 is completed ('converting internal to external setup' – also concentrating on increasing the external time status of a changeover) is undertaken. That differentiating between 'internal time' and 'external time' is central to the 'SMED' methodology is further acknowledged by Shingo: 'Mastering the difference between internal and external setup is thus the passport to SMED.'

87 Bille, J. (1989). Shigeo Shingo's concepts need more attention. *Quality Progress*, **22**, No. 3. 9. An industrial engineer's perspective of 'SMED', set out by Bille, confirms that moving tasks from internal time to external time can be viewed in industry as the major theme of the 'SMED' methodology: ' "SMED" describes how to reduce set-up times from hours to less than 10 minutes by separating inside and outside exchange of dies.' Bille's opinion reinforces our own findings of industry's perception of 'SMED'.

88 Quinlan, J. P. (1987). Shigeo Shingo explains 'single-minute exchange of die'. *Tooling and Production*, Feb., 67–71. Quinlan repeats Shingo's 'mastering the difference...' quotation (reproduced above in reference 86) word for word in a summary article on changeover improvement.

89 Baker, P. (1998). Extra minutes mean extra mints. *Works Management*, **51**, No. 3, 36–37.

90 Hall, R. W. (1983). *Op. cit.* ref. 13. Hall supports Shingo's assertion of the likely impact of moving and converting changeover tasks to external time, making a similar global claim: 'Switching as much set-up activity as possible from internal set-up to external set-up reduces downtime by as much as 50% or more.'

91 Burcher, P., Dupernex, S. and Relph, G. (1996). The road to lean repetitive batch manufacture. *Int. J. Operations and Production Management*, **16**, No. 2, 210–220. From an academic standpoint, Burcher *et al.* describe their view of activity that is required to reduce set-up (changeover) time: 'Set-up has two elements, internal and external. ... To reduce the effective set-up time you need either to remove the need for the set-up entirely (e.g. by using dedicated machines) or move from internal to external set-ups.'

92 Black, J. T. (1991). *The Design of the Factory with a Future*. McGraw-Hill, New York. Black is enthusiastic enough of Shingo's work to reproduce it directly in his book, including

figures, making similar assertions of the importance of separating and converting changeover tasks to external time.

93 De Ron, A. J. (1995). Measure of manufacturing performance in advanced manufacturing systems. *Int. J. Production Economics*, **41**, No. 1–3, 147–160. Attention is briefly given to changeover performance, where, citing Shingo, de Ron summarizes changeover improvement 'rules' as 'carry out as many set-up activities as possible during the operation of the system'.

94 Gilmore, M. and Smith, D. J. (1996). Set-up reduction in pharmaceutical manufacturing: an action research study. *Int. J. Operations and Prod. Management*, **16**, No. 3, 4–17.

95 Baker, P. (1998). *Op. cit.* ref. 89. A reduction from 23 minutes to 138 seconds is reported by direct application of the 'SMED' methodology. That some design changes needed to be made to achieve this level of reduction is noted.

96 McIntosh, R. I. (1998). *Op. cit.* ref. 3. As well as the consultant activity that is assessed within this thesis, a search of the Internet will unearth other global providers.

97 McIntosh, R. I. (1998). *Op. cit.* ref. 3.

98 Jones, J. and Griffin, D. (1993). Achieving quick change-over. IMechE Total Productive Maintenance Workshops, 25–29 October, Stratford-upon-Avon, UK. This paper is co-written by an industrial engineer at Avon Pharmaceuticals alongside a changeover improvement consultant. The application of the consultant's methodology (the 'SMED' methodology) is reported. Emphasis is placed on separating and converting internal tasks to external time. The paper is typical of many others.

99 David, I. (1997). *Op. cit.* ref. 77. For one firm of changeover consultants David describes how the 'SMED' methodology is extended by steps of 'mechanize operations where possible' and 'eliminate operations where possible'. Although mechanization and elimination both feature as assigned techniques as part of the 'SMED' methodology, neither is accorded the status by Shingo that is reported here.

100 Mather, H. (1992). *Op. cit.* ref. 22. Mather adopts the principal elements of the 'SMED' methodology but, as other experienced practitioners have done, places a greater emphasis on eliminating or reducing adjustment.

101 Monden, Y. (1981). How Toyota shortened supply lot production time, waiting time and conveyance time. *Industrial Engineering*, **13**, September, 22–30. Four changeover improvement concepts are set out: separate the internal set-up from the external set-up; convert as much as possible of the internal set-up to the external set-up; eliminate the adjustment process; abolish the set-up step itself. Also, six techniques 'to apply' these concepts are described.

102 Sekine, K. and Arai, K. (1992). *Op. cit.* ref. 32.

103 Hay, E. J. (1988). *Just-In-Time Breakthrough*. Wiley, New York. As well as highlighting attention to adjustment and clamping, Hay also describes that attention is also needed to overcome 'problems' with the existing set-up, where constantly it should be sought to get to the root cause of these problems. 'Problems', as Hay notes, is a catch-all category. We believe that many of the 'problems' that Hay refers to can be categorized more explicitly, and we have attempted to do this in our book. Attention might be directed, for example, to '*item quality*' or '*communication*' (as explained elsewhere in the current chapter).

Similarly to Shingo, Hay also draws attention to opportunities to extend the external work content of the changeover. Hay does not describe a place for Shingo's concept of 'streamlining'.

104 Zunker, G. (1995). Fifty percent reduction in changeover without capital expenditures. PMA technical symposium proc. for the metal forming industry, pp. 465–476.

105 Steudel, H. J. and Desruelle, P. (1992). *Op. cit.* ref. 37.

106 Hay, E. J. (1989). *Op. cit.* ref. 20. Although Hay pays considerable attention to organizational aspects, and indeed strongly advocates a 'low cost/no cost' approach (see earlier

Hay reference, reference 1, above), there is evidence in this subsequent text of an understanding that design can also be important. First, Hay goes some way to acknowledging a role for design when describing team composition, writing: 'With shop floor and support personnel working together, all setup reduction ideas can be implemented following sound engineering and tool design principles.' Second, Hay also writes: 'For instance, on a press, the time to get a die onto the bed of a press is often much shorter than the time to get the die squared, centred or in some other exact position ready for clamping. The die could probably be redesigned so that the first motion of getting the die onto the bed is exact enough for proper positioning before clamping.' This is an example of using design to overcome previously existing changeover tasks by elimination (in this case the procedures involved in getting the tool into the correct position). Shingo does not in any way exclude system change by design, but, as we argue later, does not sufficiently promote the use of design to eliminate tasks, or otherwise alter how a changeover is conducted (when compared to maximizing 'external' tasks).

107 Noaker, P. M. (1991a). Turning out faster setups. *Manufacturing Engineering*, **107**, No. 1, 43–46.

108 Noaker, P. M. (1991b). Pressed to reduce set up? *Manufacturing Engineering*, **107**, No. 3, 45–49. Noaker, who previously joined the chorus of 'no cost/low cost' improvement, paints a rather different picture in this article, where the use of design is acknowledged, along with the potential usefulness of external (to a shopfloor improvement team) expertise. Noaker reports the extent to which an OEM can go to improve changeover performance by design, writing: 'Danly machine (Chicago) clocks changeover at 140 seconds or less on its latest 2,700-ton twin-slide transfer press. The machine also includes automatic adjustment of shut height, cushion pressure, counterbalance pressure, and exit conveyor speed and spacing, in addition to die device timing, feed rail spacing, die clamping and unclamping, and simultaneous movement of die carriages in and out of the press. To aid setup, the part transfer feed system can run in press or feed-only modes using independent inch drives.' This is one example of what Noaker refers to as: '...other approaches to press setup reduction, none of which uses "SMED".' In fact, a survey of trade literature in many industries will indicate the extent to which capital equipment is now being marketed with the issue of changeover performance to the fore.

109 Claunch, J. W. and Claunch, J. C. (1996). *Set-up Time Reduction*. Irwin Professional, New York.

110 Imai, M. (1986). *Kaizen, the Key to Japan's Competitive Success*. McGraw-Hill, New York.

111 Monden, Y. (1983). *Toyota Production System: Practical approach to problem solving*. Industrial Engineering and Management Press, USA.

112 Schonberger, R. J. (1982). *Japanese Manufacturing Techniques – nine hidden lessons in simplicity*. Free Press, New York.

113 Sheridan, J. (1997). *Op. cit.* ref. 39.

114 Suzaki, K. (1987). *The New Manufacturing Challenge: Techniques for continuous improvement*. Free Press, New York.

115 Imai, M. (1986). *Op. cit.* ref. 110.

116 Stanton, E. S. (1993). Employee participation: A critical evaluation and suggestions for management practice. *SAM Advanced Management Journal*, **58**, No. 4, 18–23. The strength of the case for implementing incremental changeover improvement through a mechanism of team-based participation is often outlined, for example by Imai (reference 110), but the universal wisdom of this approach is questioned by Stanton. Stanton sets out the opposing alternatives of participative management (as advocated by Imai and other authors) and directive management and, by analysis of 450 articles, attempts to

determine what factors and conditions make either the more appropriate for a given situation. The analysis concludes that it is by no means clear that participative management – the route which Stanton observes management consultants normally favour – will always give the greatest benefit in a shopfloor improvement programme.

117 Poling, D. (1987). The key to single-digit-changeover. *Modern Machine Shop*, **60**, No. 6, 78–86. Incremental retrospective improvement to the way that a changeover is conducted (concentrating upon the organization of a changeover) does not feature at all in major redesign exercises, whether done in-house or by original equipment suppliers. Poling's article considers the design of lathe equipment for rapid changeover.

118 Godfrey, B., Dale, B., Marchington, M. and Wilkinson, A. (1997). Control: a contested concept in TQM research. *Int. J. Operations and Production Management*, **17**, No. 6, 558–573. Typically, for shopfloor changeover teams, devolvement of authority is encouraged in the belief that those who operate process equipment are those who best understand how to improve it. On this basis they are allocated to this task. There is also an implicit belief that those who have been so empowered have been, or will become, aligned to the goal of changeover improvement. Godfrey *et al.* investigate these assumptions (in the context of TQM); particularly that empowerment will be harnessed and will be directed effectively. It is argued that sometimes too much faith is placed in the empowerment 'process', where often constraints limit the extent to which the workforce enjoy a truly empowered status. It is also described that it can be ill-advised in the short to medium term to require the workforce to undergo what is typically a significant change in their behaviour if, for different reasons, there is notable resistance to it.

119 Anon. (1995). Teamwork is Rx for better bottling. *Packaging Digest*, **32**, No. 5, 50–56. The author describes how initially seeking improvement to changeover performance at a Burroughs Wellcome packaging line with a 'changeover reduction team' grew into a much wider project involving major equipment purchases and the involvement of external specialists. Alongside faster changeover performance, the company simultaneously took the opportunity substantially to reduce the number of line operators (from 8 down to 3), increase productivity (which was a major reason for initiating the changeover programme in the first instance) and more easily achieve compliance with FDA (US Food and Drug Agency) regulations.

120 MacDonald, J. (1992). Reasons for failure. *Total Quality Management*, August. For TQM implementation MacDonald has identified separate and contrasting approaches (which we have described can also be related to changeover improvement) of:

- Improvement through cultural change
- Improvement through a project by project approach

Changeover improvement by a 'cultural change' approach would typically be represented by devolvement of responsibility to empowered shopfloor teams. A changeover example of a 'project-by-project' approach could be design agents moving into the workplace and implementing change in response to identified problems (see also Figure 3.3). As noted by MacDonald, such an approach need not involve the cooperation of the workforce at large. In respect of the advisability of wholly pursuing one or other approach, MacDonald asserts: 'It would be unwise to select any one to be either wholly right or wholly wrong. In as much as it is possible they should be viewed as a totality.' MacDonald's conclusions are broadly in line with those of Stanton (reference 116, above) if it is assumed that Stanton's 'directive management' approach is applied when undertaking a 'project by project' improvement. MacDonald states that consultants can sometimes exaggerate a distinction between the two separate approaches to improvement identified above. One approach will be strongly favoured over the other, rather than each being presented as competing approaches, which are then assessed for each different situation on merit. Of the consultant's role MacDonald writes: 'This dangerous dichotomy is being largely fostered by consultants, each striving to find their own unique selling point.'

121 Myers, M. S. (1987). Don't let JIT become a North American quick fix. *Business Quarterly*, **51**, No. 4, 28–38. Myers, who has experience of JIT implementation in the USA, similarly has reservations in the way that industry frequently embarks upon employee-led improvement initiatives: 'In fact, the decision to foster employee participation has often been based primarily on pure faith.'

122 Radharamanan, R., Godoy, L. P. and Watanabe, K. I. (1996). Quality and productivity improvement in a custom-made furniture industry using kaizen. *Computers and Industrial Engineering*, **31**, No. 1/2, 471–474. Continual improvement is contrasted with innovation in large jumps or steps. This topic will be revisited in later chapters in the context of using design to improve changeover performance.

123 Noaker, P. M. (1991b). *Op. cit.* ref. 108. A useful overview of differing approaches is given, contrasting (although not summarized as such) design-led change and organization-led change. For example, an assessment of changeover improvement opportunities as cited by an engineering manager are: part programming; tool management and setting; first part inspection; work holding.

124 McIntosh, R. I. (1998). *Op. cit.* ref. 3. An assessment of consultant programmes is presented. There is another aspect of consultant emphasis on *kaizen*, organizational improvement that might be considered. It is only natural that consultants who take up the gauntlet of changeover performance improvement should concentrate on those issues that most closely match their own existing specialist expertise. These skills may well reside in issues of organization more strongly than in issues of design. If so, the significant work of these consultants, both in industrial initiatives and in their contribution to the literature, may distort the perception of how design might be deployed to achieve better changeover results. In our experience the bias of changeover training consultants in the area of organizational improvement is often matched by a similar receptive bias on the part of industrial production managers. It may be that production managers, or other personnel that the consultants initially have contact with, also better understand organization-based improvement techniques, expect them to be employed, and thus more readily accept them. Avoiding involving specialist design personnel might also allow independent control in the hands of these production personnel to be maintained more readily.

125 Daniels, R. C. and Burns, N. D. (1997). A framework for proactive performance measurement system introduction. *Int. J. Operations and Production Management*, **17**, No. 1, 100–116. For example (see also elsewhere), Daniels and Burns write that 'single-minute exchange-of-die' [use of the 'SMED' methodology] is one of a number of 'well known tools of *kaizen*' – which are 'relatively easy to implement and can result in large savings in a short space of time'.

126 Noaker, P. M. (1991b). *Op. cit.* ref. 108. The consultant Jerry Claunch, for example, is cited as aspiring to no cost/low cost improvement.

127 Hay, E. J. (1988). *Op. cit.* ref. 103.

128 Steudel, H. J. and Desruelle, P. (1992). *Op. cit.* ref. 37.

129 Hughes, D. and Hobbs, S. (1994). SMED – Assessing the cost effectiveness of set-up reduction. *Proc. 27th Int. Symp. on Automotive Tech. and Automation*, Aachen, 31 Oct.–4 Nov., pp. 601–607.

130 Bradbury, D. (1983). Eliminating roll tooling changeover. Technical paper MF83-544, Society of Manufacturing Engineers.

131 Tellett, D. (1993). New press strategy through to year 2000. *Sheet Metal Industries*, **70**, No. 9, 12–14. The French company Lebranchu is identified as a customer for a 2500-tonne triple axis transfer press (costing £5.5 million) equipped for 'fully automated tool changing'.

132 McIntosh, R. I. (1998). *Op. cit.* ref. 3. One company that we researched widely presented internally its '10 themes for improvement', one of which was 'Do not spend money for improvement'.

133 De Groote, X. (1994). *Op. cit.* ref. 51. De Groote might view such an insistence on low-cost improvement, especially when seeking enhanced flexibility, as shortsighted. This paper argues the importance of the relationship between the marketing function (where greater product differentiation is sought) and the production function (who strive to achieve this desired level of flexibility). It identifies that 'set-up costs' – the cost of changeovers – are so important that a significant trade-off in the cost of the technology (manufacturing systems) used to achieve low 'set-up costs' may well be justified: 'The intuition is clear: a broader product line leads to more frequent changeovers and adjustment of the machines; this justifies the choice of a technology with smaller set-up costs – at the expense of a larger initial investment.' This justification of an increased 'technology cost' to attack 'set-up costs' could be incurred equally by upgrading what exists already (allowing for a higher cost to develop changeover performance, including substantive design – the opposite of what consultants usually promote), or by buying and installing replacement equipment. As this book investigates, the whole topic of cost in relation the level of improvement that is achieved – especially sustainable improvement – is complex. There are no easy answers. It is wrong, in our opinion, to hold preconceptions that efficient, significant and sustainable improvement can always be brought about by organizational improvement alone.

134 McIntosh, R. I. (1998). *Op. cit.* ref. 3. The McIntosh thesis cites a few examples. There are others.

135 McIntosh, R. I. (1998). *Op. cit.* ref. 3. A visit was paid to a press ancillaries show in Birmingham, UK in the early 1990s. Numerous devices to aid press changeover performance were identified, manufactured by a variety of companies. Similar examples are likely to exist for other 'mature' changeover industries.

136 Iafrate, A. (1993). Quick die change concepts. *Sheet Metal Industries*, March, 22–23.

137 Mileham, A. R., Culley, S. J. and Shirvani, B. (1996). *Op. cit.* ref. 31.

138 Choi, T. Y. and Liker, J. K. (1995). Bringing Japanese continuous improvement approaches to US manufacturing: the role of process orientation and communications. *Decision Sciences*, **26**, No. 5, 589–620.

139 Johansen, P. and McGuire, K. J. (1986). *Op. cit.* ref. 2. Johansen and McGuire write: 'When management demonstrates real commitment to set-up reduction, everyone else will take the issue seriously.' A demonstration of serious management intent is stated by Johansen and McGuire as a necessary first step in the improvement process (although we would also argue that more than this – all activities of the strategic phase of the overall methodology – need to be conducted before active shopfloor participation commences).

140 Herrmann, J. W., Minis, I., Lin, E. and Schneider, P. (1997). An operator information system for parallel, off-line assembly. IEEE Symp. on Emerging Technologies and Factory Automation, ETFA, pp. 143–148. Herrmann *et al.* describe progress in communication to aid changeover performance at a US Black and Decker factory. Describing changeover associated problems, they write: 'Faced with increasing product variety, production operators need more accurate information. Currently, they waste too much time and energy seeking information that describes the product they are building. Mistakes can occur when the operators cannot easily access the right information and when they use outdated information.' The parallel between this and the case study that we will present in Chapter 10 will be apparent.

141 Carlson, J. G., Yao, A. C. and Girouard, W. F. (1994). The role of master kits in assembly operations. *Int. J. Production Economics*, **35**, No. 1–3, 253–258. The need for

information to be well communicated during changeover is similarly highlighted by Carlson *et al.* in a circuit board manufacturing scenario: 'In this flexible, small lot-size, JIT environment, there were too many variables for an operator to remember without taking time to refer to prints, specifications or other documents. The large number of variables is derived from a volume of up to 300 components per board with a mix of 10 different boards per day . . . It was essential that something be done to reduce the possibility of error and the corresponding high cost of rework, resetting and scrapped components.'

142 McIntosh, R. (1993). Attendance at factory changeover presentation, 15 February, Bath University changeover project, File note. Observation of a changeover initiative at a UK factory included attendance at an internal presentation. The presentation was given by shopfloor staff, simultaneously to other colleagues and to senior site managers. An outline of individual improvements was made. The file note reports: 'Much of the work on organizational aspects has been to do with achieving effective communication.'

143 Kobe, G. (1992). Engineer for right hand steer: how Honda does it. *Automotive Industries*, **172**, March, 34–37.

144 Lee, D. (1986). *Op. cit.* ref. 12. Lee, describing an improvement initiative at a Cummins factory at Daventry (UK), strongly identifies an involvement of maintenance personnel, and communication that needs to occur with them: 'Operating experience of the teams dictated that we needed to add a maintenance operator to the core group because of the number of occasions that problems were deemed to be caused by poor maintenance (or more correctly by poor communication between manufacturing and maintenance people).'

145 Carlson, J. G., Yao, A. C. and Girouard, W. F. (1994). *Op. cit.* ref. 141. The impact of items of poor quality (incoming product; change parts; consumables; fixed machine parts) is often not discussed. One example is supplied by Carlson *et al.*, who briefly describe a problem that this can cause: 'A major disruptive problem occurs with individual kits [of PCB parts] if there is a quality problem or a shortage. The kit may have to be returned for verification and replenishment.' In a changeover situation, as our main text describes, adjustment or other tasks might have to be of an extended duration. The problem of shortage that Carlson *et al.* refer to is one of item logistics but, if discovered only once the changeover is underway, its impact upon the changeover similarly may be significant.

146 Silbereis, J. (1996). Creating a good impression. *The Canmaker*, May, 97–100.

147 Noguchi, quoted in Schonberger, R. J. (1982). *Op. cit.* ref. 112. While Schonberger clearly shows his support for team-based improvement activity in respect of changeovers, he recognizes that a wider responsibility when seeking quality improvement is necessary, quoting Noguchi: 'Workers and foremen can solve only 15 percent of all quality control problems. The rest must be handled by management or the engineering staff.' In other areas beyond item quality, where an improvement team may enjoy only a limited influence or otherwise possess only a limited capability to get things done (for example, possibly in issues of mechanical design, planning control or implementing major new logistics systems), our view is that Noguchi's perspective similarly might be appropriate.

148 Ghobadian, A. and Gallear, D. (1997). TQM and organization size. *Int. J. of Operations and Production Management*, **17**, No. 2, 121–163. Reference is made to a TQM case study, where 200 personnel were trained, each for a period of 2 weeks, by a well-known consultant team. The observation is made: 'The company's willingness not only to sanction but also to encourage a substantial time investment puts the message across that senior management was serious about its commitment to quality.'

149 Melcher, A., Acar, W., Dumont, P. and Khauja, M. (1990). Standard-maintaining and continuous-improvement systems – experiences and comparisons. *Interfaces*, **20**, No. 3, 24–40. Melcher *et al.* present very forceful arguments for adopting a CI regime.

150 Imai, M. (1986). *Op. cit.* ref. 110. Imai's text is often cited, where he asserts that continual, incremental improvement is perhaps the key reason for the rising dominance of Japanese manufacturing industry since the 1950s.

151 Schonberger, R. J. (1990). *Building a Chain of Customers*. Free Press, New York. We have often focused on cost during this chapter, where, frequently, the pervading opinion that changeover improvement can and should occur with little expenditure has been presented. This view is often expressed, as noted in the chapter, by changeover improvement consultants (with the important exception of consultants that directly advocate design improvements to the existing manufacturing system). A more balanced view of the use of substantive design – hence greater expenditure for changeover improvement – will be made in subsequent chapters. Similarly, examples of buying replacement equipment are presented through the book. Many writers, both academics and industrialists, have argued that low-cost improvement of existing equipment is often a better option. This book debates this issue, arguing that a predetermined position on cost should not be adopted, and that all possible improvement options should be evaluated on merit.

152 Bessant, J. and Caffyn, S. (1997). High-involvement innovation through continuous improvement. *Int. J. of Technology Management*, **14**, No. 1, 7–28. Some of the problems (principally cultural ones) of getting a meaningful programme of continuous improvement underway are described.

153 Shapiro, D. L. and Kirkman, B. L. (1999). Employees' reaction to the change to work teams: the influence of 'anticipatory' injustice. *Journal of Organizational Change Management*, **12**, 51–68.

4

Financial benefit analysis

This chapter sets out the many different ways by which better changeovers can gain advantage for a business. A classification of potential benefit opportunities is presented. This classification may be used when making a financial assessment of the impact of better changeover performance.

A financial assessment serves several highly important purposes. First, it allows the business to decide which of the sometimes conflicting potential benefit opportunities it should pursue. Second, it allows informed resource allocation decisions to be taken (to achieve the improvement targets that the business has set itself). Third, a financial benefit analysis also allows the business to assess a changeover improvement initiative more objectively relative to other initiatives that it may also be considering.

Major changeover improvement initiatives should be assessed as a strategic investment, based on a strategic evaluation of business needs. It should be understood what level of improvement is being targeted and why.[1–3] Even if a company is imbued with pursuing low-cost improvement,[4] resources (people and time) still have to be allocated to undertake improvement work. Business processes, both in the immediate shop-floor environment and elsewhere, may need to be restructured.[5,6] Additionally, the improvement effort needs to be managed and results need to be measured. Ill-judged decisions can be taken at all stages of an initiative if it is not known in financial terms what the initiative is likely to deliver. Primrose[7] makes observations in respect of investment in high-technology equipment that are equally valid for changeover improvement:

> An enormous amount has been written about manufacturing strategy in an attempt to provide some form of scientific basis for making investment decisions. Unfortunately, if the investment can not be evaluated in financial terms, neither can the strategy, which means that, although it may be called a strategic decision, the investment decision is purely subjective. Not only are such decisions prone to personality bias, they can be strongly influenced by current management fashion.

A financial benefit analysis thus places the business in a far stronger position to take decisions in respect of changeover improvement: whether to proceed in the first instance; where to focus improvement effort; what improvement target to set; what resources can be justified; how to exploit the changes that are made.

Parallels with TQM experience may be drawn. TQM has been described as one of the most important of recent business philosophies.[8] Implementing TQM is likely to be a major undertaking[9] yet, despite this, its benefits are by no means always quantified to determine if this investment is worth while.[10] For changeover improvement the same considerations apply. A business is advised to determine how better changeovers can contribute to its 'bottom-line' performance. This chapter is presented to show how this may be done.

4.1 Benefit arising from better changeover performance

Most authors writing on the topic will cite one or more benefits attributable to better changeover performance. Often one potential benefit will be singled out and discussed in detail (often inventory reduction or enhanced productivity). Alternatively, benefits arising from changeover improvement might be more broadly described in terms of attaining a JIT manufacturing capability or a lean/agile manufacturing capability. Despite a widespread understanding that an array of benefits are potentially available,[11,12] little work has been done to classify these benefits.

As well as the contribution of better changeovers to a JIT environment, little work has similarly been done to date to assess the financial impact of better changeover performance in its own right.[13,14] Typically any empirical assessments that have been undertaken have been restricted to quantifying the benefit of reducing machine downtime (hence increasing productivity if the same number of changeovers are conducted).[15] More theoretical assessments similarly can be restricted in this way.[16,17] The benefits of better changeover performance can be far wider than this.

Classifying and quantifying benefit

We present our benefit classification in the next section. The classification will be supported by examples (within numerous exhibits) and by reference to the literature.

Our classification can be used, if chosen to do so, when conducting a financial benefit analysis, for which a procedure is subsequently presented (see section 4.3).

Particular reference is made in our exhibits to two separate UK factories. Each factory manufactures commodity products. The two factories, between them, demonstrated the considerable scope for benefit that can be available (and demonstrated as well some of the pitfalls of not conducting a complete, quantified analysis).

Engaging an internal management accountant

It is recommended that organizations undertaking a thorough benefit analysis appoint a suitably skilled accountant to conduct data-gathering and analysis tasks. This work may also require the involvement of a production manager or engineer. It should be noted that some estimates may have to be made, where data are incomplete or unreliable.[18] Alternatively, it might be necessary to record current changeover times, or otherwise ensure that data are representative.

Effort might also be made to determine the internal systems, or business processes, that the company employs in respect of changeovers, with a view, possibly, to upgrading these to exploit any improved changeover capability to maximum advantage.

Data gathering

Recommendations will be made as to data that a business might be advised to acquire (upon which to base a benefit analysis). It is deliberately left to individuals within the organization to decide how this is to be done, according to the company's own internal accounting and information systems, and its own specific market circumstances.

Constraints

Different constraints can significantly influence the benefit analysis. Whoever conducts an analysis needs to be aware of constraints that can apply. Constraints are often imposed by limited resource (labour, material and equipment). Constraints can restrict how benefit can be taken, or affect how changeover operations occur.[19] For example, factors to do with personnel skills and the organizational structure might limit the number of changeovers that can be conducted during a 24-hour period. The topic of constraints will be addressed in the final section of this chapter. Case study examples will again be presented to illustrate the points that we make.

Better changeovers: speed and quality

Care is taken throughout this book to describe that better changeover performance has two components: a changeover that is faster and a changeover that is of a higher quality.[20,21] Either or both may occur as a result of changeover improvement activity. Most writers concentrate upon the former aspect, but a business would be wise not to lose sight of the potential advantages that higher-quality changeovers can also bring. Similarly, the business should be conscious that focusing exclusively on conducting a faster changeover might compromise the changeover's quality.[22]

The influence of changeover frequency on changeover quality

While conducting an individual changeover more quickly might compromise its quality, there is evidence that conducting changeovers more frequently (manufacturing in smaller batch sizes) can be a major step in itself in improving changeover quality.[23–25] In this case higher quality stems both from practice that is intrinsic in repeatedly conducting changeover tasks, and from the discipline that is also required more generally of other personnel to maintain a successful high-frequency changeover operation.

4.2 A classification of where benefit may be derived

As discussed, changeover improvement typically makes available a wide range of possible benefits.* Figure 4.1 presents our benefit classification. This classification is

* At the same time as improving changeover performance, other wider improvements to the manufacturing operation are often also sought. In these cases care needs to be taken to isolate the benefits specifically attributable to better changeover performance.[26]

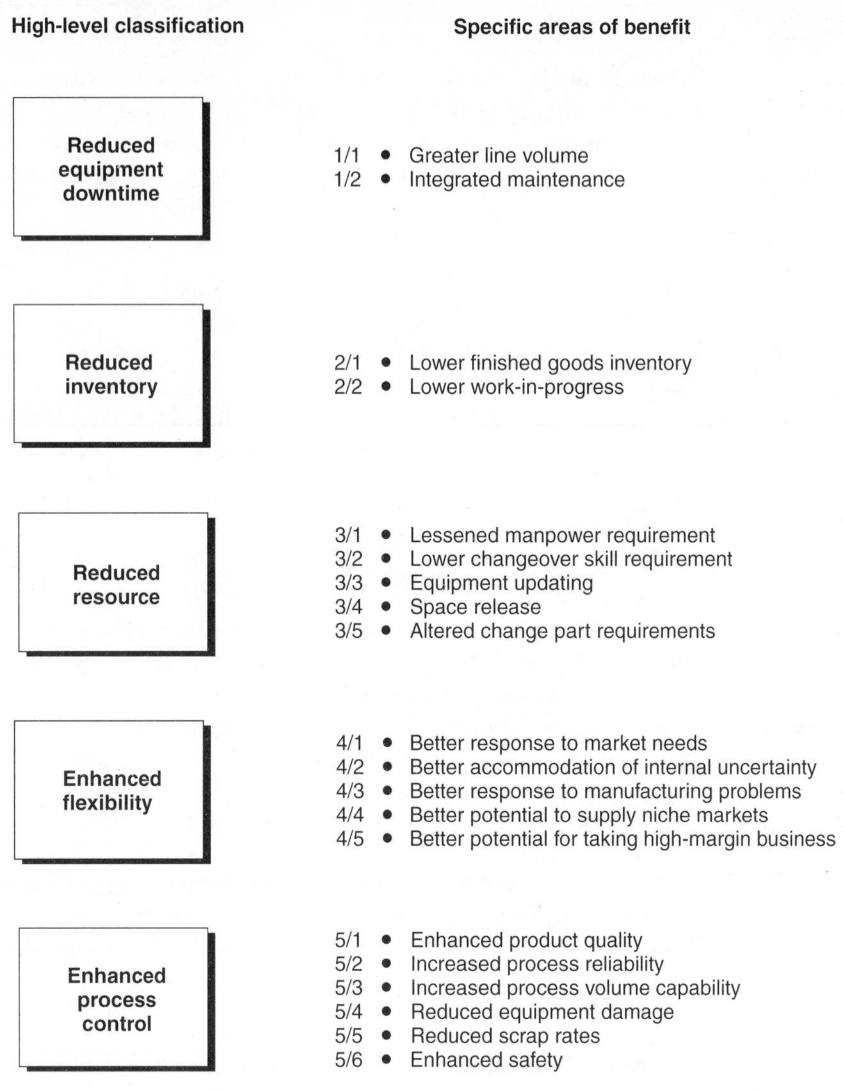

High-level classification
Specific areas of benefit
Reduced equipment downtime
1/1 • Greater line volume
1/2 • Integrated maintenance
Reduced inventory
2/1 • Lower finished goods inventory
2/2 • Lower work-in-progress
Reduced resource
3/1 • Lessened manpower requirement
3/2 • Lower changeover skill requirement
3/3 • Equipment updating
3/4 • Space release
3/5 • Altered change part requirements
Enhanced flexibility
4/1 • Better response to market needs
4/2 • Better accommodation of internal uncertainty
4/3 • Better response to manufacturing problems
4/4 • Better potential to supply niche markets
4/5 • Better potential for taking high-margin business
Enhanced process control
5/1 • Enhanced product quality
5/2 • Increased process reliability
5/3 • Increased process volume capability
5/4 • Reduced equipment damage
5/5 • Reduced scrap rates
5/6 • Enhanced safety

Figure 4.1 Classifying benefit arising from improved changeover performance

for the manufacture of products from a specified product range. Two levels to the classification are presented: a generic 'high-level classification' and, in more detail, 'specific areas of benefit' (for each high-level area).

Some potential benefits typically will be considerably more difficult to assess. Likely problems in this regard are indicated within our text. Problems of assessing benefit will apply particularly within the classification of 'enhanced process control' which, for this reason, is discussed separately. Importantly, benefit will not be possible in all categories simultaneously because the way benefit is taken is dependent on the frequency with which changeovers are conducted (and, correspondingly, the batch sizes that are manufactured).

Primary benefit classification (high-level classification)

Figure 4.1 presents five primary areas of benefit:

1 **Reduced equipment downtime** refers to the ability to manufacture more products in a given time period in which changeovers occur.
2 Benefit accrued by holding **reduced inventory** arises from an ability to manufacture products more frequently in smaller batches.*
3 The high-level classification of benefit arising from **reduced resource** includes both manpower and equipment usage.
4 Benefits attributable to **enhanced flexibility** arise through being better able to commence manufacture of any quantity of a given product from the specified product range at any chosen time. Ensuing improvement in manufacturing flexibility can be used, for example, to enhance responsiveness to customer needs, hence possibly yielding opportunities for greater market penetration.
5 The category of **enhanced process control** refers to benefit arising from conducting higher-quality changeovers. As described below, this is potentially the most difficult area of benefit to quantify.

Secondary level benefit classification (specific areas of benefit)

The secondary level classification describes potential individual benefit opportunities in more detail:

1/1 **Greater line volume** in a nominated time period becomes possible due to increased equipment availability, arising from changeovers being conducted more quickly, but at a substantially unaltered frequency.[27,28] Exhibit 4.1 presents an example. Belief is sometimes mistakenly expressed that this always represents the best use of an improved changeover capability.[29]

1/2 Overall equipment downtime might be reduced by **integrating maintenance** more closely with changeover activity. For example, the preventive replacement of part-worn components might occur simultaneously, while equipment is out of production during changeover.[30,31] More generally, changeover activities could also include condition monitoring, thereby assisting maintenance and ensuring that equipment condition does not decline to the extent that changeover performance is compromised[32,33] (see also Chapter 3). Maintenance attention for greater product quality might also be an objective.[34] By seeking better integration of maintenance activity with changeover activity, the business stands to gain a further reduction in overall equipment downtime, beyond what it could achieve by implementing both programmes entirely separately.† The similarity between what occurs during a changeover and what can occur during maintenance (effectively, changeover of the same part) also should not be missed. The

* Finished goods inventory is sometimes split as safety stock (to cover poor forecast reliability and poor supply reliability) and cycle stock (dependent on scheduling time and batch size). Better changeover performance can have an impact on both 'safety stock' and 'cycle stock'. We have chosen here not to distinguish between them.

† This is a topic of on-going research.[38]

techniques that can lead to improvement in either of these disciplines can often be similarly effective in the other[35–37] – in other words, there can be substantial overlap between them.

Exhibit 4.1 Changeover improvement to enhance productivity

A UK commodity product manufacturer operated a 24-hour production facility. The product was an established one to which little significant technological change had been made for many years. Other major manufacturers were competing in the same market and margins were tight.

The plant recognized that it had problems with its changeover performance.* As described in other exhibits in this chapter, the line's changeover performance was significantly influencing factory operations and customer service.

At the time of our study the line's changeover time had been reduced from in excess of 72 hours (3 days) to approximately 24 hours. During our research the company was engaging TPM consultants, who advised that better changeover performance should be sought, but that this – as required under TPM – should be dedicated to reducing downtime losses by maintaining the current changeover frequency.

We assessed financially the impact of reducing changeover time from 24 hours to 4 hours. By maintaining the current changeover frequency, an additional profit potential to the business of approximately £950 000 per annum was calculated. The analysis assumed that all additional output could be sold. However, our study concluded that greater benefit still would arise if the company were to maintain productivity at existing levels; instead conducting more changeovers, significantly reducing inventory and improving aspects of customer service (see later exhibits in this chapter).

The study also determined that changeover performance was of such significance to this business that any extra manpower that could usefully be added to aid changeover operations could easily be financially justified.

* If the factory had benchmarked the changeover performance of the press equipment it was using (where the changeover bottleneck existed) with what other manufacturers in other industrial sectors were achieving with similar equipment, the true extent of their problems – and of the opportunity for improvement – would have been recognized.[39]

2/1 By providing for manufacture in frequent small batches, rapid changeover can lead to a significantly **lower finished goods inventory**, without loss of overall productivity (see Exhibit 4.2). Penalties attributable to high levels of inventory can extend beyond direct inventory holding costs and, in total, can be substantial.[40] Indeed, as well as incurring fewer penalties, holding low inventory in itself may also have significant positive benefits in terms of responsiveness to customers and being able quickly to identify manufacturing problems. In markets of rapid technological advance or where raw material prices are falling, holding low inventory can afford a business significant competitive advantages that would not otherwise be available, allowing it more readily to stay abreast of – and take advantage of – changing market conditions.[41,42] With such potential to influence its performance, the business should decide how it wants to manage inventory, considering particularly to what extent it wants to seek to minimize it. Chapter 2 described that inventory decouples delivery flexibility from manufacturing flexibility. If delivery demands exceed a company's short-term manufacturing capability it might be appropriate to supply in part from stock.[43,44] Even under such circumstances the levels of finished goods inventory that are to be held still need to be very carefully assessed, where cost considerations should determine that no more finished goods than are absolutely necessary are to be kept in store.[45] A particularly dramatic saving arising from inventory reduction can sometimes arise by eliminating the need to use separate warehouse facilities.[46]

2/2 Small-batch manufacture can also bring about significantly **lower work-in-progress**.[47–49] Any form of inventory might be regarded as expensive insurance. In turn it can lead to complacency in manufacturing operations.[50] Again, it needs to be assessed whether the reduction of WIP inventory to an absolute minimum is always wise, where carefully considered use of WIP inventory may enhance line performance.[51,52]

Exhibit 4.2 Changeover improvement to reduce finished goods inventory

A simplistic analysis determines that where changeover time is halved and changeover frequency is simultaneously doubled, finished goods inventory will be halved.[53] Such an analysis makes a number of basic assumptions (see below) that are unlikely fully to apply in a real manufacturing situation. Nevertheless, frequent, small-batch production can still very markedly diminish finished goods inventory. To test the extent to which rapid changeover applied for this purpose can benefit a company we conducted a study at a UK factory engaged in the manufacture of a family of commodity products. The products all conformed to an industry standard, whereby competition was limited in terms of technical innovation and product enhancement. Competition thus was more focused on market responsiveness and cost.

There were significant operational differences between the plant featured in Exhibit 4.1 and the plant of the current exhibit. The factory in Exhibit 4.1 operated two flow lines full time, and each line manufactured one of just two alternative products. The factory of this study had a significant level of surplus capacity, where 12 separate flow lines were operated to manufacture 40 discrete product sizes, as described below in Table 4.1:

Table 4.1 Line utilization

Flow line number	*Utilization (%)*	*Product sizes*
Line 1	50	4 Sizes
Line 2	48	5 Sizes
Line 3	40	3 Sizes
Line 4	74	6 Sizes
Line 5	33	4 Sizes
Line 6	41	3 Sizes
Line 7	12	Dedicated (one size)
Line 8	48	10 Sizes
Line 9	40	Dedicated (one size)
Line 10	12	Dedicated (one size)
Line 11	48	Dedicated (one size)
Line 12	81	Dedicated (one size)

Exhibit 4.2 (*Continued*)

Each flow line comprised nine machines. The machines from line to line were near-identical. The overall utilization of the facility was 42 per cent.

The factory had identified a number of ways that better changeover performance could help it to become more competitive (although these opportunities had not been quantified). As it was operating in a mature market, factory managers were conscious that customer service, particularly on-time delivery and meeting demands for ever smaller batches, was an important issue. Internally, the factory had identified that it could dispose of surplus lines, thus freeing factory space and reducing manpower. The factory also expressed a desire for better control of WIP and a desire to reconfigure the remaining lines (which the space freed by eliminating some lines would allow it to do) into a 'U' format (for purposes of better line control and further manpower reduction).

The impact of improved changeover performance on finished goods inventory alone is investigated below. The analysis is based on the company's forthcoming stock–make–sell forecast. The impact of differing changeover regimes on finished goods inventory was modelled based on this information.

The initial analysis was undertaken based on the following assumptions:

- No minimum stock levels are held
- Dedicated flow lines remain dedicated
- Short-term capacity constraints are not exceeded
- Changeovers do not clash (i.e. manpower constraints are avoided)
- Changeover frequency is doubled
- Consistent batch sizes are retained

The results of the analysis are shown in Figure 4.2.

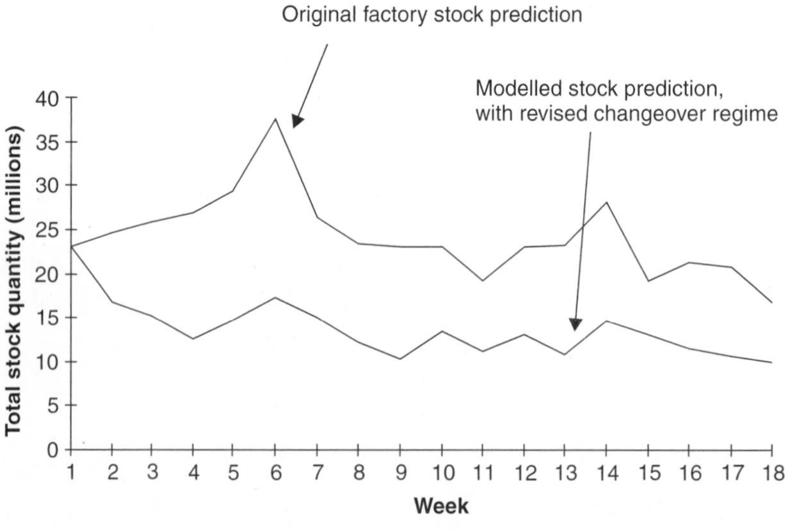

Figure 4.2 Stock predictions

The results indicate that over the time period shown the company can reasonably expect its finished goods inventory approximately to halve. In this instance the saving equates to a goods value of approximately £750 000. Savings for the revised changeover regime fall away rapidly from the start point until stability is established (the effects of the previous stock–make–sell regime have been overcome). The model, of course, is dependent on the accuracy of the original forecast data.

The savings noted above would result from very simple conceptual changes – essentially that batch size is halved and the frequency of manufacture is doubled. Importantly, the revised analysis assumes that identically sized smaller batches are manufactured at identically spaced intervals. A revised manufacturing schedule whereby production becomes more exactly matched to the customer's demands (batch size and delivery date) will have a greater impact still on finished goods inventory.

Figures 4.3 to 4.6 show the effect that revisions to manufacturing practice can have. To show this, one particular product has been isolated (product 17). Respectively the figures show: the original factory stock–make–sell prediction for this product; doubling the frequency of manufacture, using equal-sized batches (the scenario that yielded Figure 4.2); quadrupling the frequency of manufacture, again using equal-sized batches; manufacturing in batch sizes and at intervals that closely match the customer's delivery demands. Particularly for Figures 4.5

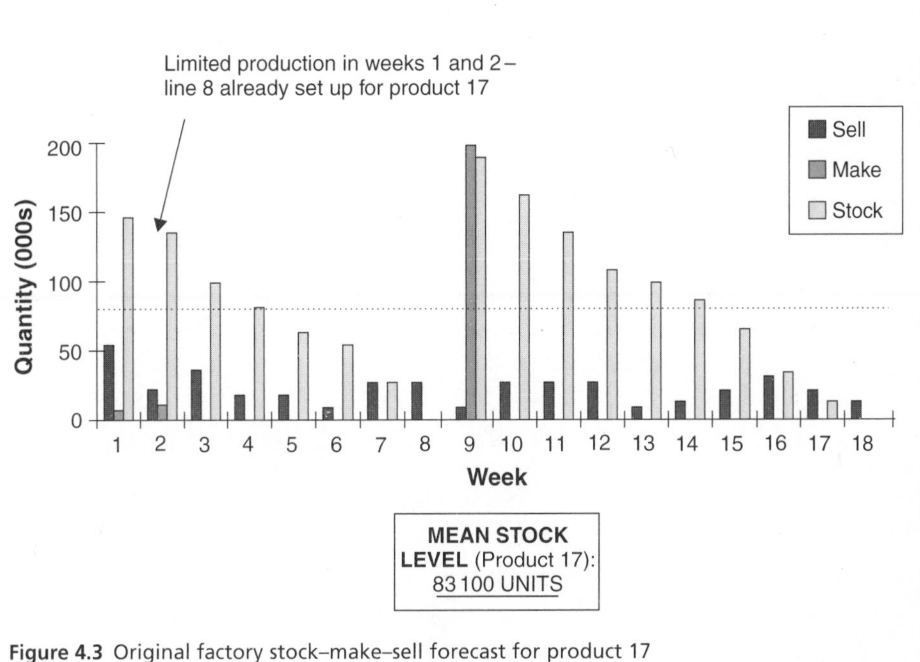

Figure 4.3 Original factory stock–make–sell forecast for product 17

Exhibit 4.2 *(Continued)*

and 4.6 the mean stock level that is presented is inflated by inclusion of the finished goods inventory during transition to the new manufacturing regime. Indeed, for Figure 4.6, it is this alone that contributes to a mean stock level of 27 200 units.

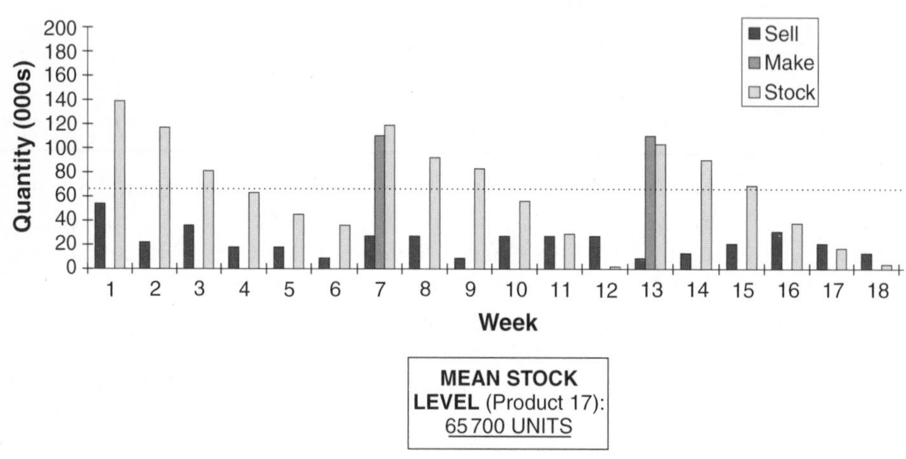

Figure 4.4 Product 17 manufactured in batches of 110k units at 6-week intervals

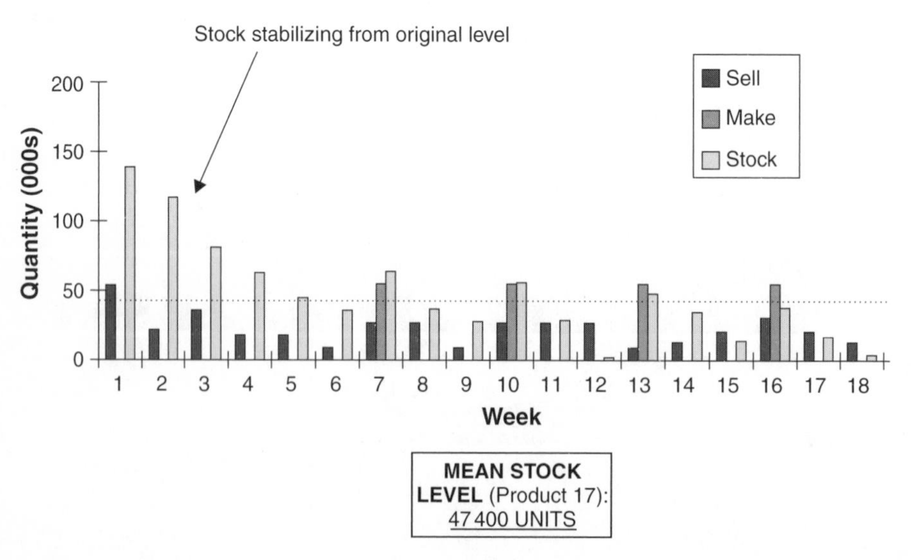

Figure 4.5 Product 17 manufactured in batches of 55k units at 4-week intervals

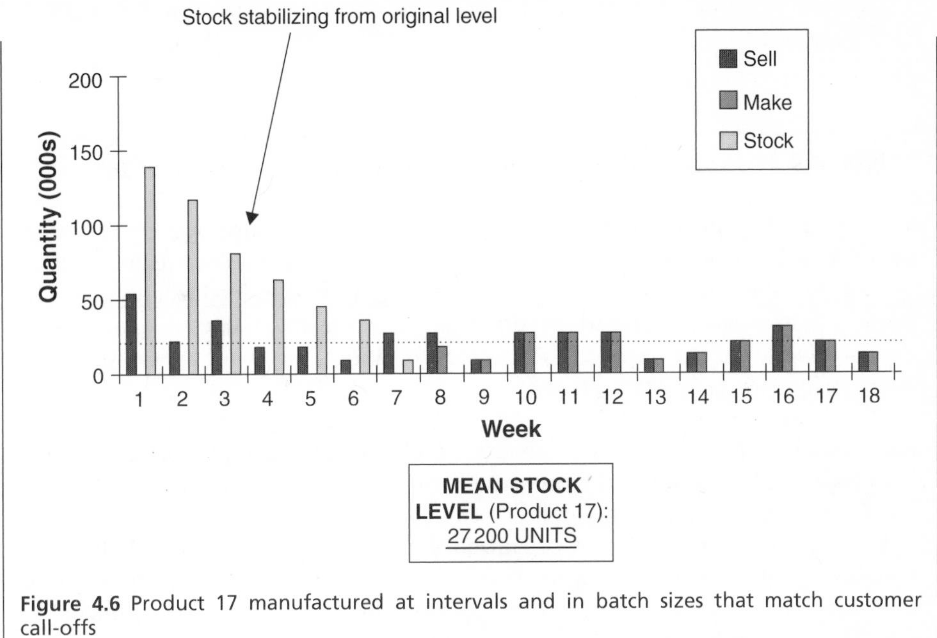

Figure 4.6 Product 17 manufactured at intervals and in batch sizes that match customer call-offs

3/1 Objectively a **lessened manpower requirement** is desirable. This applies both to changeover activity and to normal line operation. Savings might be sought in the number of personnel that conduct a changeover, but care should be taken that the time the changeover takes to complete is not disproportionately compromised by removing personnel from it.* Many different opportunities to change working practices and eliminate personnel from the changeover may be available, often arising because of design changes that reduce work content. Similarly, if new equipment is installed, the number of personnel that are required to conduct a changeover might fall dramatically.[55–57] As detailed in the previous chapter, even ensuring that equipment is maintained in good condition might reduce the number of people required to conduct a changeover. Use of specialized changeover teams (see Exhibit 4.3) might also be considered. Moreover, manpower savings need not arise only in terms of direct changeover labour. If the opportunity can be taken to use better changeover performance to concentrate manufacture on a reduced number of parallel manufacturing lines, as detailed in Exhibit 4.4, the manpower associated with the day-to-day running of the manufacturing facility might also be significantly reduced. Inventory

* The desirability of reducing the number of personnel that conduct a changeover must be balanced against the possible benefit of assigning additional personnel to conduct more tasks in parallel, and hence to reduce the total elapsed time for the changeover. Financial data will greatly assist in making this decision. The cost of employing additional personnel needs to be set against the financial return of being able to execute the changeover more quickly. In analyses that we have conducted, the use of additional personnel, where applicable, has easily been justified.

savings might also lead to manpower savings (in areas of inventory storage and management).[58]

Exhibit 4.3 Possible limitations imposed by specialist changeover teams

The factory described in the previous exhibit elected to change over its manufacturing lines by using specialist changeover teams. These teams (two teams, each of five people, alternating between the two shifts) comprised trained technicians. The primary responsibility of personnel on these teams was to conduct changeovers. If their services were not required for changeover duties, they were then assigned to alternative work within the factory.

The reasoning behind this use of specialist teams was that skilled personnel were required to conduct the changeover tasks, and that, by regularly practising them, the tasks could be conducted more quickly and to a higher standard. No-one else on-site was permitted to conduct a line changeover.

For the most part the teams organized their activities well. Each of the team members undertook specific tasks within the overall changeover (involving nine machines, plus associated trackwork). Where possible, tasks were conducted in parallel and interdependencies (see later chapters) were taken into account in the work sequence. Where necessary, team members paired up to conduct certain tasks where, by virtue of the current equipment design, two people were needed.

Specialization within the team, however, could lead to some difficulties. It was recognized that certain team members had acquired an expertise significantly beyond that of their colleagues for some of the tasks. When these individuals were sick or on holiday those tasks needed to be conducted by other team members. This, though, was a comparatively minor problem, and one that could be overcome by training. More significant was the fact that, because a typical changeover took approximately 5 hours to conduct, only one complete line changeover could be undertaken during a shift. There was a factory policy not to change shifts with a changeover only partly completed. Moreover, changeovers were of significantly variable duration, both because of the type-specific nature of individual changeovers, and because unexpected problems (often concerning item quality) frequently arose. In some instances, as described elsewhere for changeovers at other factories, changeovers could not be completed and had to be abandoned.

These inconsistencies made planning difficult. On the one hand, it was difficult to assign work for the team members beyond their changeover duties, because it was by no means certain that they would be available when expected to do this work. Similarly, manufacturing sequence planning was made more difficult by the constraints that use of a changeover team imposed. Only one changeover could be conducted at any one time and during any one shift. This limited the flexibility that the factory was able to achieve, and hence the benefits that the factory could expect to realize. If continuing to work the current team changeover system, the factory acknowledged that average changeover time would have to fall to below 3 hours before, because of the variability in changeover performance that was still likely, two line changeovers per shift could be contemplated.

The use of changeover teams (or 'pit-stop' teams) is sometimes described in the literature.[59] The alternative is equipment operators conducting changeovers themselves. Numerous different factors will influence the decision as to which approach to adopt. The purpose of this exhibit is not to prescribe that changeover teams represent either good or bad practice. Rather, deliberate focus has been given to difficulties that can arise.

Exhibit 4.4 Equipment reduction made possible by changeover improvement

The same factory (see also Exhibits 4.2 and 4.3) was conscious that better changeover performance might afford it the opportunity to dispose of some of its 'surplus' equipment. At the time of the study just 42 per cent of the available capacity was being used (see Exhibit 4.2). The retention of twelve lines was strongly influenced by the existing changeover performance: although their productive time frequently was low, line 'activity' levels, when including changeover and maintenance, were far higher.

The factory assessed how sub-10-minute changeovers might allow equipment to be removed.* By also making other major assumptions it was calculated that, at just 42 per cent current overall utilization, the same capacity ultimately could be achieved with just six lines. On a line-by-line basis, each line that could be eliminated would mean that two fewer operators would need to be employed. In addition, other benefits would be likely in terms of space release, WIP inventory and equipment updating. Possibly, also, utilizing fewer lines would reduce the factory's total maintenance task. As noted previously, the opportunity would also be afforded to change the existing line configuration.

* The factory at this time, like the factory studied in Exhibit 4.1, was engaged in a TPM initiative. Line OEE data were influential when considering which lines should be disposed of (factory OEE data were being used within the group as a key performance indicator). A dilemma was that those lines with the best OEE figures were not the ones that had the highest speed (volume) capability.

3/2 A **lower changeover skill requirement** may be engineered by changes in the way that the changeover is conducted.[60] Eliminating the need for specialized skills could mean, for example, that all staff working on-site could be trained to conduct any changeover, hence giving added flexibility.[61] An objective that a business might set itself, if it is not already occurring, is to provide for all changeovers to be conducted by those staff that directly operate the equipment.[62] Other benefits may also come about by reducing the skill required to conduct a changeover. It is likely, if simplified procedures are employed, that individual changeovers will become more consistent with one another.[63] In addition, lower-skilled personnel are likely to be less costly to employ. Changeover 'pit stop' teams might have previously been used, whose more highly skilled members can now be alternatively deployed on other duties[64] (see also, for example, Exhibit 2.2). In some cases savings in labour costs alone can be sufficient to justify considerable changeover improvement expenditure.

3/3 The ability to concentrate manufacture on less equipment, described above in terms of its effect on manpower, might additionally lead to other substantial benefits. Exhibit 4.4 presented a situation where a significant reduction in manufacturing equipment could be contemplated. Exhibit 4.5, taken from the same manufacturing plant, describes that significant savings can become available under these circumstances when there is a need to **update equipment**.

3/4 Manufacturing equipment takes up space, and in many businesses space is a prized asset. Better changeover performance can effect **space release** by reducing inventory and by concentrating manufacture on fewer parallel lines. By releasing space the business will have an opportunity to 'unclutter' its manufacturing operation, and hence to control it better.[65] Alternatively, the newly found space could be dedicated to other purposes. Freeing up existing space, by improving changeover performance, might even mean that new premises need not have to be acquired to accommodate an expanding manufacturing operation.[66]

Exhibit 4.5 Reduced costs of updating/overhauling equipment

Many businesses undertake on-going improvement to their existing plant. The factory described in the current series of exhibits was no exception. Indeed, circumstances were such that equipment updating was a major undertaking within the group, where dedicated central teams of engineers were charged with this task. In particular the performance of two specific machines was constantly under review.

One of these machines was involved in the changeover described in Exhibit 4.2. This machine is complex and highly expensive. While exact cost data were not available, it is known that the average spend per machine per year on performance-enhancement modifications is significant. Notable savings would be available if fewer machines were used.

Another machine worked in an alternative department, where as many as twenty such machines operated in parallel. Some were dedicated to specific products, but many of them underwent slow changeovers. There was thus the opportunity, as before, if changeover performance could be appreciable improved, to pare down the number of machines that needed to be used. These machines, on average, were taken out of service and given a major overhaul/update once every ten years. The estimated cost for this work (in 1993) was £65 000 per machine.

3/5 A significant benefit can sometimes be available in **altered change part requirements**. Alteration to production equipment can mean that cheaper change parts are required (a case study in Chapter 10 will provide an example). In some cases a previous need for change parts can be eliminated altogether, particularly when changeover newly involves programming permanently attached, variable position systems – rather than engaging physical change elements.[67,68]

4/1 **Better responsiveness to market needs** typically is one of the most important benefits of better changeover performance, and is often a major reason why improvement is sought.[69–73] An expression of better responsiveness would be continued delivery of ever more exacting repeat orders in terms of the batch sizes sought by the customer and at the time requested by the customer.[74–76] Alternatively, the customer might require delivery of an order with as short a lead time as possible.[77–81] Better changeover performance can contribute to opportunities such as these, without incurring the many penalties associated with holding high levels of inventory.[82–84] Product diversification in itself is also likely to place new pressure on equipment changeover times.[85–87] Exhibits 4.6 and 4.7 describe some ways that better changeover performance can enhance responsiveness to market needs. There are other possible opportunities, both offensive and defensive in nature, as set out in the current exhibits and in the references, that can come about once the business has divested itself of excess inventory. Opportunities to respond better to customer needs will vary significantly from business to business. Potential opportunities should be under constant review (supply of the right product; supply at the right quality; supply in the right format; supply at the right place; supply at the right time; supply in the right quantity; supply at the right price; supply in a professional manner). By benchmarking performance it will be known what needs to be done to maintain a competitive edge,[88] and hence ways that better changeover performance might be exploited. Similarly, as outlined in Exhibit 4.2, benchmarking current changeover performance standards, including devices and improvement methodologies that might be applied, will keep the company abreast of levels of performance to which it can legitimately aspire.

Exhibit 4.6 Poor responsiveness

This exhibit and Exhibit 4.7 relate to the factory that was introduced in Exhibit 4.1. The description of factory operations highlights a number of significant opportunities for benefit.

Both production lines within the factory undertook manufacture of two standard-sized products (four products in total). With both poor changeover performance and tight margins the business was reluctant to conduct more changeovers than, it believed, it absolutely had to. Typically, with the factory perceiving of changeover costs as a fixed expense, production runs for any one product were scheduled for 4–5 weeks' duration.[89] The manufacturing strategy sought to minimize these 'necessary' costs, rather than try to reduce them by improving changeover performance.

To further minimize its costs the factory was concerned that it should strictly control its finished goods inventory. This represented a real problem because the product sizes not currently in manufacture had to be supplied from stock. As a high-speed, 24-hour production facility, the output that could be put into store for this purpose over the previous 5-week period was large.

To alleviate this problem the factory operated what it termed a 'windows' manufacturing system, whereby customers were made aware in advance of the forthcoming production schedule. Price discount inducements were offered to encourage the customers to place orders at periods that were advantageous to the company. The system, inevitably, was not watertight. When major customers

Exhibit 4.6 (*Continued*)

made demands for specific products that were not currently in production, nor could be supplied from stock, then the forward-planning schedule needed to be torn up and production changed over. Whether or not to do this was a decision that was strongly influenced by the customer's status. When production was disrupted in this way, delivery to alternative customers was frequently compromised.

Discussion

Some important opportunities for benefit become available if rapid changeover is used to reduce batch sizes from the current 4–5-week production runs. A conscious decision needed to be taken to exploit these opportunities. If, instead, by maintaining the current changeover frequency, faster changeover performance is used only to reduce equipment downtime (as the company was anticipating under its TPM programme) these benefits could not be realized. It was not possible to exploit these contrasting opportunities to maximum effect simultaneously.

Some benefits of greater responsiveness arising from frequent, small-batch manufacture could be quantified. Not least of these was the price discounting that was being used,[90] whereby a more appropriate pricing structure might be applied. More significantly, achieving a better supply capability should engender a better perception of the company across all its customers. An indication of the importance of this was provided by one customer who had recently rescinded sole supplier status from the company, largely because of delivery problems. What was previously a £13.1 million per annum business was now being shared with an alternative European supplier.

Such events indicate that internal measures, such as lead-time performance, should be used to audit the improvement that is being made. The marketing function should always be fully aware of the current manufacturing capability.

Benefits of greater responsiveness, of course, are but a part of the overall portfolio of opportunities that the company might expect. Also highly important was the projected inventory reduction which, as with the inventory reduction demonstrated in Exhibit 4.2, constituted a very considerable saving (included in which was the likely ending of renting a large external warehouse unit, and the associated local transportation costs thereto).

Exhibit 4.7 Sub-contracting work

Another consequence of poor changeover performance was the need, sometimes, to sub-contract work. In part this need arose to overcome short-term capacity problems. On other occasions sub-contracting of short runs was used because production in-house – with the existing poor changeover performance – was not viable.[91] The factory indicated that this work was often done at a loss, and was accepted from its customers usually as part of a larger package.

Significantly, manufacture of the products of this discussion, either at the company or at the sub-contractor's premises, would be undertaken on broadly similar equipment. The fact that small runs were viable elsewhere was at least indicative that other manufacturers were enjoying substantially better changeover performance.

4/2 Market requirements represent external influences upon the business. By contrast, as discussed in Chapter 2, a business should also be able to adapt to **accommodate internally generated uncertainty**. Better changeover performance enhances the ability of a business to change its manufacturing operation responsively to accommodate, for example, equipment failure, inadequate resource management (including ensuring material supply), poor forecasting, poor communication, confused scheduling or slow decision taking. The better the responsiveness of the business, the less impact such poor internal control is likely to have. Exhibit 4.8 describes how better changeover performance can alleviate the disruption caused by changeable scheduling. Likewise, Exhibit 4.9 describes how poor quality of incoming material can disrupt factory operations. In either case the impact of better changeover performance is relatively easy to quantify. This is not always the case, and in many instances the assessment that is made, because of its inherent inaccuracy, will be used only to qualify decisions that are made. Often (as we describe in more detail below for the high-level classification of 'enhanced process control') an assessment will need records of failings or difficulties that have occurred in the past, coupled with an understanding of better changeover performance's contribution to overcome such problems should they occur in the future.

Exhibit 4.8 Changeable scheduling

The situation of changeable scheduling at a mainland European factory was described in the previous chapter.[92] Production jobs were being broken down before completion to accommodate changed scheduling priorities. Each job affected in this way thus required two changeovers to occur. The financial contribution of better changeover performance to alleviate these losses can be estimated. To enable a calculation to be made, records of when jobs were broken down (interrupted) because of scheduling changes need to be available. The approximate cost of non-production additionally has to be known. Different factories and different circumstances will prompt different internal measures to be used. In this case the factory was unable to keep pace with demand for its products. Profit per unit manufactured was known, as was the average production rate. Lost production could thus be equated to line downtime arising from changeover. From this the approximate financial loss to the company from changeable scheduling could be calculated.

More than two jobs per week were being broken down because of scheduling problems. The average changeover time per job was 79 minutes. Thus (approximating) two additional changeovers had to be conducted, adding 160 minutes per week to line downtime (for a line that was being operated 24 hours a day). For these high margin products, in excess of £115 000 was being lost annually.

Exhibit 4.8 (*Continued*)

Discussion

Faster changeover performance could help alleviate these losses by two mechanisms. First, if the same level of interruptions were to occur, any reduction in changeover time will lead to a commensurate reduction in downtime – and hence to a reduction in these losses. If, for example, changeover time was to halve, a reasonable approximation is that savings of £57 500 per annum would be experienced. Second, if better changeover performance was being used to manufacture in appreciably smaller batches, then the need to break jobs down before completion might in any case be eliminated. If so, larger savings still would be expected.

Changeable scheduling is only one representation of internally generated uncertainty. The next example, taken from a printing operation at another mainland European factory, considers a different problem.

Exhibit 4.9 Poor item quality

We have often witnessed failure to commence manufacture of a scheduled batch because of quality problems with changeover items. Item quality deficiencies – which can be apparent in a number of areas (see the previous chapter) – may be classified as internal control failings, whereby it has not been ensured before the commencement of the changeover that all items are present to the quality standard that is required. Quality problems, in our experience, can match logistics problems (actually getting items to be present in the first place) in terms of the frequency at which they occur. Their impact can be greater, though, as in many cases quality problems are only determined once the changeover is well underway. Even when problems have been identified, there often follows a lengthy period of fruitlessly attempting to make adjustments to compensate for them – hence significantly extending the changeover before it is finally abandoned.

Estimating the cost of these problems can be straightforward – as long as adequate failure records have been logged. For example, in one instance, line records and scheduling records indicated the frequency and duration of abandoned print changeovers because of ink problems. This one failure reason alone, on just a single print line, was estimated to be losing the company £27 000 per annum.

If anything, this figure represents an underestimate of the extent of the ink problems. What was not separately recorded was instances of the ink being poor enough to disrupt the changeover but not, once extensive adjustments had been made (including sometimes reformulating the ink by adding other inks of different colours), leading to the changeover being abandoned. In such cases, which occurred comparatively frequently, the changeover would be significantly extended. Again, ink problems are impacting upon the print operation. With sufficient data, such problems could be quantified.

4/3 Vagaries of the manufacturing operation might be such that sometimes quality or production rates will fall below what is sought. Minimal inventory (WIP and finished goods inventory) can allow any such problems to be more readily identified. Hence **response to manufacturing problems** can be enhanced.[93,94] As previously, those determining benefit in this area might need to make assumptions based on historical data to allow an assessment to be made.

4/4 Responsive small batch manufacture may allow a business to consider new work that previously it has been economically obliged to turn away. Better changeover performance thus can allow **better potential to supply niche markets**.[95–97] By doing so market presence is also increased. In other cases, for many different reasons, and even though changeover capability is poor, a company might anyway commit itself to manufacture widely divergent batch sizes. In such instances significant changeover time losses are likely to encroach upon the profitability of the smaller batches.[98] Here, sharper changeover performance can alleviate the losses to which the company's obligations might otherwise expose it.

4/5 Often small batch work can command premium rates, and one example of improved changeover performance allowing such high-margin business to be taken is presented in Exhibit 4.10. Such work might also have the benefit of providing an introduction, which can later be built upon by more substantial business. Batch size considerations are not the only way that higher-margin business might be gained. Better changeover performance leading to improved lead time or better product quality might equally allow premium rates to be commanded.

Exhibit 4.10 High-margin, short-run business

The factory first described in Exhibit 4.1 was engaged in the manufacture of industry-standard sizes of its products. As previously noted, this site manufactured four standard sizes. Elsewhere within the group a separate factory was engaged in the dedicated manufacture of the final industry-standard size variety of the product.

This consumer packaging market was undergoing a period of transition. Increasing calls were being made by customers for non-standard 'promotional' sizes. These calls for change were unlikely to recede. Competing manufacturers thus had an opportunity for competitive advantage by responding first to this demand. The prices that customers were willing to pay for short-run promotional size products were 25 per cent above what would be paid for the nearest standard-sized equivalent product. With such small margins for the standard products, this was potentially a very attractive opportunity.

If the manufacturer in our study was to seize this opportunity two significant hurdles had to be overcome. First, tooling costs had to be considered. These costs were high, although it was likely that the tooling would be re-used for future orders – either for the same customer or for other customers. Second, the manufacturing equipment for this product family had not originally been designed with good changeover performance in mind. The original design concept for the equipment went back to the late 1960s. Improvements to aid changeover performance had been made since, but changeover performance was still poor. The situation was faced where changeover to the new size would take 24 hours to complete (at the

Exhibit 4.10 *(Continued)*

then-current level of performance), to be followed by just a 6-hour period of production, to be then followed by a further 24-hour changeover. Until the changeover time could be significantly improved, the company was far adrift from being able to justify this new small-batch promotional work.

Enhanced process control – the difficulty of quantifying benefit (5/1 to 5/6)

Potential benefits within the classification of 'enhanced process control' will be difficult to quantify. Better discipline, design changes, better integration of maintenance, better equipment layout, better changeover procedures or greater changeover frequency might all have contributed to higher-quality changeovers taking place, and hence to benefit within this classification. Evaluating of the contribution that better changeovers provide, however, is likely to be problematical. In each case – for **enhanced product quality, increased process reliability, increased process volume capability, reduced scrap, reduced equipment damage** or **enhanced safety** – the analysis can be made significantly more difficult by not being able to predict accurately in advance the level of improvement that better changeover performance is likely to bring about. Historical data (should historical data of sufficient quality exist) will identify the extent of existing problems, but that still leaves unresolved improved changeover's contribution to overcome them. This difficulty of evaluating the possible level of improvement can be compounded further, for example, for **enhanced product quality**, by it being highly difficult in turn to assess what the improvement itself will contribute to the business.

As researchers we have excluded this final benefit category from previous formal benefit analyses that we have conducted. We have simply adopted the stance that a positive contribution towards process control will be made as changeover performance is improved – for which both evidence exists[99,100] and which our experience tells us will come about. Each specific area of benefit, for example **enhanced safety**, might separately make a major contribution (even if only on a one-off basis).[101,102] Many businesses expect scrap to be generated as part of the changeover process, whereby the potential savings can be significant if **reduced scrap** can be brought about.[103–105] Similarly, for example, the impetus to **enhance product quality** should always be present, for which considerable potential rewards can also be available.[106]

For each of these categories there is real difficulty in precisely quantifying potential benefit. This is not to say that a cautious limited assessment cannot be made, at least estimating the contribution that might ensue. Historic data will normally need to be available to enable an assessment to be made. Additionally, some underlying assumptions will need to be made which either correlate what has occurred in similar circumstances elsewhere, or which concentrate on highly likely outcomes as changes to existing practice are made. We give an example now concerning tooling damage/refurbishment costs. To highlight how a benefit analysis might take place, three possible scenarios are briefly described:

Scenario 1: Reduced equipment damage (*analysis assuming a likely outcome*)
Tooling damage is attributed to poor manual alignment procedures during changeover. The use of precision location devices might be expected to eliminate, or at least significantly reduce, the extent and frequency of occurrences of damage. Under such circumstances a benefit approximation could be made, for example, if adequate evidence exists to do so, by halving annual die set refurbishment costs.

Scenario 2: Reduced equipment damage (*correlation with similar outcomes*)
There are possible alternative ways of assessing the situation of tooling damage/refurbishment costs outlined above. If there is evidence from previous corresponding situations, then this evidence might be used in the current analysis. For example, the company itself might have used the same improved location technique at other sites. It will thus have evidence of the improvement's effectiveness. This evidence can be brought to the current assessment. Depending on the similarities that exist, the validity of this evidence can be judged and allowances given in the assessment that is made. Alternatively, similar improvements in tooling location might previously have been reported, for example in the trade press, which can be cautiously correlated with the company's own situation.

Scenario 3: Reduced equipment damage (*analysis assuming a likely outcome*)
A third way of accessing evidence to make an assessment of tooling damage/refurbishment cost savings might be to take account of device manufacturer's claims. For example, there might be a manufacturer of tooling location devices who will present claims of reduced damage during changeover that use of his devices will yield. Caution will need to be exercised when using manufacturer's data as these may be either overstated or not directly applicable.

Existing literature on the extent and impact of higher-quality changeovers, either separately or *vis-à-vis* faster changeovers, is not extensive. Potentially this is an area of further research.

Changeover frequency

Not all potential benefits outlined above (Figure 4.1) may be taken simultaneously. The single most influential factor here is the frequency with which changeovers are conducted. For example, conducting changeovers of one quarter their previous duration at four times their previous frequency by itself will have little or no impact on total downtime (and hence line productivity). The same production regime, though, can contribute significantly towards greatly reduced inventory and enhanced manufacturing flexibility. Alternatively, maintaining the same frequency of changeovers, but now of a significantly reduced duration, could have a substantial impact on line output.

Notwithstanding changeover quality considerations, there is inherent conflict as to how faster changeovers might be employed. This conflict will become of ever less significance as changeover times continue to improve. In a situation of near-instantaneous changeovers, a business might realistically start to seek benefit in all available areas simultaneously.

Improved changeover performance and TPM

Figure 4.1 shows the extent to which benefit potentially is available to a business beyond the confines of changeover's application under a TPM programme. TPM can contribute much to a business, but the way that TPM seeks to use faster changeover performance (specifically, to reduce line downtime) can constrain its impact (see also Chapter 2).

Undertake a full benefit assessment

It should be noted that the benefits that are described within the classification of Figure 4.1 are often highly interwoven. For example, low work-in-progress inventories are typically strongly related to manufacturing lead time.[107–109] Alternatively, if the number of manufacturing lines can be reduced, a beneficial knock-on effect on WIP may arise. In addition, inventory reduction and plant reduction can both lead to more space becoming available and less people being engaged.

The classification nevertheless gives focus to a benefit-appraisal exercise. Ideally the business should attempt to quantify benefit in all areas, or at least in all areas that are deemed possible given the extent of the available information with which to conduct an analysis. We have determined, for example, that when a business seeks to exploit better changeover performance only to achieve enhanced output, this may be simply because the benefit available in other areas has not been assessed. Alternatively, management pressure can oblige changeovers to be exploited in a predetermined way (Exhibit 4.1). On more than one occasion during our field studies we have concluded that significantly greater financial benefit would have been accrued had the business in question decided to exploit better changeover performance in an alternative way.

Goodwill

Most potential benefits within the high-level classification of **enhanced flexibility**, by their nature, will be experienced directly by the customer. In other categories – **reduced downtime; reduced inventory; reduced resource; enhanced process control** – better changeover performance typically only directly benefits the manufacturing organization.

Service that meets or exceeds customer expectations can enhance customer goodwill. The lost opportunity for promotional work noted above in Exhibit 4.10, for example, may have been compounded by a less easily quantified loss: a reduced customer perception of the manufacturer, diminishing any goodwill that might exist.[110,111] Because of the difficulty in assessing it, goodwill has deliberately been omitted from the classification presented in Figure 4.1.

Improved working environment

Like goodwill, changed worker attitudes might be anticipated,[112–114] but will be largely intangible and hence, again, highly difficult to quantify. We have additionally omitted changes of this type from our classification.

Group technology

Previous authors have described better changeover performance in relation to group technology.[115–118] The benefits that are described in this chapter apply.

4.3 A financial benefit analysis procedure

The purpose of the benefit classification (Figure 4.1) is to direct the business to the many potential ways that benefit might be available. For a financial analysis to be conducted it will be necessary to have access to various data, some of which might need to be recorded first-hand. A comprehensive benefit analysis might include a sensitivity model, so the benefits at different levels of changeover reduction can be calculated (which can significantly help a business decide what level of improvement it should aspire to*). A thorough financial benefit analysis will also need to take full account of different constraints that might apply.

As discussed above, a wholly accurate financial analysis is unlikely to be possible for all potential areas of benefit. At times, using the best available data, some well-judged estimates will be necessary. In all, once it is complete, the benefit analysis will enable implementation decisions (whether to go ahead with an initiative and where to direct the improvement effort) and exploitation decisions (how best to make use of the newly improved changeover facility) to be taken with greater authority. The wider topic of a benefit analysis procedure is now briefly considered.

Methodology

Three principal elements of a financial benefit analysis are briefly described:

- Data gathering
- Data analysis (using a classification of where benefit might be available)
- Data modelling (conducting a 'what-if' sensitivity analysis)

These elements are now considered. Constraints will need to be assessed concurrently as the analysis is undertaken.

Data gathering

The amount of information that needs to be collected for a comprehensive financial analysis can appear daunting. By the same token, the financial reward that a business might expect from significant improvement to its changeover performance can be

---

* While the financial benefit for differing levels of changeover improvement can be calculated (although this is not necessarily simple), assessing in advance the likely cost of achieving these levels of improvement will not be at all easy. Some rules of thumb can be employed, for example that a 50 per cent reduction in changeover time can be achieved without substantive design changes being made,[119] but this needs to be done with extreme caution (see also Chapters 3, 5 and 7). Perhaps the most certain way to assess the extent of improved performance against the cost of achieving it is through the purchase of alternative, more changeover-proficient equipment. Even here problems may arise because, as we have explained previously, manufacturers' performance claims in respect of changeover can be highly optimistic.

substantial. Analyses that we have conducted (see also previous exhibits in this chapter) have run to benefit in excess of £1 million per annum for a single line. To understand how to harness the potential of changeover improvement to best advantage, data gathering, in all its complexity, has to be confronted.

Typically information will need to be gathered from finance, planning, commercial and production functions. Data will be acquired as necessary to conduct individual assessments as separate benefit opportunities are identified. Data as noted might be required (although this should not be taken as an exhaustive list, nor that similar data need be collected in all instances):

- **Finance**
  (Manpower costs; product costs; pricing structure; margin variance for differing customers or for differing volumes; inventory costs; scrap costs; non-delivery costs; cost of buying in surplus product; cost of sub-contracting work)
- **Planning**
  (product range; manpower use; performance indicators; lead times; order profile; line selection; scheduling tools/methodology; perceptions of changeover; planning horizon; capacity; batch sizing; inventory; seasonal factors; liaison with marketing/commercial functions)
- **Commercial**
  (product sales profile; accepting business; customer agreements; small batch business; inducements; customer base; service criteria; competition benchmarking; seasonal variability; market maturity; emergency demands; delivery failure histories; lost business)
- **Production**
  (existing changeover data; work shop space; capacity; decision authority; minimum order size; manpower allocation; changeover documentation; change parts; product scrap histories; production equipment/tooling damage/wear and refurbishment costs (including downtime losses); maintenance structures and procedures)

Data analysis

Data analysis is the task of using the data accessed above and using it, for each of the classification areas of Figure 4.1, to determine the benefit that the business could accrue. It will be necessary to have established a new target changeover time for this analysis to be conducted. As noted, a thorough understanding of all potential benefit opportunities is required if a full analysis is to be made.

Data modelling

Data modelling represents a refinement to the data analysis step, above, by allowing a sensitivity analysis to be performed. In the past we have constructed models using personal computer spreadsheet packages. By varying changeover time targets in these data models it was possible quickly to assess how potential benefit to the business varied as changeover times varied. By performing a sensitivity analysis additional information is made available which may be used when judging what changeover performance to aim for.

4.4 Constraints

Opportunities for benefit in the areas described by Figure 4.1 must be tempered by an understanding of possible market (external) and operational (internal) constraints. In normal manufacturing operations, resource in both manpower and plant needs to be balanced effectively to achieve the required manufacturing capacity to fulfil customer orders.[120] These two components of capacity and resource are now discussed briefly in relation to improved changeover performance.

Manufacturing capacity/resource availability

Faster changeovers can be used to increase overall manufacturing capacity, without the need to increase levels of manpower or equipment. This option, though, is unlikely to be taken when existing lines are already underutilized and where surplus capacity already exists. The situation of existing surplus capacity has been relatively common within the businesses we have studied. In the case of there being existing surplus capacity there would be little point, in the short/medium term at least, in assessing benefit in this area.

Exhibit 4.4 describes how faster changeovers can be used to reduce equipment levels (and consequently, staffing levels). An imbalance again exists between what the market demands (in terms of product volume) and what the business is geared to provide. In this case the decision is taken to pare down existing resources, whilst still meeting set performance targets. The exhibit described the minimum possible level of resource that was needed to meet existing demands. The exhibit did not take into consideration:

- possible increased future demand
- fluctuating short-term demand
- breakdown history
- planned maintenance

These issues are described further elsewhere in our book. Their effect is to constrain the changes that it is wise to make to the manufacturing facility. These and other similar issues should be understood by anyone undertaking a financial benefit analysis.

Constraints are not concerned solely with issues of balancing equipment levels to market requirements. Manpower considerations similarly can significantly affect the conclusions of a financial benefit analysis. The factory described in Exhibit 2.2, for example, experienced short-term seasonal production peaks. Equipment was taken from store and set up to meet this increased demand when it occurred. As well as there needing to be equipment, so too there needs to be manpower to operate the equipment, coupled with an additional on-site commitment to manage these staff. More directly, in respect of changeover activity, the use of additional equipment, which similarly requires changeovers to occur, imposed significant extra loads on the limited pool of skilled engineers on site who conducted changeover tasks. If such staff simply are not available, then the changeover plans, and hence the financial benefit analysis, need to be revised accordingly. Alternatively, if additional staff are to be taken on, employment and training costs will need to be factored into the assessment. In such circumstances there is also likely to be a delay before the desired changeover capability is achieved.

These constraints on changeover activity are imposed by specialist abilities limiting who can do what during a changeover. The need for physical strength to conduct a changeover in one instance has already been given (Chapter 3), where one particular technician, in a predominantly female workforce, needed to be present for the changeover to proceed. If, for example, this changeover ideally needed to occur on the night shift, the factory had the dilemma of amending this man's working hours or postponing the changeover until he next came on-shift. Other potential manpower constraints should also be allowed for. Exhibit 4.3 described how the use of a specialist changeover team actually limited the number of changeovers that could take place during a day – rather than, as was hoped, concentrating expertise, reducing the average time that a changeover took to complete, and hence increasing the number of changeovers that occurred.

The staff constraints noted above present a strong argument for simplifying (de-skilling) changeover activities, so as to make it easier for any personnel on-site to conduct a successful, rapid, high-quality changeover. By job demarcation, by lack of skill, or for other reasons, restricted manpower flexibility can significantly impact upon the level of changeover performance that is achieved.

Overcoming constraints

Like changeover improvement itself, an important step in overcoming internal constraints is the adoption of a 'can do' attitude. For example, design changes might be applied, or a training programme might be instigated. By design or otherwise, significant revisions to existing tasks and procedures may be made. All the internal constraints described above could be alleviated if it was decided that it was appropriate to do so, coupled with a will to see changes through; although in some cases major revisions to current practice may be involved.

4.5 Conclusions

A changeover improvement programme should not be entered into simply on the perception that it is 'a good thing'; perhaps commenced on the basis of current fashion. Instead, changeover improvement should be undertaken only once it has been determined strategically in what ways improvement will assist the business, and further, once this advantage has been quantified. The benefit classification and the procedure outlined in this chapter can greatly assist in allowing this to be done.

Opportunities for operational advantage and commercial advantage are both likely to exist. It is important that all possible ways of exploiting improved changeover performance are assessed and that no preconceptions of how better changeover is to be used are entertained. Such preconceptions may well be misplaced.

When conducting a financial benefit analysis the business should also be aware of, and take into account, constraints that can influence how better changeover performance might be exploited. Alternatively, constraints might limit the level of improvement that may be achieved and hence, once again, the benefit that the business is able to derive from a changeover improvement initiative.

References

1 Leschke, J. P. (1997). The set-up reduction process: part 2 – setting reduction priorities. *Production and Inventory Management Journal*, **38**, 38–42. It is noted that changeover time reduction down to 'single minutes' is not always justified: '. . . "SMED" implies that set-up times should be driven into the single-digit range without an explicit examination of the costs. By way of contrast, the academic literature emphasises the trade-off between the cost of reducing set-up and its benefits and, thus, it may not be optimal to drive set-up times arbitrarily to single-digits.' Thus there are two fundamental financial questions: what performance is required by business needs; what level of improvement can be financially justified?

2 Vörös, J. (1999). Lot sizing with quality improvement and set-up time reduction. *European Journal of Operational Res.*, **113**, No. 3, 568–574.

3 Myers, M. S. (1987). Don't let JIT become a North American quick fix. *Business Quarterly*, **5**, No. 4, 28–38. Much has been written recently on measuring business performance, where it is noted that assessment purely in financial terms can be unwise. Myers, for example, sets out some non-financial internal performance measures that might be used. Improvement in one measure will often lead indirectly to improvement in other respects. Nevertheless, a financial assessment of any improvement measure remains an important indicator of its potential contribution.

4 Zunker, G. (1995). Fifty percent reduction in changeover without capital expenditures. *PMA technical symposium proc. for the metal forming industry*, pp. 465–476. Zunker has experience as a US changeover consultant. He makes generalized claims (as do many others) about the levels of reduction that a typical changeover improvement programme can expect to achieve. The low-cost aspect of improvement is also emphasized. Unlike some others, however, Zunker writes that latter improvements are likely to require equipment modifications, which may be costly. It is asserted that reductions in changeover time from 100 per cent to 50 per cent of the previous time can come about by low-cost 'method improvements'. Subsequent improvement beyond 50 per cent will involve 'press and tool modifications': 'The last 40 to 50 percent reduction in set-up time requires major press and tool revisions. This could result in major expenditures and long lead time for implementation.' The pattern of a typical improvement programme passing from low-cost method improvements through to equipment modifications will be discussed later in this book. The pattern is not always followed, but in many instances (ignoring the 50 per cent figure that Zunker apportions) can represent a sound way to proceed. Stating this, the practitioner still has to decide when each of these two approaches is the most appropriate. A methodology for identifying improvement options will be described later in Chapter 7.

5 Suzaki, K. (1987). *The New Manufacturing Challenge*. Free Press, New York. Suzaki sets out an accomplished overview of the many changes that can be made to a manufacturing organization to raise its competitiveness. Changeover improvement is featured (drawing heavily on Shingo's 'SMED' methodology), but Suzaki makes it clear that changeover improvement out of context – without commensurate improvement elsewhere in the organization – will not make as much of a contribution as it could otherwise do.

6 Beard, T. (1995). A system for quick change. *Modern Machine Shop*, **68**, No. 12, 100–107. Changes made in a machine shop to increase its changeover capability are described. The changes extend beyond the process equipment and the operators, where wider system

changes are also necessary. In respect of the cutting tools, Beard writes: '[A system was required]. . . to support quick change both on and off the machine. Its a system that's roughly one third hardware, one third information and one third discipline.'

7 Primrose, P. (1991). The appliance of science. *Manufacturing Engineer*, **70**, No. 9, 42–43.

8 Sohal, A. S., Samson, D. and Ramsay, L. (1998). Requirements for successful implementation of total quality management. *Int. J. of Technology Management*, **16**, No. 4–6, 505–519.

9 Camisón, C. (1998). Total quality management and cultural change: a model of organisational development. *Int. J. of Technology Management*, **16**, No. 4–6, 479–493.

10 Taylor, W. A. (1997). Leadership challenges for smaller organisations: self-perceptions of TQM implementation. *Omega*, **25**, No. 5, 567. Taylor presents the results of a questionnaire survey in which perceptions of the financial impact of TQM were investigated. Ninety-six per cent of respondents offered an opinion on whether TQM was contributing to their company's 'bottom line'. Of these 96 per cent, 63 per cent believed that TQM was yielding financial benefit, 36 per cent believed it was too soon to make a judgment and 1 per cent reported that TQM had had no effect so far. It is significant that this study is founded on *perceptions* of the financial impact of TQM, rather than on a quantified assessment. In respect of TQM we cannot comment on the ease with which a proper financial assessment could be made. From other research we can state, however, that the same is not true for a changeover improvement initiative, where a financial benefit analysis may be undertaken. Doing so puts the business in a far stronger position to decide whether or not changeover improvement should be conducted and, if so, how this should be done and what improvement targets to apply.

11 Productivity Consulting Group, Portland, USA. http://www.productivityconsulting.com/bodies/body_quick_changeover.html (site accessed 20 March 1999). A changeover consultant's advertisment (advertising a changeover workshop based on Shingo's 'SMED' methodology) typically sets out a range of potential benefits: 'Learn a proven methodology for reducing changeovers (make-readies) by at least 50%. You'll increase capacity, cut inventory, reduce lot sizes, respond faster to customer demands, clear floor space devoted to inventory, and lay the foundation for true just-in-time production.'

12 Steudel, H. J. and Desruelle, P. (1992). *Manufacturing in the Nineties – how to become a lean, mean world-class competitor*. Van Nostrand Reinhold, New York. A wide variety of potential benefits are set out.

13 Arinze, B., Kim, S. L. and Bannerjee A. (1995). A multicriteria model for supporting set-up reduction investment decisions. *Production Planning and Control*, **6**, No. 5, 413–420. A model to assess the impact of better changeovers is presented by Arinze *et al.* It is theoretical in nature and is built on a number of significant assumptions. The model uses weighted factors affecting likely set-up improvement (seven factors are described), which the model user must assess. The paper's authors acknowledge that this approach is somewhat subjective. The method of assessing the financial impact of better changeover performance described in this book is instead empirically based. For each area of benefit that is described it should be possible to acquire data with which to derive a figure for 'bottom-line' contribution to the business.

14 Fabry, R. and Gelders, L. (1994). Cost/benefit analysis of a JIT supply system in a car-assembly plant, *Logistics Information Management*, **4**, No. 4, 32–35. Benefits of introducing JIT manufacturing or lean/agile manufacturing are sometimes quantified. For example, Fabry and Gelders, among other criteria, assess inventory reduction. The study shows that in-factory inventory of car components can be significantly reduced (by an amount of DM8.9 million in 1994) by improvements to part supply and throughput. Caution is needed to differentiate such inventory reduction to that which can be achieved by better changeover performance alone.

15 Rozema, H. and Travaglini, V. (1995). Quick mold change systems for high volume stack molds. *Proc. Annual Tech. Conf. ANTEC 1995*, Soc. of Plastics Engineers, Vol. 1, 1011–1015.

16 Gallego, G. and Moon, I. (1995). Strategic investment to reduce set-up times in the economic lot scheduling problem. *Naval Research Logistics*, **42**, No. 5, 773–790.

17 Boyer, K. K. and Leong, G. K. (1996). Manufacturing flexibility at plant level. *Omega*, **24**, No. 5, 495–510.

18 Dale, B., Boaden, R., Wilcox, M. and McQuater, R. (1998). The use of quality management techniques and tools: some key issues. *Int. J. of Technology Management*, **16**, No. 4–6, 305–325. Observations are made in relation to gathering quality improvement data: 'The communication barrier that separates the professional accountant from the non-accountant is a major hurdle. There is also the precision and attention to detail employed by accountants which means that they are loathe to use assumptions and make estimates, even where this is the only means of obtaining cost data.' The same observations might apply equally to collecting data to conduct a changeover analysis. As discussed later in the chapter, it is sometimes necessary to make estimates based on the best available information.

19 Mapes, J. (1992). *Manufacturing Strategy – the strategic management of the manufacturing function*. MacMillan, London. Mapes states a task of the production manager is: '. . . to achieve high levels of customer service using low levels of resources while coping with unpredictable levels of demand.'

20 Oishi, O. (1990). Total quick mold change. *Proc. 48th Annual Tech. Conf. of Soc. of Plastics Engineers*, Dallas, USA. Reports of changeover time improvements dominate within changeover improvement literature. A proportion of the literature, for example Oishi, also records that higher-quality changeovers have simultaneously been achieved.

21 Takami, K. and Nakagawa, Y. (1997). Advanced electrical equipment for hot rolling mills. *Mitsubishi Electric Advance*, **79**, June, 2–4. Mitsubishi claim that design changes made to their equipment (particularly, the use of a parallel adjustment system) improves changeover time and, by enhancing its control capabilities, also improves the new on-line product's quality.

22 Smith, D. A. (1991). *Quick Die Change*. SME, Dearborn, USA. Higher-quality changeovers should give rise to better process consistency (see also to the quotation by Smith provided in Chapter 1/reference 35).

The accuracy of parameter settings typically is intrinsic in attaining better-quality changeovers. Ways that parameter settings may be accurately and swiftly achieved will be investigated later in this book, especially in Chapter 5. For the majority of changeovers, adjustments need to take place to achieve the product-specific settings that are required. This can be a time-consuming process, not least if it is done on a 'trial-and-error' basis. Fast changeovers that are fast simply because due attention has not been given to aspects of adjustment will be paid for in sub-standard quality (of the product and, quite possibly, of other line performance measures). Alternatively a 'fast' set-up period, when parameters are not correctly set while the line is stationary, can be exposed by the subsequent need for greater attention to adjustment, leading to an overlong run-up period and, possibly, a longer total elapsed changeover time than would otherwise have occurred.

23 Ebrahimpour, M. and Schonberger, R. (1984). The Japanese just-in-time/total quality control production system: potential for developing countries. *Int. J. of Production Research*, **22**, No. 3, 421–430. Ebrahimpour and Schonberger state: 'Japanese experience shows that when setup time and lot size are reduced, there will be a significant improvement in quality, scrap and worker motivation.'

24 Womack, J. P., Jones, D. T. and Roos, D. (1990). *The Machine that Changed the World*. Rawson Associates, New York. Womack *et al.* similarly report that conducting more

changeovers can lead to better quality: 'The discipline imposed in the plant by manufacturing small lots is one of the key steps to greater efficiency and quality in lean production.'

25 Wantuck, K. A. (1984). Set-up reduction for quality at the source. *APICS Seminar Proceedings – Zero Inventory Philosophy and Practices*, pp. 179–183. Wantuck's article places emphasis on changeover activity contributing to product quality. Wantuck asserts that: 'Almost universally overlooked is the fact that set-up reduction is the keystone to Quality at the Source...' Wantuck also writes: 'Dedicated tooling, automatic die positioning, standard shut heights, duplicate tool holders, etc., will more than pay for themselves in better product quality. This is also the most expensive step, because most improvements in this category require tooling or capital monies.'

26 Sriparavastu, L. and Gupta, T. (1997). An empirical study of just-in-time and total quality management principles implementation in manufacturing firms in the USA. *Int. J. of Operations and Production Management*, **17**, No. 12, 1215–1232. Improvement in changeover speed and changeover quality will often be part of a wider improvement initiative: '...we conclude that companies which implemented JIT observed significant improvements in quality and productivity without changing the production lot size. Specifically, companies observed significant improvements in lead time, delivery cycles and rejection rates. A substantial reduction in stocks of finished goods and WIP as well as a reduction in set-up time were also reported. These positive results are attributed to the use of work-cell concepts such as uniform workload, kanban, and group technology principles, which are very important for a successful implementation of JIT.'

27 Duplaga, E. A., Hahn, C. K. and Watts, C. A. (1996). Evaluating capacity change and setup time reduction in a capacity-constrained, joint lot sizing situation. *Int. J. of Production Res*, **34**, No. 7, 1859–1873. Duplaga *et al.*, concentrating only on changeover improvement to enhance effective capacity, develop a model that can be used to evaluate the effectiveness of line capacity change. Changeover improvement is contrasted with other alternatives to achieve this objective.

28 Klospic, A. R. and Houser, W. F. (1997). Increased throughput with rapid changeover at Tenneco. *National Productivity Review*, **17**, No. 1, 59–65. Additional productivity is the driving motive behind the improvement in changeover performance reported at a 24-hour continuous production facility. Changeover, in this case, relates to coil feedstock changeover. The product under manufacture remains constant.

29 Suzaki, K. (1987). *Op. cit.* ref 5. Suzaki reminds his readers of the folly of this stance. A summary of other possible opportunities is made, wherein it is noted: 'Increased capacity is not the primary reason to reduce set-up time.' This assertion still needs to be treated with some caution: there may be occasions when increased capacity does indeed represent changeover improvement's most advantageous use. However, the simple point not to be blind to other opportunities is well made.

30 Kelly, A. and Harris, M. J. (1978). *Management of Industrial Maintenance*. Butterworth, London. Kelly and Harris' book is a useful overview of factory maintenance, including preventive maintenance.

31 Feely, R. and Menard, J.-P. (1997). Increased PCB throughput via reflow oven design. *SMT Surface Mount Technology Magazine*, **11**, No. 6, 52–54. A need to increase maintenance effort in PCB reflow ovens is described. To do so with minimum adverse effect on line downtime, a situation is described of preventive maintenance occurring only during product changeover.

32 Hall, R. W. (1983). *Zero Inventories*. Dow Jones-Irwin, New York. Hall notes that preventive maintenance's role is to preserve and enhance the capability of the equipment – which, as we discuss elsewhere, can itself contribute to better changeover performance.

33 Arinze, B., Kim, S. L. and Bannerjee, A. (1995). *Op. cit.* ref. 13. This paper, unusually, cites preventive maintenance as a way to improve set-up performance (by helping to overcome the need for excessive adjustment). It is a view with which we concur.

34 Baker, P. (1998). Extra minutes mean extra mints. *Works Management*, **51**, No. 3, 36–37. In this particular example the bounds of changeover activity and maintenance activity are extremely blurred. Another way that changeover activity can contribute to process quality might be identified.

35 Productivity Consulting Group, Portland, USA. http://www.productivityconsulting.com/bodies/body_autoliv_asp_uses_tpm_page_3.html (site accessed 20 March 1999 – see also 'page_1', 'page_2' and 'page_4'). A TPM newsletter demonstrates notable similarities between the disciplines of changeover improvement and maintenance: 'In another operation, two rocker arms control the flow of product through the machine. Between the rocker arms is a cam follower that needs to be replaced once a month. Previously, every time the cam follower needed to be changed, a maintenance technician had to remove a shaft, remove a spacer between the rocker arms, remove the cam follower, replace it, then put everything back together. This procedure took 105 minutes. "The team realized this procedure alone was eating up a tremendous amount of time," said Whitley, "because we have ten of these machines." The team modified each spacer by cutting it in half and drilling two holes for Allen bolts. Now, changing a cam follower simply involves removing the Allen bolts that hold a spacer together. It's no longer necessary to remove the shaft and realign it after the new cam follower is installed. The new procedure takes just 15 minutes, which eliminates 180 hours of downtime per year on this machine. This translates into potential sales of an extra $4.7 million annually.'

The next chapter will investigate design changes that can be made to assist changeover performance, where some of the techniques described above for maintenance will be shown to be directly applicable. The excerpt also describes a financial assessment of the impact of the improvements that have been made, in this case (as the improvements have been conducted as part of a TPM programme) using them to reduce line availability (downtime) losses. It indicates what changeover improvement can achieve in this regard: if the same line availability gains had been achieved by changeover improvement (perhaps by employing similar techniques), equally large benefits should ensue.

36 Baker, P. (1998). *Op. cit.* ref. 34. The use of the 'SMED' methodology for maintenance purposes is described.

37 Petronis, T. J. and Krause, L. R. (1984). Innovative robotic workcell performs tool set-up. *Robotics Today*, **6**, No. 6, 37–39. A robotic workcell is described that performs what was previously a time-consuming task of replacing cutters in an automated aerospace application. In a highly automated process the previous dull cutter condition is monitored. A new, identical, sharp cutter is accurately installed to replace it. The operation is effectively preventive maintenance. The operations that are performed – and the way the problem is tackled – may be likened to what occurs during automated CNC tooling changeovers and, more generally, other highly automated changeover operations (see next chapter).

38 McIntosh, R. I., Culley, S. J., Mileham, A. R. and Owen, G.W. (2000). Changeover improvement: a maintenance perspective. *Int. J. Production Economics* (accepted for publication 13 November 2000). A more complete assessment is provided.

39 Anon. (1991). JIT a pressing matter. *Appliance Manufacturer*, **39**, May, 59. A brief article indicates the improvement in changeover performance that can be achieved when equipment designed to facilitate changeover is installed and used. The changeover time that is cited is just 3½ minutes. This is just one of many possible examples. Press system developments for changeover have been on-going for many years. The topic will be discussed further in the next chapter.

40 Suzaki, K. (1987). *Op. cit.* ref. 5. Suzaki claims that many manufacturing problems are 'covered' by excess inventory: poor scheduling; machine breakdown; quality problems; long transportation; vendor delivery; line imbalance; long set-up time; absenteeism; lack of house-keeping; communication problems. Suzaki further discusses 'inventory waste', writing that excess inventory increases the cost of the product. Some problems of excess inventory are identified: 'It [inventory] requires extra handling, extra space, extra interest charges, extra people, extra paperwork and so on.' Suzaki cites manufacturing small lots by reducing set-up times as one way to reduce inventory (also citing: disposing of obsolete materials; not producing items unless required by the subsequent process; not purchasing or bringing in items in large lot sizes).

41 BBC Radio 5 (1998). Late Evening Financial News Service, 19 August 1998. A short broadcast explained an offensive competitive strategy reliant on low inventory at Dell Computer. Approximately one third of Dell sales were being made via the Internet. All computers were stated as being made to order, and manufactured and shipped quickly to the customer, hence allowing the company to minimize inventory. A situation of rapidly falling component prices was described, particularly for the computer chips because of world overcapacity. Simultaneously, chip-processing capability was rapidly improving. Dell was strongly placed to pass on these advantages to its customers, which was cited as a successful strategy to increase market share. Further, in a market of rapid technological change, the company were not being saddled with stock that was becoming obsolete almost by the day. In time, the selling of this stock, for companies retaining high finished goods inventory, might have necessitated discounting. Doing so would have compounded the comparatively higher prices paid for components in those machines in the first place.

42 Gavirneni, S. and Morton, T. E. (1999). Inventory control under speculation: myopic heuristics and exact procedures. *European J. of Operational Res.*, **117**, No. 2, 211–221. A converse argument (see above, reference 41) favouring inventory might also be made – that speculative purchases in situations of increasing product/material cost might be justified. Such a stance assumes that savings made will exceed the costs of holding this purchased inventory.

43 Jones, T. (1998). Welcome to Britain's biggest fridge. *Independent on Sunday*, London, 11 October, 4. The building of a more responsive, centralized new distribution centre for Bird's Eye Walls frozen foods is described. The warehouse is a giant refrigerator 34 m high × 66 m wide × 140 m long, which operates at −28 °C. It is Britain's largest frozen single-chamber container. The building is big enough to store 5 ice-cream bars for every person in Britain, or to store a fish finger for every person in China. It cost approximately £37 million to build. Operating costs are comparatively high. The site is fully automated, partly for safety reasons because of the extreme conditions. The company's project manager is cited: 'We're building the fridge because of customers. We had to become more high-tech to get stock out to them very quickly. Where before we had a lead time of two days between orders and delivery, we will now be able to turn things around within 24 hours. Demands for things like ice-cream can be very whimsical. When the sun shines, you sell more, so we've got to be able to respond.'

44 Cunningham, R. (1998). Balancing inventory and service levels. *APICS – The Performance Advantage*, **8**, No. 8, 42–46. A measure of a successful company is asserted to be management of the conflicting demands of inventory and customer service levels. Cunningham asserts that excellent communication within the organization is necessary, particularly between sales/marketing functions and production planning (see also Chapter 3).

45 Malucci, L. J. (1999). Communications with the University of Bath – April 1999, National Research Analyst, American Production and Inventory Control Society (APICS), Alexandria, USA. Louis Malucci discussed the complexity of determining inventory carrying cost, explaining how assessment methods have changed where, particularly, attempting to

measure carrying cost as a percentage of inventory value has declined. Readers may choose to pursue this topic further. It suffices here to note that inventory costs are typically found to be substantial.

46 Robins, R. M. (1989). Quick changeovers – fast paybacks. *Manufacturing Systems*, **53**, No. 3, 53–55.

47 Krajewski, L., King, B., Ritzman, L. and Wong, D. (1987). Kanban, MRP and shaping the manufacturing environment. *Management Science*, **33**, 39–57. A simulation is conducted, in which the impact on WIP of using reduced set-up times to manufacture more frequently in smaller batch sizes is determined.

48 Cho, K., Moon, I. and Yun, W. (1996). System analysis of a multi-product, small-lot-sized production by simulation: a Korean motor factory case. *Computers and Industrial Engineering*, **30**, No. 3, 347–356. A further simulation by Cho *et al.* is used to confirm that changeover time is a key factor to maintain small buffer sizes.

49 Bergman, I. (1999). Handling reform. *Manufacturing Engineer*, **78**, No. 5, 194–197. Reduced changeover times are by no means the only way that inventory reduction can be achieved. An account is given of improvement at a Swiss automotive component manufacturer.

50 Smith, D. A. (1991). *Op. cit.* ref. 22.

51 Owen, G. W. (1994). *The Buffering of Transfer Lines*, PhD thesis, University of Bath, UK.

52 Caputo, M. (1996). Uncertainty, flexibility and buffers in the management of the firm operating system. *Production Planning and Control*, **7**, No. 5, 518–528. Caputo argues in favour of considered use of WIP buffers as an economic way of maximizing line capability.

53 Slack, N., Chambers, S., Harland, C., Harrison, A. and Johnston, R. (1995). *Operations Management*. Pitman, London.

54 Gunasekaran, A., Goyal, S. K., Martikainen, T. and Yli-Olli, P. (1993). Determining economic inventory policies in a multi-stage just-in-time production system. *Int. J. Production Economics*, **31**, No. 7, 531–542. By manufacturing in this way (Figure 4.6) the company would be achieving what Gunasekaran *et al.* describe as a JIT ideal of producing 'the required item at the time and in the quantities needed'. The ideal, described another way, is a stockless production system. To manufacture in such a way will also require, among others, highly competent scheduling and logistics (for changeover) capabilities.

55 Anon. (1995). Teamwork is Rx for better bottling. *Packaging Digest*, **32**, No. 5, 50–56.

56 Anon. (1995). Lead times decimated. *Work Study*, **44**, No. 8, 28–29. One possibility (see also Chapter 5) can be to eliminate a series of machines and replace them with a 'single set-up' machining centre. This can have the effect of dramatically reducing changeover times (eliminating previous set-ups on successive machines for successive machining operations), lead times and, perhaps, manpower requirement.

57 Remich, N. (1990). Uptime now at 95% vs. 60%. *Appliance Manufacturer*, **38**, November, 42–43.

58 Souloglou, A. (1992). Scroll shear changeover. Bath University Changeover Group, internal case study report, July. Souloglou describes change at a research site: 'Over the last 3 years the finished goods inventory has halved from approximately 4 million to 2 million units. In addition, with the introduction of JIT, storage methods are better organised and as such the distribution staff have a much easier job. There has, in fact, been a reduction in staff from 25 to 15. Although space has not been freed for other uses the same space is now used more cost-effectively and with less frustration.'

59 De Ron, A. J. (1995). Measure of manufacturing performance in advanced manufacturing systems. *Int. J. of Production Economics*, **41**, No. 1–3, 147–160.

60 Hollingum, J. (1995). Quick change acts for robotic systems. *Industrial Robot*, **22**, No. 5, 31–33. Improvements may be 'engineered' both to existing work practices and, physically, to the equipment itself. The next chapter will consider in greater detail the use of

design to de-skill and otherwise alter changeover operations, including automated and semi-automated changeovers. Though not written about changeover *per se*, Hollingum strongly identifies automation's role (which, typically, is expensively acquired) to change tasks that previously existed: 'In a manual operation much depends on operator skill, whereas with automation, in most instances, the skill of the operator is less important as the machine takes over and gives not only a better and faster result, but also a more consistent result.'

61 Suzaki, K. (1987). *Op. cit.* ref. 5. Suzaki describes that increasing the skill of factory personnel enhances flexibility, allowing them to conduct additional tasks. Here in our text the same increased ability to conduct new tasks is not being accomplished by raising individual's skill levels, but rather by reducing the skill needed to undertake a task.

62 Thomas, A. (1996). Double whisky. *Packaging News*, October, 20–21. It is identified that United Distillers sites in Scotland are working hard to reduce changeover times. Revised design of a change assembly is cited for one machine as reducing changeover time to approximately 10 minutes and, further, allowing female line operators to conduct the changeover: 'The whole size change assembly now comes as a unit on trolleys we developed, which are simply wheeled up to the line, slid into place and locked up.'

63 Hollingum, J. (1995). *Op. cit.* ref. 60.

64 Smith, D. A. (1991). *Op. cit.* ref. 22. The availability of fork-lift trucks and skilled die setters is noted by Smith as a problem. This mirrors our own experience. The time that these problems can contribute to the total elapsed changeover time is noted in one example: 'By eliminating the delays for the forklift truck and delays due to interruptions to service production, the set-up time was reduced from 111 minutes to 71 minutes.' In the situation described in Exhibit 2.2, the fitters who conducted the changeover could sometimes be absent from the changeover for 2 or more days.

The converse situation could also occur. Entirely freeing skilled personnel from changeover duties (by eliminating the need for their services) could mean that line downtime because of breakdown, or for other reasons, will be reduced – as these personnel become more available to tackle production problems.

65 Suzaki, K. (1987). *Op. cit.* ref. 5

66 Slack, N., Chambers, S., Harland, C., Harrison, A. and Johnston, R. (1995). *Op. cit.* ref. 53.

67 Ueno, K. (1998). New guideless CNC shaper for helical gears. *Gear Technology*, **15**, No. 2, 17–19. A number of references can be found that describe use of programmable, variable-position systems.

68 Doyon, P. (1999). Changeover time minimized. *Connector Specifier*, **15**, No. 6, 14–15.

69 Schonberger, R. J. (1990). *Building a Chain of Customers*. Free Press, New York. Schonberger attributes significant benefits to responsiveness to customer needs: 'The main benefits of quick response are growth of sales and market share.'

70 Lam, K. and Xing, W. (1997). New trends in parallel machine scheduling. *Int. J. of Operations and Production Management*, **17**, No. 3, 326–338. Lam and Xing describe JIT firmly in terms of responsiveness – in terms similar to those we use when describing changeover capability (Chapter 2): 'The JIT philosophy is to produce the necessary products in necessary quantities at the necessary time.'

71 Cooper, M. (1998). Britain's best factories 2003. *Management Today – Guide to Britain's best factories* (supplement), **5**, 3–5. The opinion is sometimes passed that flexibility is a more valuable commodity than productivity.

72 Ayers, R. and Butcher, D. (1994). The flexible factory revisited. *American Scientist*, **81**, No. 5, 448–459.

73 Forth, K. D. (1994). Quick die change helps auto stamper produce to order. *Modern Metals*, **50**, No. 9, 30–35.

74 Lankford, E. (1995). Making short runs profitable on existing packaging presses. *Gràvure*, **9**, No. 1, 26–27. In many industries there is a marked trend towards ever smaller, ever more frequent batch deliveries. The example here relates to the print industry, where a company has undertaken internal improvement to its equipment and work practices. These changes have been in response to changing customer requirements, and a need to remain competitive to meet these changing requirements. It is stated that the company has increased the changeovers on its gràvure press stations by 185 per cent in less than 3 years. In that time it is reported that the average run size has diminished by 65 per cent, with only 16 per cent of jobs in 1995 being classed as long runs. Mechanical changes to aid changeover performance undertaken include an optical alignment system. Start-up scrap is stated to have been reduced by 30 per cent since the improvements were made. The JIT philosophy is to produce the necessary products in necessary quantities at the necessary time. Empowerment of the operators is also cited as an important change that has taken place.

75 Bungert, W. (1993). New solutions for ERW pipe mills in the size range from 0.5″ to 4″ O/D with quick-change over system and computerised mill settings. *Proc. 35th MWSP Conf. Mechanical Working and Steel Processing*, ISS-AIME, Vol. 31, pp. 199–213. Some information is given on changing batch size requirements in one particular market. The reasons for this change are briefly discussed, and implications on manufacturing operations described.

76 Miller, W. H. (1992). Chesebrough–Ponds. *Industry Week*, **241**, October, 43–44. Citing factory staff, changes to nail varnish batch sizes are discussed: 'We used to run a shade for two days on our nail polish line. Now we change eight to twelve times a day.'

77 Piroird, F. and Dale, B. (1990). The importance of lead time control in the order fulfillment process. *Production Planning and Control*, **9**, No. 7, 640–649. Better changeover performance (in this case, faster changeovers) can contribute to better lead times. Other revisions to working practice equally can contribute to better lead time performance. Piroird and Dale describe that the manufacturing component of lead time comprises: set-up time; process time; move time; wait time; queue time. It is noted that shortening the 'set-up time' impacts also upon WIP, hence additionally contributing to reduced lead time: 'One means for reducing the lead time is to shorten the set-up time, and this action directly decreases in-process inventory.'

78 Robins, R. M. (1989). *Op. cit.* ref. 46. Many authors cite that reduced changeover times can impact upon lead time. Robins briefly summarizes a case study, where he quantifies the lead-time reduction that has arisen. An 80 per cent decrease in changeover time (from 150 minutes to 30 minutes) is claimed as resulting in a 67 per cent reduction in manufacturing lead times.

79 Souloglou, A. (1992). *Op. cit.* ref. 58. Souloglou further describes change at one of our research sites: 'Over the last 2 to 3 years the factory's lead time has been reduced from 6–12 weeks to 15–25 days, with a 4 week guarantee. This, coupled with the fact that the minimum order now accepted has gone down from 50,000 units to 25,000 units, has given the factory the opportunity to increase the number of customers they serve by accepting more smaller orders. They have not taken this opportunity. However, this year has seen a 17% increase in volume, and this is attributed to the factory's fast response to the needs of existing customers. More small orders are being taken from these existing customers. 60% of all orders are now less than 50,000 units.'

80 Little, J. D. C. (1961). A proof for the queuing formula $L = \lambda W$. *Operations Research*, **9**, 383–387. It is stated that lower work-in-progress inventory will result in lower throughput time, hence reducing customer response time.

81 Yang, J. and Deane, R. H. (1993). Set-up time reduction and competitive advantage in a closed manufacturing cell. *European J. of Op. Res.*, **69**, 413–423. A feature of faster changeover performance described by Yang and Deane, specifically for a closed

manufacturing cell, is that variance in job flow time is reduced. Hence queuing is improved and flow time is improved. In each respect, the business gains better control of its internal manufacturing operation.

82 Pratsini, E. (1998). Learning complementarity and setup time reduction. *Computers and Operational Research*, **25**, No. 5, 397–405.

83 Anon. (1995). Lead times decimated. *Work Study*, **44**, No. 8, 28–29.

84 Anon. (1997). Quick change key to just-in-time delivery at Nissan. *Manufacturing Engineering*, **118**, No. 2, 81–83.

85 Ayers, R. and Butcher, D. (1994). *Op. cit.* ref. 72. Product diversification has been widely written about. The Ayers and Butcher article provides an overview, discussing product diversity and the market. In some cases product diversification can be taken as a survival strategy.

86 Hammond, D. (1997). Changing demands in the world of beverage preparation. *Beverage World*, **116**, No. 1636, 144. Market changes in beverages are very briefly summarized, where their impact on batch size, and the pressure they place on changeover performance, are noted. Hammond describes how market changes have not necessarily been anticipated by process equipment manufacturers: 'Most bottling plants were designed in an era when the number of products was small and the product runs long. Times have changed. Bottlers now produce up to 10 times the number of product types and flavours as they were originally set up for.'

87 Kobe, G. (1972). Engineer for right hand steer: how Honda does it. *Automotive Industries*, **172**, March, 34–37. Changes in market share by provision of right-hand drive are discussed.

88 Camp, R. C. (1989). *Benchmarking: The search for industry best practices that lead to superior performance*. Quality Press, Milwaukee. A Chinese general is quoted: 'If you know your enemy and know yourself, you need not fear the result of a hundred battles.'

89 Coates, J. B. (1974). Economics of multiple tool setting in presswork. *Sheet Metal Industries*, **51**, No. 2, February, 73–76. The thinking that changeover costs should be minimized by avoiding small batches is illustrated by Coates: 'Results show that, for batch-press-working, where tools are loaded and unloaded manually, tool setting time decreased in all cases with multiple tool set-ups. Costs vary, but can be particularly unfavourable where small batch sizes are involved. The correct costing approach for the recovery of overhead is to calculate total productive time as the number of hours available for both running and setting presses. Setting cost then becomes a fixed expense for each batch run and short production runs are penalized accordingly.' Future developments in the presswork industry (discussed elsewhere in this book) show the extent to which the notion of changeover as a 'fixed expense' is flawed.

90 Suzaki, K. (1987). *Op. cit.* ref. 5. Suzaki cautions that entertaining volume discounts might not always yield expected benefit to the business: 'Savings achieved through volume discounts may be more than offset by inventory waste.'

91 Esler, W. (1996). Fast track math: $2 \times 2 = 4$. *Graphic Arts Monthly*, **68**, No. 3, 49–50. New equipment, incorporating automation to aid changeover performance (including greatly improved registration), is described as enabling a small print operation to take on work that it previously brokered out.

92 Bergman, I. (1999). *Op. cit.* ref. 49. Similar splitting of batches at a Swiss factory because of out-of-control scheduling is identified.

93 Suzaki, K. (1987). *Op. cit.* ref. 5. Suzaki describes that lower WIP inventory allows manufacturing problems to be identified more easily, and allows for a swift response to them to be possible.

94 Belbin, R. M. (1970). Quality calamities and their management implications. OPN8, British Institute of Management, London. Identifying product quality problems is one matter. *Acting upon them* is another: 'In many cases calamities result from failures of a very

simple nature. A typical instance is that of the company manufacturing raincoats. Trials had established that a new silicone treatment had excellent waterproofing properties. The new method was then put into full-scale operation and appeared to go well but for the problems created in the very last process which was that of imparting the final finish to the garments by ironing. At this stage the silicone treatment was found to set the creases. The difficulties created by removing them were waived aside as teething troubles and mass production of the garments continued. In due course after a large stock had been built up, the prospective customer maintained that the final finish of the garments fell below the minimum standard of acceptability. The finish could not be improved and the whole stock had to be written off.'

95 Schonberger, R. (1982). *Japanese Manufacturing Techniques – nine hidden lessons in simplicity*. Free Press, New York.

96 Rebsamen, W. (1997). Dash to the finish. *American Printer*, **219**, No. 2, 60–62. Automated rapid changeover of book-binding equipment permitting entry to the market for small runs on demand is described. While this may represent a niche market at the time of writing, the author observes that there is a strong market trend towards small batch work to cut book publisher's warehouse inventory. In time, small batch work might represent not a niche, but the norm.

97 Anon. (1997). Small runs, big business. *Graphic Arts Monthly*, Nov., 79. Again in the printing industry, the success of a business is described that set out to capitalize on a niche market for short-run print work. The ability to change over rapidly between successive products is strongly identified as a key ingredient of the firm's success. Further, the need to have suitable equipment to do this is also identified: 'We sought to provide excellent service and fast turnaround at a fair price.... We realised right away that hard work and the right equipment were crucial to success.' In this case, suitable changeover-proficient equipment was purchased, rather than upgrading the performance of existing equipment.

98 Slack, N., Chambers, S., Harland, C., Harrison, A. and Johnston, R. (1995). *Op. cit.* ref. 53. A good example of the range of batch sizes that might be required is provided. The Royal Mint at Llantrisant operates presses that manufacture coins at a rate of up to 750 coins per minute. Batch sizes vary from 1000 million coins to just 5000 coins for an order for a small island.

99 Lankford, E. (1995). *Op. cit.* ref. 74. Many commentators assert that 'higher quality' will result from changeover improvement. It is less common to set out specific attributes of 'higher quality', let alone quantify them, particularly in terms of process improvement. Lankford indicates, for the study in question, that better changeover practice contributes to a reduction of start-up scrap by 30 per cent.

100 Garvin, D. A. (1988). *Managing Quality*. Macmillan, New York. That product quality problems can arise from changeover operations is noted by Garvin: 'In the room air conditioning industry, for example, changeovers were cited frequently as a source of quality problems.' Garvin, identifying a trend to improved product quality as the production run for a given product continues, later writes: 'From a quality standpoint, the worst manufacturing environment is one with bumping, short runs, and frequent changeovers.' Probable reasons for such quality problems (arising from conducting a changeover) will be described in more detail in the next chapter. Changeover improvement that addresses the reasons for these problems – either directly or indirectly as a consequence of seeking faster changeovers – is likely to lead to better post-changeover process control.

101 Bradbury, D. (1983). Eliminating roll tooling changeover. Technical paper MF83-544, Society of Manufacturing Engineers. Safety is a commodity that is extremely difficult to put a financial value on. To some companies, perhaps, the incremental benefit of enhanced safety arising from revised changeover practice would seem inconsequential, and be dismissed from a financial benefit calculation. Every country, though, has its own

history of compensation for industrial accidents, including the possibility of substantial legal costs or large fines. Bradbury is one of relatively few who highlights safety in respect of changeover activity: 'When multiple parts are roll formed, roll tooling changeover occurs. This creates problems. First, changeover is time consuming and expensive; second, a set-up technician is required to adjust and prove the tooling. Third, exposure to accidents is greater during set-up, when machine guards are removed for accessibility to the rollformer. Finally, scrap losses are high during set-up. These and other problems cause late orders, reduce part quality, decreased productivity and even lost sales.'

102 Klospic, A. R. and Houser, W. F. (1997). *Op. cit.* ref. 28. It is wrong, of course, to associate improved safety only with design changes. The Klospic and Houser case study concentrated upon changeover method improvements. It is reported that upon completion of the improvements the factory's managers believed that they had a safer process and a higher-quality product. Nevertheless, by implication, improved safety under such conditions will only be maintained if the improved work procedures are maintained. There is scope, under difficult operating conditions, or by time degradation, for more exacting procedures to lapse.

103 Beard, T. (1995). *Op. cit.* ref. 6. Sophisticated aerospace component manufacture is described. In a situation of complex machining on expensive material, it is shown that systems were set up within the company to avoid having to scrap any components during changeover. For each component that is not scrapped the savings can run into thousands of dollars.

104 Anon. (1978). Fast tool change speeds transfer line. *Tooling and Production*, **44**, No. 5, 91–92. A redesign for faster changeover undertaken by the DeVlieg Company reduced changeover times per spindle on a transfer line from 12 minutes to 30 seconds. The design changes also contributed to a higher-quality changeover taking place: 'In practice, we found that our quick change system also contributes to improved tool life, improved part quality and greater scrap reduction.'

105 Souloglou, A. (1992). *Op. cit.* ref. 58. Scrap can often be associated with trial-and-error adjustment activity. Eliminating occurrences of trial-and-error adjustment can have a dramatic impact on scrap production. Souloglou reports a specific instance where scrap production was almost entirely eliminated.

106 Garvin, D. A. (1988). *Op. cit.* ref. 100. Just one potential cost of poor product quality – call-out service cost – is described for the room air-conditioning industry (for which poor changeover performance is partly blamed): 'During the first year of warranty coverage [1975] there was one service call for every nine units in the field.'

107 Vastag, G. and Whybark, D. C. (1993). Global relations between inventory, manufacturing lead time and delivery date promises. *Int. J. of Production Economics*, **30–31**, 561–569.

108 Piroird, F. and Dale, B. (1990). *Op. cit.* ref. 77.

109 Krajewski, L., King, B., Ritzman, L. and Wong, D. (1987). *Op. cit.* ref. 49. Krajewski *et al.* describe the effect of simultaneously reducing changeover times and lot sizes to improve WIP and customer service.

110 Anon. (1985). Accounting for goodwill – A commentary on SSAP 22. Deloitte Haskins and Sells, London. Goodwill is defined as the difference between the value of a business as a whole and the aggregate of the fair value of its separate net assets. This difference can be positive or negative.

Our use of the term 'goodwill' assumes value has been engendered through previous professional, competent trading between the manufacturer and the customer. This additional value is highly intangible. Goodwill is likely to lead to continued support – at least in the short term – from that customer in the future. Similarly, any transactions that are poorly conducted – that do not meet the customer's requirements – will, depending on their severity, detract from the likelihood of continued customer support in the future.

Some accounting characteristics of goodwill are set out:

- The value of goodwill has no reliable or predictable relationship to any costs that the company may have incurred.
- It is impossible to value individual intangible factors that may contribute to goodwill.
- The value of goodwill may fluctuate widely according to internal and external circumstances over relatively short periods of time.
- The assessment of the value of goodwill is highly subjective.

111 Taguchi, G. and Clausing, D. (1990). Robust quality. *Harvard Business Review*, **68**, No. 1 (Jan.–Feb.), 65–75. The cost of deficient product quality (and thus not including other ways that the customer may be poorly served) is summarized. The more measurable associated costs are described – cost of fixing/remake, transportation and tracking – to which the cost of 'loss of reputation' may be added. Taiichi Ohno is cited: 'Whatever an executive thinks the losses of poor quality are, they are actually six times greater.'

112 Ebrahimpour, M. and Schonberger, R. (1984). *Op. cit.* ref. 23. (Refer to previous Ebrahimpour and Schonberger text in reference 23.)

113 Souloglou, A. (1992). *Op. cit.* ref. 58. Souloglou passes an opinion of a more intangible benefit of changeover improvement: '. . . the frustration of long and awkward change-overs has been greatly reduced.'

114 Monden, Y. (1981). How Toyota shortened supply lot production time, waiting time and conveyance time. *Industrial Engineering*, **13**, September, 22–30. Monden similarly reports that teamwork commenced in an improvement exercise will improve the working environment.

115 Flynn, B. (1987). Repetitive lots: the use of a sequence-dependent set-up time scheduling procedure in group technology and traditional shops. *J. of Operations Management*, **7**, No. 1, 203–216.

116 Kannan, V. R. (1996). Incorporating the impact of learning in assessing the effectiveness of cellular manufacturing. *Int. J. Production Research*, **34**, No. 12, 3327–3340. In a brief summary passage, Kannan observes: 'Cellular systems appear to perform well relative to comparable job shops only under a limited set of conditions, in particular when set-up and material handling times are high relative to processing times, and when demand patterns are stable.'

117 Choi, M. J. (1996). An exploratory study of contingency variables that affect the conversion to cellular manufacturing systems. *Int. J. Production Research*, **34**, No. 6, 1475–1496.

118 Steudel, H. J. and Desruelle, P. (1992). *Op. cit.* ref. 12.

119 Zunker, G. (1995). *Op. cit.* ref. 4.

120 Hill, T. J. (1983). *Production/operations Management*. Prentice Hall, London.

5

The role of design

This chapter investigates the potential role of design to improve changeovers. Design alters the existing manufacturing system, often allowing substantial revision to the tasks that comprise a changeover.[1] Design change – to physically alter the manufacturing system – can be applied beyond the immediate shopfloor environment and the hardware undergoing changeover. For example, more widely applied design can impinge upon both the quality of change parts and, perhaps by devising racks or handling devices, their delivery. Also, the manufacturing system is inclusive of the product that is being manufactured: just as physical changes may be made to equipment, so too physical changes can be made to the product under manufacture.

Here we will investigate design issues. Organizational improvement (where no physical change to existing manufacturing hardware occurs) will be contrasted with the use of design to enhance changeover performance. The chapter also investigates how design solutions might be identified, introducing a tool we term as the 'Reduction-In' strategy. We conclude by presenting design rules that can be applied either for retrospective improvement or for the design of new equipment.

5.1 Acknowledgement of a place for design

Previously we have written that design is often accorded scant formal recognition when changeover improvement is approached as a low-cost, *kaizen* initiative. We have argued that many changeover consultants promote an organization-led approach. For any company to ignore that design can make a significant contribution, however, is unwise.

That there is a place for design is acknowledged in current practice and in recent literature. We identified in Chapter 3 that companies exist whose *raison d'être* is to make design changes to manufacturing hardware for changeover purposes (typically, for press systems).[2,3] We also gave limited examples of devices that have been designed to aid changeovers.[4]

In the literature many authors similarly describe, by use of examples, how design might by applied. Not least of these is Shingo, who additionally states that changeover improvement comprises of two separate components: method improvement and design.[5] Such a distinction is made elsewhere.[6] More frequently, however, rather than stating that design is an important improvement option, only tacit acknowledgement of the use of design is made in the description of how improvement has been undertaken.[7,8]

The extent of design improvement

Fellow researcher Graham Gest undertook an analysis of Shingo's book *A Revolution in Manufacturing: The SMED system.*[9] Gest's analysis shows that the case studies in Shingo's book belie the low formal emphasis that is given to design as part of the 'SMED' methodology. This research indicates that purely organizational solutions account only for approximately 36 per cent of all solutions that Shingo describes. This is not to say that these solutions should be dismissed in any way – far from it – rather, that design warrants greater prominence in a structured appraisal of how improvement might be achieved.

Design change: mechanisms of *kaizen* improvement and 'step change' improvement

A further assessment is warranted of the use of design as part of a *kaizen* improvement exercise. Focused incremental improvement by shopfloor teams is a recognized feature of modern manufacturing philosophies.[10–12] The team approach that is advocated is one of constant appraisal of the manufacturing environment by those who work in that environment, continuously seeking incremental improvement to working practice.[13] It is an approach that we have expounded for retrospective changeover improvement[14] (Chapter 3).

Kaizen improvement may be contrasted with 'step-change' innovative improvement.[15] Step-change improvement is not necessarily contributed to by those who work on the shopfloor, is often costly and can have a comparatively lengthy evolution. For example, an original equipment manufacturer (OEM) might undertake design improvements leading to the replacement of existing shopfloor equipment with more changeover-proficient alternative equipment. Alternatively, a programme of substantive retrospective design by a company's internal professional design engineers might effectively achieve the same results.

It is described in other texts that a *kaizen* approach should not represent the sole means by which a company should seek to alter its manufacturing operation: that *kaizen* improvement should be considered alongside options for step-change improvement.[16,17] This concurs with the different improvement strategies that we describe in Figure 3.3. These two approaches of continuous, incremental improvement and step-change improvement might be thought of as opposing options. However, it needs to be questioned whether these two approaches in reality can be so clearly differentiated in respect of applying design. With reference perhaps to examples given in this book, when is a particular design change a step change, and when is it incremental improvement? Does the impact of a design change affect its categorization as a *kaizen* improvement, or is such a categorization more correctly determined by the manner in which the design change is sought? Considered another way, if a *kaizen* approach to changeover improvement is deliberately adopted, how should potential design changes, if at all, be accommodated? For example, should there be a limit to the cost of any one improvement that is proposed? Should a workshop team be augmented by personnel with a degree of proven design expertise so that design solutions might be more readily sought?

Design as part of a *kaizen* improvement programme

Some of the above questions are perhaps academically relevant – but need not necessarily be of consequence to a shopfloor, retrospective improvement team. *Kaizen* improvement in no way intrinsically precludes seeking design changes. It is not necessary at all to attempt to predetermine what 'level' of design a team should seek. Rather, the team should be encouraged simply to seek design change alongside organizational change. If the team is familiar with design-biased improvement options, then wider-ranging, more beneficial use of design is likely. The key here is that team members are able to identify a full range of possible design-led opportunities. As part of the improvement exercise all proposals will subsequently need to be evaluated against one another (see Chapter 7), where the decision to proceed with any particular option, be this design biased or otherwise, will finally be taken based on a number of criteria. Or, in simple terms, the decision as to which improvement proposal to pursue will ultimately be taken according to what an idea can contribute, set against what is involved in implementing it (Chapter 7). One notable issue will be cost. Other issues might include line disruption during installation and ease of sustaining the improvement. Substantive design can be a legitimate team-improvement option.

The foregoing discussion indicates that structure is required of the *kaizen* improvement effort if it is to be of maximum use. We will later describe in detail (Chapter 7) how design improvement might be sought alongside organizational improvement. An improvement team should be equipped to do this (Chapters 6 and 7) – for example, employing suitable documentation and appraisal tools. The composition of the team (Chapter 6) and training that is undertaken should also all be considered in this light.

OEM design

Design, as noted, has a significant role in step-change improvement options, for example as undertaken by OEMs.[18–21] Indeed, there are historic precedents of OEM design for changeover. Some examples can be singled out. Early in the 1960s the Danly Machine Corporation took out a patent relating to the changeover capability of its press systems.[22] More recently it has been reported that this company has maintained a focus on the changeover capability of its products.[23] Examples of other companies in the metal forming industries that have maintained a long-term interest in changeover capability (by design) are reported.[24,25] Notable OEM design focus is also being applied by companies in various other industries, such as injection moulding[26,27] and machine tool systems.[28–30]

These instances of applying design can be studied. What is learnt from them will be shown to be useful, not only for OEM step-change improvement, but also for team-based, retrospective, *kaizen* improvement. The machine tool industry in particular is singled out later in this chapter for this purpose.

5.2 Defining design changes

It is necessary to define what constitutes design improvement if design is to be assessed in relation to organization-led improvement of a changeover. In a sense this is very easy to do, for design can be taken to comprise any activity where purely organizational changes do not occur. Thus, where the process equipment is modified, design changes have been made. Design changes can include modification by permanent addition or replacement, for example replacing bolts with quarter-turn, quick-release devices. Design here relates to hardware changes: it does not include the design of new working methods.

Take, for example, a desire to repeat setting positions for a changeover to a previously manufactured product. If those conducting the changeover repeat, say, a machine slide position by newly lining up a feature of the slide with a static feature on the body of the machine, then what has occurred is an organizational improvement. If, conversely, the operators place a scale and a marker on the slide and frame respectively, then a design change will have occurred. Strictly, by our definition, a design change will have occurred even if a pair of scribed marks had been applied instead.

Design to alter the way that changeover tasks are conducted

The examples of simple design changes that are described above show two ways that design may influence changeover operations. In the one instance, of adding scribed marks or a scale, the changes made to the existing changeover are arbitrary: if it is decided not to make use of them, the changeover would proceed exactly as before. This particular design change does not impose change to the way that the changeover is conducted.

The same is not true of the example of the quick-release devices referred to above. In this case change has been imposed on the existing changeover procedure. The previous way of conducting the changeover cannot be reinstated unless the design modification (adding the new devices) is withdrawn.

This latter case is far more common. For the most part design changes will impose change on the way that a changeover has to be conducted. Ever more significantly changed procedures typically will be imposed as the design changes become more involved.

Substantive and non-substantive design solutions

We will make reference to substantive and non-substantive design solutions. This is an arbitrary distinction, where non-substantive design may be regarded as comparatively trivial solutions, of the type of the bolt/quick-release device example described above. Gest's previous analysis of Shingo's case studies indicated that non-substantive solutions are common in typical changeover improvement programmes. In contrast, substantive design refers to significant mechanical changes to the existing manufacturing system. These solutions will probably be costly and will have a lengthy evolution. Substantive design usually would require the experience and skill of professional design engineers if it is to be concluded satisfactorily.

Primary design and secondary design

A further distinction is possible between design changes made directly to the equipment undergoing changeover – as opposed to remote design changes (modifications to other, remote equipment) that can also improve the same changeover.

The great majority of design changes made directly to the machine undergoing changeover effect a unique change to the way the changeover is conducted. Where these exact changes to changeover tasks cannot occur unless a design change is made, we nominate that primary design modifications have taken place. Often design modifications to the product experiencing the changeover will similarly have a unique impact upon the way that the changeover is conducted. These modifications too thus constitute primary design.

Conversely, design changes that are made to other machines beyond the immediate changeover environment typically have the effect of altering the delivery of items for changeover (whether for logistics or for item quality). These aspects usually can be identically altered by alternative organizational means: in other words, the design changes do not uniquely alter the changeover. For example, consider lithographic printing plates. In one case study we determined that it was difficult to load the plates satisfactorily during changeover. A likely cause for this poor loading was the precision of the plates' punched location notches. If so, an option existed to improve the plate changeover by improving – by design – the cutting of these notches. This is an example of design applied to aid changeover performance away from the immediate changeover machine. The improvement that it yields (providing higher-quality plates) could be achieved by alternative means – for example, by sourcing the plates from another supplier, or by instigating a rigorous inspection 'gate' to reject poor-quality plates. These latter options represent organizational change.

5.3 Design for changeover in the machine tool industry

It is useful to reflect on the development of computer aided machining (NC, DNC, CNC, FMS). Changeover is prominent in this development. Unlike much retrospective improvement of existing practice, this development has relied heavily on design improvement. The development of changeover-proficient machine tools provides good historical evidence of the way that design for changeover has been applied within one particular industry. These design developments have largely been conducted by OEMs. We believe the use of design within this industry to be wholly separate to the influence of the 'SMED' methodology. Also, design improvement has been undertaken, to our knowledge, without the aid of formal design rules for changeover.

Improving machine tool operations

One characteristic of the Industrial Revolution is the widespread development and use of machine tools,[31] including metal-cutting tools. Some machine tools in use today are little different from those used one hundred years ago. For example, for a lathe, a hardened steel tool is ground to the profile required by the user and brought against the workpiece. Cutting commences, where the part is frequently inspected by the operator until the desired finished dimensions are achieved. When a new cutting operation is required, for example parting off, a different tool frequently will be ground up and used.

Although this description can still be applied today to hobby machining, it does not necessarily represent manufacturing practice at commercial workshops in industrially developed countries. Simply, this way of machining metal has been superseded. Although apparently uncomplicated, this conventional way of machining has considerable scope for improvement.

Changeover and machine tools

Changeover is inherent in almost all metal-machining environments. It is required when a different product (a new batch of one item or more) needs to be processed. Changeover might also be thought of in terms of replacement of tooling as required, as the machine's tools become worn or when alternative tools are needed.

Machine tool development frequently seeks to reduce the overall processing cycle, from loading the workpiece in the first place through to machining it, and finally removing it. Machine tool development can additionally impact upon issues of storage and transportation.[32] The overall processing cycle time can be reduced in many different ways and some opportunities are described below. Further, simultaneously or otherwise, improvement can be sought in different facets of the capability of the machining operation, for example increasing the rate of metal removal or achieving more exacting dimensional tolerances.

Some areas in which improvement has been sought include:

- tool storage
- position monitoring
- precision, uniform shape tool inserts (qualified tooling)
- presetting of tools
- rapid movement of tools
- rate of metal removal
- location (workpiece and tool)
- clamping
- pallet loading

Much of the time that is saved in the overall processing cycle occurs through greatly improved changeover performance – between different batches, and for the tooling that is used. In turn, excellent changeover performance significantly contributes to the widespread use of computer-aided machining for responsive manufacture[33] and small-batch manufacture.[34] The improvements in respect of changeover individually permit tasks to be conducted with greater speed and precision. Taken together, these improvements can allow the need for manual trial-and-error adjustments during changeover to be eliminated. Ultimately a fully automated system becomes possible, where both the tool and workpiece are loaded, cutting is completed and the workpiece, tool and swarf are removed.[35] For modern computer-controlled machines, multiple tool changing occurs while the same workpiece remains mounted in position, again fully automatically.[36,37]

The literature confirms the developments that have been made.[38–46] Many of the developments since the 1960s will have added considerably to the equipment's cost.

That a buoyant market for computer-aided machine tools exists indicates that the expense incurred in achieving significantly improved changeover performance, among other benefits, can be justified.[47] Overall, upon cursory investigation (see references), cost strictures do not appear to have been allowed – within limits – to restrict how improvement is sought. This can be contrasted with some of the cases highlighted previously within this book in which consultant-led retrospective improvement of existing equipment is contemplated.

Augmenting the definition of a changeover

In Chapter 1 we argued that a changeover should not be defined simply as the time that the line is stopped (a 'good piece to good piece' definition). Instead we asserted that the definition should be broadened; that it should be the interval when the line or machine is being modified, including the run-up period, when stable condition production is not occurring. This time-based definition is important in terms of assessing the company's manufacturing capability.

This definition is still limited, however. It does not, for example, define the interval when changeover tasks are being conducted because many tasks, typically for a well-organized changeover at least, will occur before the line is actually stopped. Neither, as we will discuss, does a definition that only describes an elapsed changeover period give any meaningful insight into requirements of a changeover, serving instead simply to highlight a need to reduce 'internal' time. The development of changeover in the machine-tool industry indicates that other important improvement opportunities are available.

To investigate design issues an alternative assessment of a changeover might be made: that a changeover is the bringing together of all changeover items, in the condition required of them, including the product, into the positions necessary to enable successful processing operations upon the new product to occur. A rapid changeover is one where these objectives are achieved as quickly as possible. Thus improvements can be sought that enable these objectives to be achieved more quickly. Consideration of a changeover in this way – in ensuring that all items are present, correctly located and in acceptable condition – encompasses the manufacturing system in its entirety. Figure 5.1 portrays this interpretation of changeover activity. Time is still important, but this perception of a changeover is not reliant upon (or constrained by) notions of 'internal' and 'external' time; instead focus is placed upon what is actually required to happen during a changeover. Improvements are sought that engender quickly and accurately achieving prescribed processing parameters for the new product.

A changeover as described by Figure 5.1 assists in seeking improvement options beyond moving and converting tasks into external time. Suzaki,[48] describing improvement generally in a manufacturing environment, writes:

> The basic idea of improvement is simple. We want to do our work easier, faster, cheaper, better and safer. To do so, a basic approach to improve our operations is to simplify, combine, and eliminate.

While 'simplify', 'combine' and 'eliminate' might not represent all options, a broader, non-time-dominated approach to changeover improvement is identified. The role of design to improve changeovers is now discussed in this context.

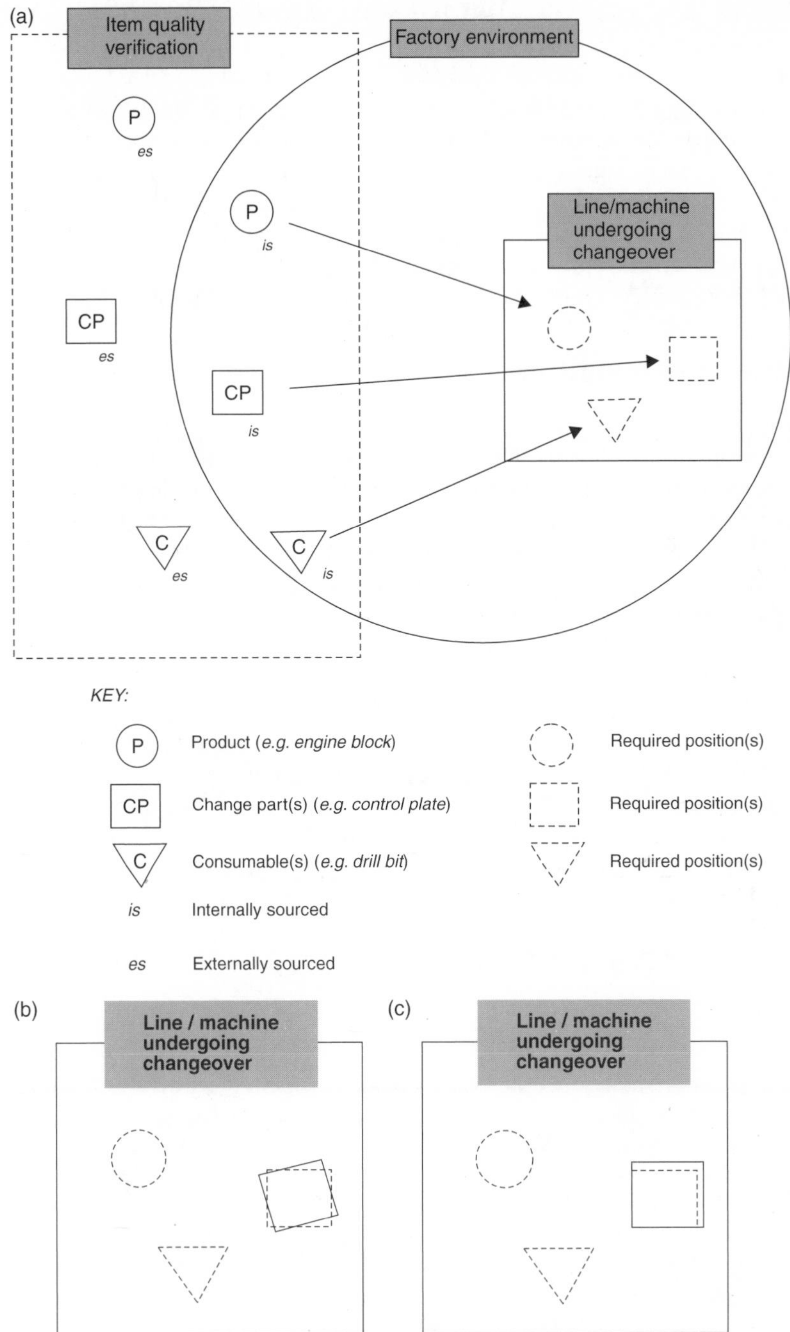

Figure 5.1(a) A design-orientated view of changeover activity: ensuring that all items are present, correctly located and in acceptable condition, (**b**) Change part incorrectly positioned (incorrect relative position of machine and change part) and (**c**) Oversize – out of specification – change part (again, incorrect relative position of machine and change part). *Note*: Incorrect size only is shown. Other parameters, for example shape and hardness, might also apply

5.4 Some issues applicable to using design

A study of design to improve changeovers for computer-assisted machine tools indicates issues that are also likely to be relevant when applying design in alternative industrial changeover environments. Our purpose here is to investigate some issues that we perceive to be significant. The discussion at this stage is intended only to give an insight into what design might achieve, including ways that good design can eliminate, or at least minimize, the need for adjustment. Our proposed design rules for changeover, which represent a structured application of design, are presented at the end of this chapter. Design is also discussed further in Chapters 7 and 9.

The issue of elimination

Elimination of changeover tasks currently can assume secondary importance as a means to gain improvement. This is exhibited, for example, within the 'SMED' methodology where, until its final phase is reached, attention is concentrated upon procedural change to the way that the changeover is conducted.[49] Elimination is only featured as an 'improvement technique' as part of the 'SMED' methodology's final 'streamlining' phase.[50] We assert that, because of its potential contribution, elimination of existing changeover tasks should be sought more earnestly throughout an initiative.

For a typical changeover there will be a limit to the level of improvement that can be achieved by attention to work practice alone,[51] within the confines of existing equipment.[52] To get beyond this point, as described in Exhibit 5.1, and as we shall also investigate later, fundamental change to existing changeover tasks will need to take place. Typically, some tasks will need to be eliminated if radical improvement is to be achieved. Other tasks will need to be simplified, reduced in duration or otherwise changed. As described in Exhibit 5.2 and particularly in Exhibit 5.3, the impact of even simple design changes can be very considerable. As is also described in these exhibits, changeovers where certain tasks have been eliminated can typically expect new tasks to be introduced.

Exhibit 5.1 Structural change to existing changeover tasks by design

The changeover of a large press tool included changeover of its coil feed equipment, where it was required to alter the fed coil length between successive products. The changeover of this assembly as originally witnessed took slightly longer than 6 hours to complete. Figures 5.2(a)–(d) highlight particular features of the changeover that are described below.

It was apparent that some immediate time savings could be made to this changeover at very little cost. In order to conduct work upon it the assembly was removed from the back of the press and taken to a separate anteroom, where a table was built up to receive it. The table was made as required from loose pallets. Once placed on this make-shift table there was typically a significant delay while hand tools were located.

Figure 5.2(a) The 'Schmitt' drive coupling to the feed rolls

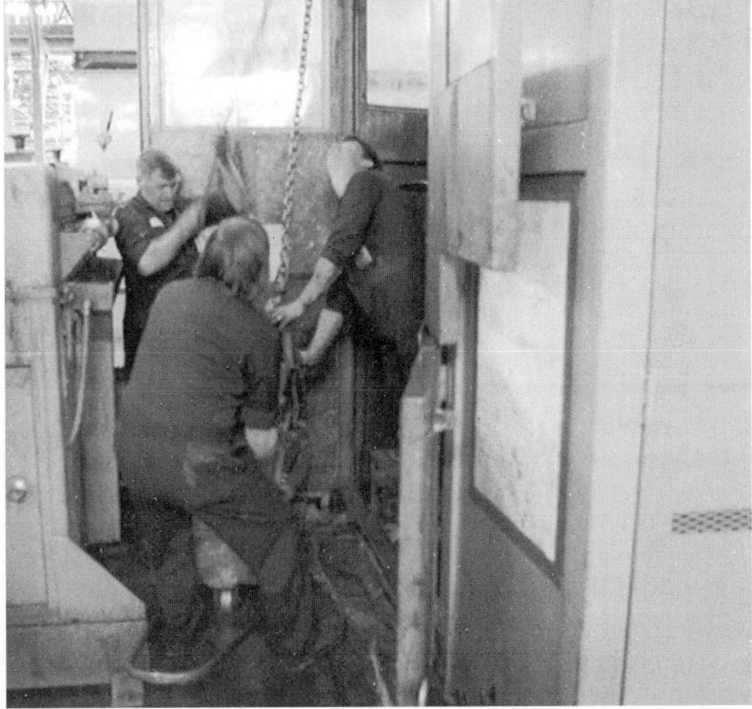

Figure 5.2(b) Removal of the coil feed assembly from the rear of the press

Exhibit 5.1 (*Continued*)

Figure 5.2(c) The assembly on a makeshift work table in a separate anteroom

Figure 5.2(d) Changing the feed rolls

These tasks are candidates for moving/converting into external time, without in any way changing their content. Doing so would improve the elapsed changeover time. Another candidate for low-cost organizational improvement was preparing for three people simultaneously to be at the rear of the press to extricate the assembly in the first place. As described by the changeover operatives, and as subsequently observed, awaiting this manpower often caused significant delay.

As other exhibits also describe, however, there was a limit to the improvement that could be gained by low-cost organizational improvement alone.[53] The coil feed changeover necessitated exchanging a pair of feed rolls for those of a different diameter. The complete assembly is large, heavy and awkward to manipulate. With the new feed rolls in place, for example, the assembly has to be transported back to the rear of the press, moved into rough position, precision aligned and connected. Options were discussed to conduct these tasks more efficiently, but until their content could be significantly altered there was unlikely to be dramatic and sustainable improvement to the total elapsed changeover time.

A wholly different approach involving a design change was suggested. The drive to the feed rolls was from a shaft on the press via a 'Schmitt' coupling. It was suggested that instead motive power should be provided by a servo motor directly coupled to the feed rolls on the side of the assembly.[54,55] By altering its drive program, the servo motor could be used to provide different length coil feeds. This solution completely eliminated the need to remove the assembly from the press, along with all associated tasks (as outlined above). At a stroke it overcame the organizational improvements that were being contemplated, rendering them superfluous. Further, only one man was needed to effect the change, which, as a simple reprogramming task, was also estimated to take just one minute to complete. The total cost of the modifications was estimated at less than £25 000. A financial benefit analysis indicated that the payback period for this investment was very comfortably less than 1 year.

Exhibit 5.2 The potential of simple design modifications

A large, 24-station machine of rotary design required changing over as part of the overall changeover of the manufacturing line in which it was situated. On average the changeover for this machine was taking approximately 7½ hours to complete.

A large proportion of this time was dedicated to gaining access, twenty-four times over, to a single securing bolt on each station. To gain this access additional non-changeover components from each station needed to be removed and subsequently, once the changeover items had been substituted, replaced.

A simple modification – machining a small hole – permitted access to these previously inaccessible securing bolts without having to remove and replace non-changeover elements of the machine. The modification was conceived by the line operatives and cost less that £1000 to implement. It represented a single, simple design improvement idea that shaved 3½ hours off the previous machine changeover time, hence reducing it by more than 50 per cent. The modification eliminated at a stroke all the changeover tasks that were previously associated with removing and replacing the non-changeover assemblies.

Exhibit 5.3 Design change eliminating previous changeover tasks

This exhibit details proposed design modifications to the adjustment of roof-mounted tracking between process machines to enable different-sized products to be transported. Programmable servo motors were suggested to move the track components accurately between preset positions. The proposal was estimated to reduce the time for this changeover from 120 minutes to 1 minute at a cost of approximately £20 000. If adopted, this changeover essentially would be simplified to setting a dial on a console.

The original changeover tasks may be compared to the tasks that occur after the design changes have been made. This is shown respectively in Tables 5.1 and 5.2. It is seen that the new design fundamentally impacts upon individual changeover tasks where, excepting communication with a supervisor, no task is carried forward into the new procedure. The simpler, more automated, new procedure has made the changeover inherently more repeatable. It has also significantly deskilled it, made it safer and, because of the greater accuracy and repeatability of settings, is likely to contribute to better product quality and line performance (for example, breakdown, scrap and line speed).

Table 5.1 Identifying tasks eliminated from original changeover procedure

Original changeover task (for changeover of roof-mounted tracking)	*Present in servo motor changeover?*
Locate step ladder	× (Eliminated)
Locate hand tools	× (Eliminated)
Locate dummy components	× (Eliminated)
Dummy components in position for trial run	× (Eliminated)
Receive permission to proceed (line stopped)	✓
Move ladder (repeated)	× (Eliminated)
Ascend ladder (repeated)	× (Eliminated)
Release adjustment rail bolts (repeated)	× (Eliminated)
Adjust rail to dummy component (repeated)	× (Eliminated)
Secure adjustment rail bolts (repeated)	× (Eliminated)
Descend ladder (repeated)	× (Eliminated)
Trial run with dummy components	× (Eliminated)
(possible fine tuning of adjustments = redo many of the tasks indicated above)	× (Eliminated)
Confirm completion to supervisor	✓

Table 5.2 The tasks comprising the revised changeover

Servo motor changeover task (for changeover of roof-mounted tracking)	*Present in original changeover?*
Receive permission to proceed (line stopped)	✓
Read appropriate servo console dial setting	New task*
Set dial to new setting	New task
Confirm completion to supervisor	✓

*Task indicated may be conducted in external time

Tables 5.3 and 5.4 show how tasks might have been altered under a more conventional, sequential, 'SMED' improvement programme (which was not done). First, Table 5.3 shows how the original changeover could be modified if it was only sought to separate existing tasks, where possible, into external time. This was only possible for comparatively few tasks. The changeover time saving that this would have made was commensurably small. Also, even though tasks might be moved into external time, they still have to be conducted.

Table 5.3 Assessing the impact of conducting existing tasks in external time

Original changeover task (for changeover of roof-mounted tracking)	*Possible to conduct task in external time?*
Locate step ladder	✓
Locate hand tools	✓
Locate dummy components	✓
Dummy components in position for trial run	✓
Receive permission to proceed (line stopped)	×
Move ladder (repeated)	×
Ascend ladder (repeated)	×
Release adjustment rail bolts (repeated)	×
Adjust rail to dummy component (repeated)	×
Secure adjustment rail bolts (repeated)	×
Descend ladder (repeated)	×
Trial run with dummy components	×
(possible fine tuning of adjustments = redo many of the tasks indicated above)	×
Confirm completion to supervisor	×

Table 5.4 The impact of adopting low-cost 'streamlining' improvements (quick-release fasteners and setting pieces)

Original changeover task (for changeover of roof-mounted tracking)	*Present if employing quick-release fasteners and setting pieces?*
Locate step ladder	✓
Locate hand tools	× (Eliminated)
Locate dummy components	✓ (Needed for trial run)
Dummy components in position for trial run	✓ (Probably)
Receive permission to proceed (line stopped)	✓
Move ladder (repeated)	✓
Ascend ladder (repeated)	✓
Release adjustment rail bolts (repeated)	× (Eliminated – replaced by releasing quick-release fasteners)
Adjust rail to dummy component (repeated)	× (Eliminated – replaced by adjusting to setting piece)
Secure adjustment rail bolts (repeated)	× (Eliminated – replaced by releasing quick-release fasteners)
Descend ladder (repeated)	✓
Trial run with dummy components	✓ (Probably)
(possible fine tuning of adjustments = redo many of the tasks indicated above)	✓ (Possibly)
Confirm completion to supervisor	✓

Exhibit 5.3 (*Continued*)

Some improvements are possible that might allow tasks to be 'converted' to external time. An example (although not a very useful one) might be the provision of multiple step ladders, which are all set in place before the changeover commences. 'Streamlining' the original tasks was also considered by the factory-improvement team (the meaning of which was open to their interpretation), and a number of other possible options were identified:

- Use standard bolts throughout the assembly to reduce the number of hand tools that are used.
- Construct a walkway alongside the overhead tracking for use during changeover.
- Use specially constructed setting pieces in place of the damage-prone dummy setting component.
- Set the tracking by means of marked settings on the assembly itself (for example, use scales to position the adjustable rail).
- Replace the bolts with quick-release fasteners.

Table 5.4 shows the effect of adopting two of these changes: using quick-release fasteners and specially constructed setting pieces (which are both relatively low-cost options). Here the suggested changes tend to have the effect of substituting easier or more accurately conducted tasks in place of the original tasks. It should also be possible to conduct these tasks more quickly. It should be noted that each of these changes would be rendered obsolete if it was subsequently decided to adopt the servo motor solution (Table 5.2).

The issue of mechanization

Like elimination, mechanization has been described as a potentially useful technique within previous changeover improvement methodologies.[56–59] Mechanization can be applied to a single changeover task, as described for example in Exhibit 5.4 or, in larger measure, can contribute towards the full automation of an existing changeover.[60–63] Mechanization of a single task can also have considerable impact on associated tasks, sometimes eliminating them altogether. For example, remotely actuated hydraulic clamps are commercially available. The need for hand tools can be removed when such clamps are installed. Moreover, the mechanized clamping operation is likely to be simplified and completed in a more repeatable fashion. Where manual clamping previously occurred in confined spaces the advantage can be greater still. Here, perhaps, prior changeover tasks associated with gaining access might be eliminated.

Exhibit 5.4 Mechanization to alleviate bolt access problems

Access to one bolt head during an involved changeover was difficult. The thread concerned adjusted the stroke of a crank by moving the crank pivot in relation to its rotation axis. The machine was nearly new, and the thread was in good condition. Typically, before improvement, this operation was being done by hand – an operator simply reached across to grasp and turn the thread. A spanner was rarely used because access to the hexagonal head was poor. Even so, the task still remained time consuming. In time it was expected that the thread would become contaminated or damaged; that the task would become more difficult and hence more time consuming to complete.

An idea was conceived to mechanize this task. A battery-powered, hand-held screwdriver was left on charge adjacent to the machine. Instead of a blade, its chuck was used to mount a 13 mm socket. When this new tool was used the time saving for this single task exceeded 90 per cent.

This exhibit hints at the widespread application of this idea, either using this hand tool or others that are available on the market that are specifically designed to hasten removing or attaching bolts. Even so, the improvement practitioner should not lose sight that use of such a tool represents just one improvement alternative. It might be more appropriate, in general, to make use of quick-release couplings instead of bolts, particularly quick-release devices that do not require the use of separate hand tools. To decide which option is the best, as we describe in later chapters, the practitioner needs to be able to identify and then evaluate as many competing improvement options as possible.

Figure 5.3 Illustration of the hand-held powered screwdriver in use

The issue of simplification

Simplification has been discussed previously in terms of its impact on flexibility and changeover quality. While simplification can be instigated by organizational change, the use of design, potentially, might have a greater impact still.[64,65] Indeed, simplification is apparent in earlier exhibits in this chapter.

Simplification can infer that less skill is required to complete the changeover. In the same manner, simplification might also be taken to mean that a changeover requires less mental or physical effort or, perhaps, eliminates the need to use tools, which otherwise have to be available.[66,67] Each of these can apply, either separately or in combination. Simplification is inherent when existing changeover tasks are eliminated (assuming that these original tasks are not substantially substituted by other tasks of equal complexity). More than this, existing tasks can be retained essentially unaltered but become simpler to conduct, often by applying standardization (see next section). An example would be using uniform 8 mm bolts with 13 mm hexagonal heads throughout a changeover assembly. This might be in place of a previous mix of 6 mm and 8 mm bolts and a combination of hexagonal head and Allen head types.

Exhibit 5.5 demonstrates how, by using design in a way that ensures tasks are completed without the possibility of error[68] and without the need for adjustment, a changeover can sometimes be simplified/deskilled to great advantage. Specifically, the exhibit investigates how mounting components with careful attention to location and with careful attention to constraining their movement can contribute to these aims. This topic is investigated further in Exhibit 5.6.

Exhibit 5.5 Datums, reference dimensions and *poke-yoke*

The way that components are designed to be assembled can significantly influence a changeover's speed and quality. For example, provision of quick-fitting devices can be highly beneficial. Our intention here is to assess datums, tolerances and geometrical features of components. Such features can include those that prevent components being assembled in anything other than their correct position.

As discussed later in this chapter, design solutions will be seeking to fulfil often-conflicting objectives. Thus, in general, seeking good changeover performance will be competing with seeking to fulfil wider design objectives. For example, high precision, which might be applied to eliminate aspects of adjustment is likely to be incompatible with low cost manufacture. The datums that are employed can be particularly important in terms of achieving high precision at low cost. When considering datums it is also often desirable to reflect upon mounting components as simply as possible, compatible with fully securing them or permitting their desired movement (see also exhibit 5.6).

These issues were relevant in the design of a plastic blow moulding tool for the manufacture of shallow pots at a Zimbabwean factory. Figure 5.4 shows views of the assembled tool.

The shallow mould cavity features a wide neck. This configuration permitted a novel mould design that enabled adjustment-free assembly at an isolated factory. In particular there was limited immediate technical assistance that this isolated factory could draw upon. The design also permitted replacement tooling to be dispatched from England in the event of wear or damage, with confidence that this replacement tooling too could be fitted 'out-of-the-box', again without the need for adjustment.

Manufacture was undertaken as shown in Figure 5.5. Critical dimensions are indicated. Much of the machining that took place was done with the mould cavity blocks clamped together. This work included machining the pin and bush locations, plus turning the cavity and turning the neck thread profiles. The pins and bushes were not used to locate the mould bodies to the backplate – instead a step feature across the backplate was used to perform this function, via the use of the precision-machined base component (Figure 5.5(c)). Precision manufacture largely depended throughout upon surface grinding (to achieve high geometrical and dimensional tolerances), which can be done with comparative ease and at relatively low cost.

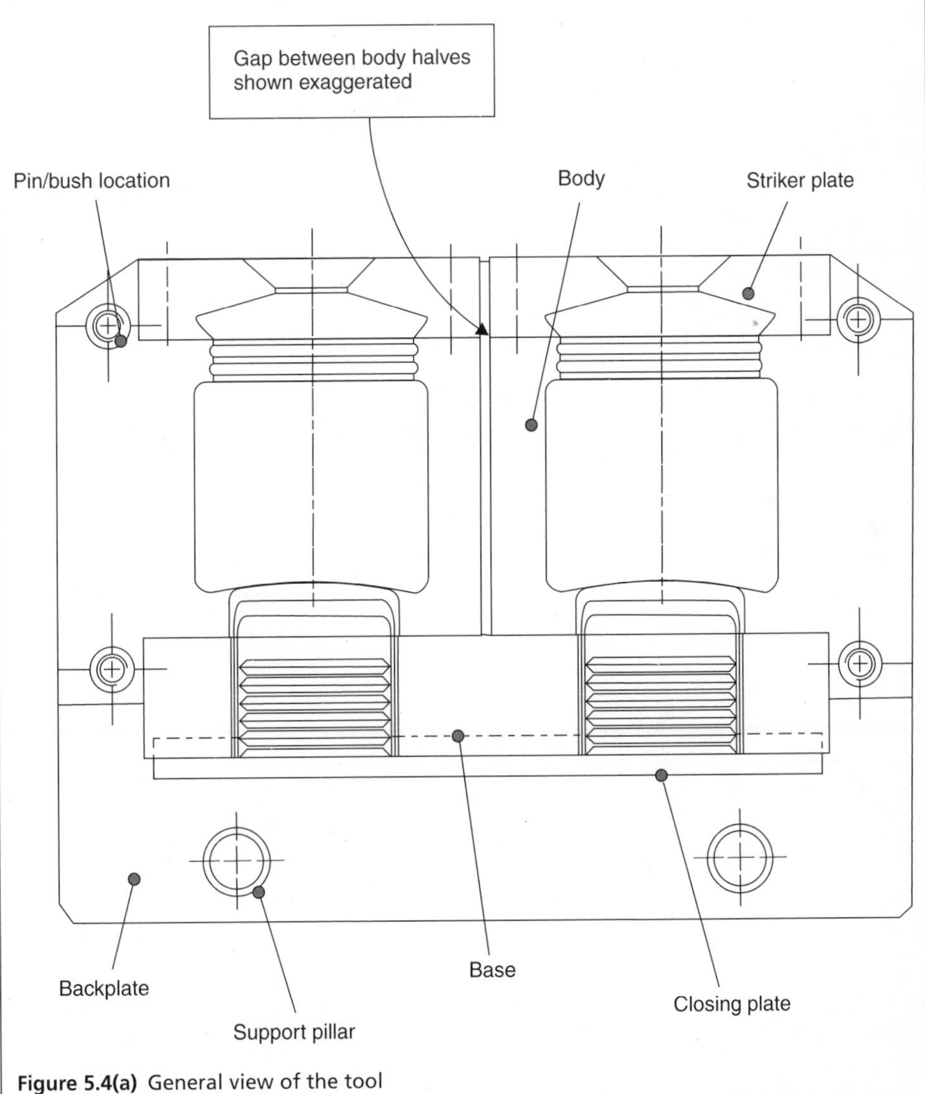

Figure 5.4(a) General view of the tool

Exhibit 5.5 (*Continued*)

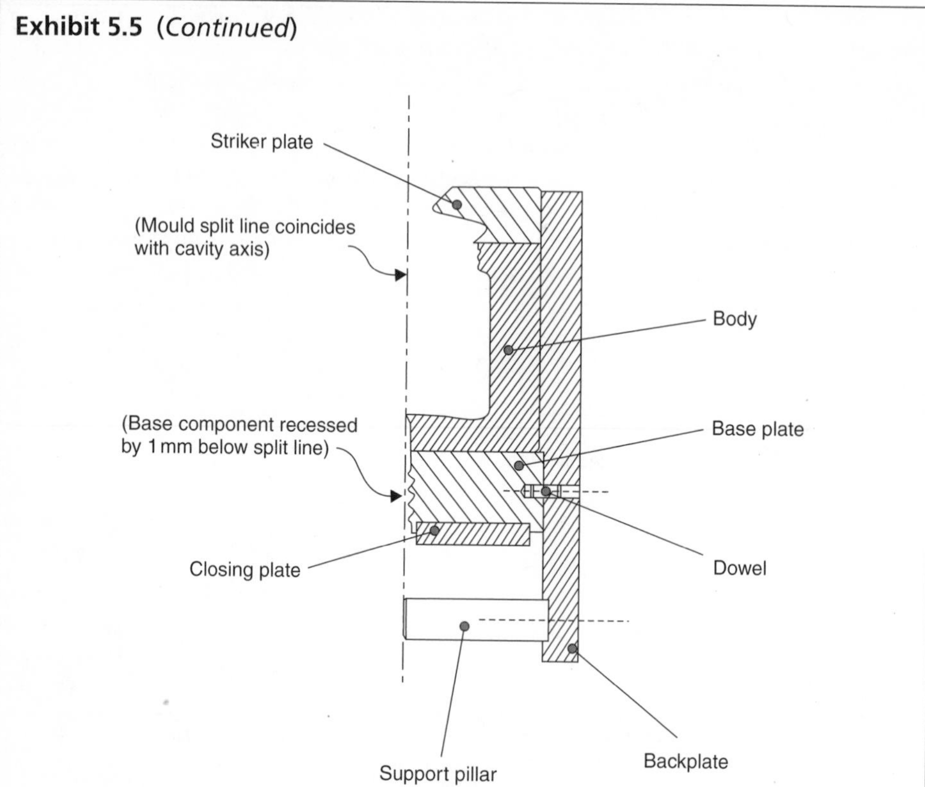

Figure 5.4(b) Sectional view of the tool (omitting screw and water channel details)

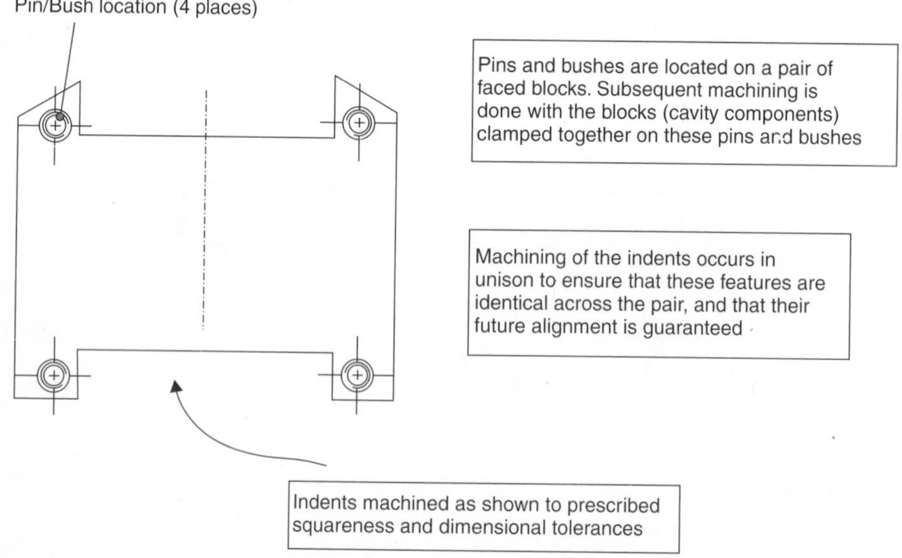

Pin/Bush location (4 places)
Pins and bushes are located on a pair of
faced blocks. Subsequent machining is
done with the blocks (cavity components)
clamped together on these pins and bushes
Machining of the indents occurs in
unison to ensure that these features are
identical across the pair, and that their
future alignment is guaranteed
Indents machined as shown to prescribed
squareness and dimensional tolerances

Figure 5.5(a) Manufacture of the body cavities: 1 (not to scale)

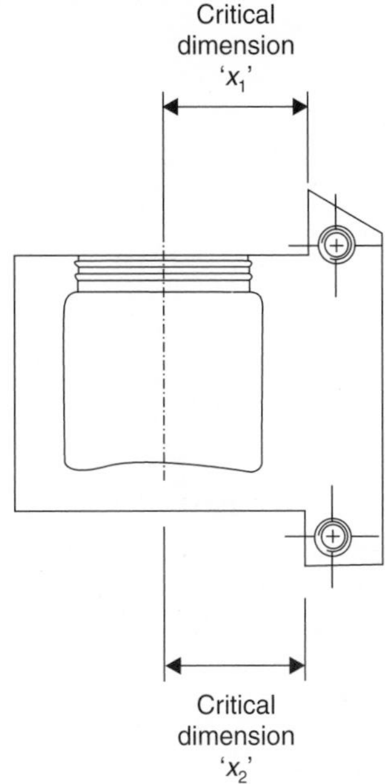

Figure 5.5(b) Manufacture of the body cavities: 2 (not to scale)

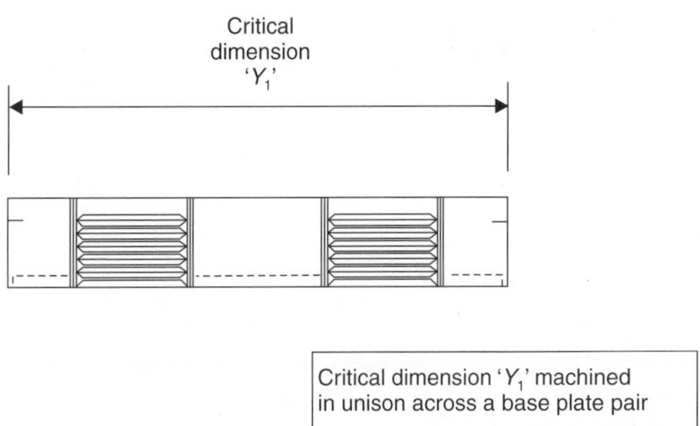

Figure 5.5(c) Base component manufacture (not to scale)

In the mould backplates, as superimposed on Figure 5.6, pairs of pins were inserted where the mould cavity bodies were to be mounted. These pins matched clearance holes on the rear surface of the bodies. For the four different positions (cavity elements), four separate pin patterns were used. This meant that it was impossible to mount a mould body component in any other but its correct position.

Exhibit 5.5 (*Continued*)

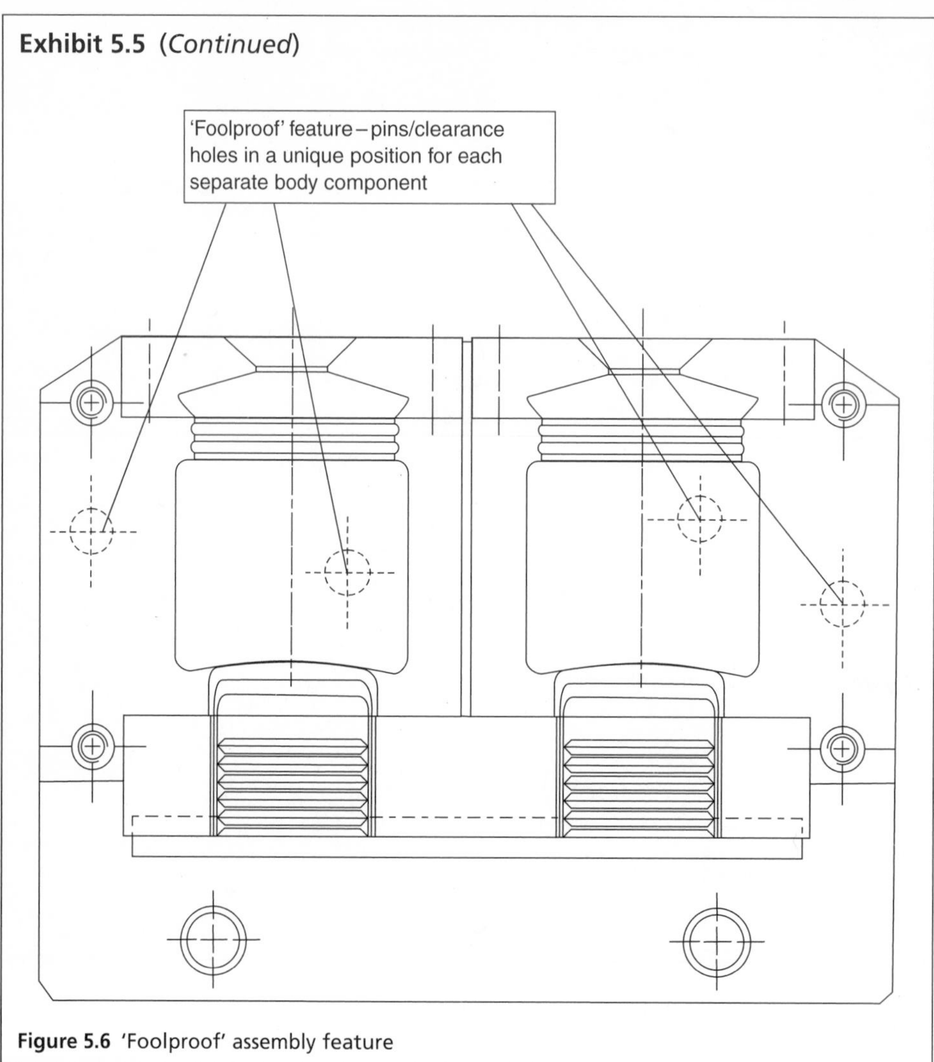

Figure 5.6 'Foolproof' assembly feature

Exhibit 5.6 Constraining movement[69]

As indicated in the previous exhibit, how components are located in position and constrained can be important. A static component needs to be fully constrained simultaneously in three linear axes and three rotational axes. As suggested in the previous exhibit, the constraint system that is used typically affects the datums that are employed and the ease (and hence the cost) of component manufacture. In Exhibit 5.5 the selection of datums was substantially influenced by manufacturing considerations.

Figure 5.7 presents a few possible methods to locate and constrain a flat plate. For each of these methods, for simplicity, a securing bolt is employed in conjunction with a 'keyhole' slot. The bolt's contribution should be recognized and distinguished from features that are additionally concerned with location. It is seen that some location/constraint/rapid changeover features can conflict with one another.

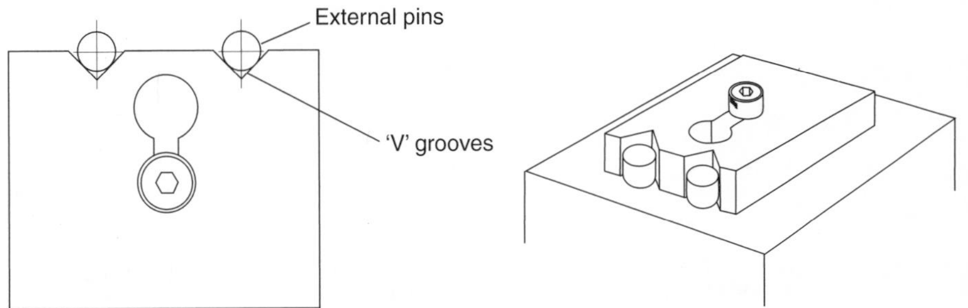

Figure 5.7(a) Possible method to locate and constrain a plate on a flat surface: 1

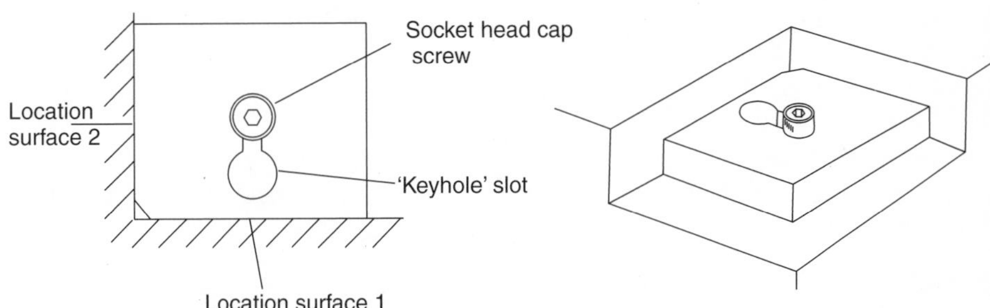

Figure 5.7(b) Possible method to locate and constrain a plate on a flat surface: 2

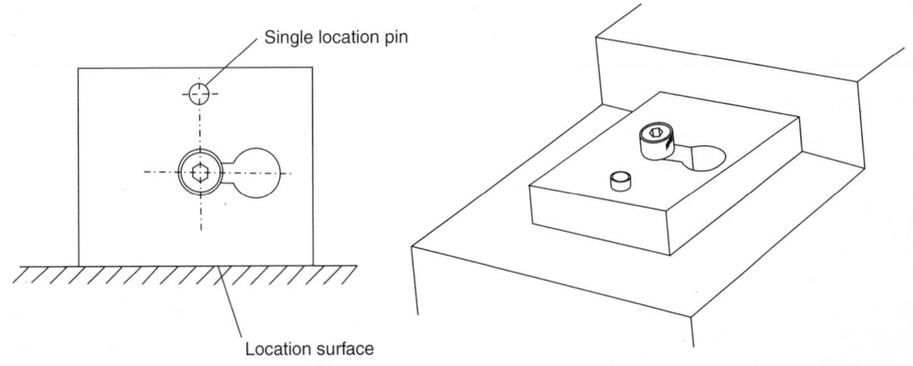

Figure 5.7(c) Possible method to locate and constrain a plate on a flat surface: 3 (keyhole slot no longer serves as a quick-release feature)

Exhibit 5.6 (*Continued*)

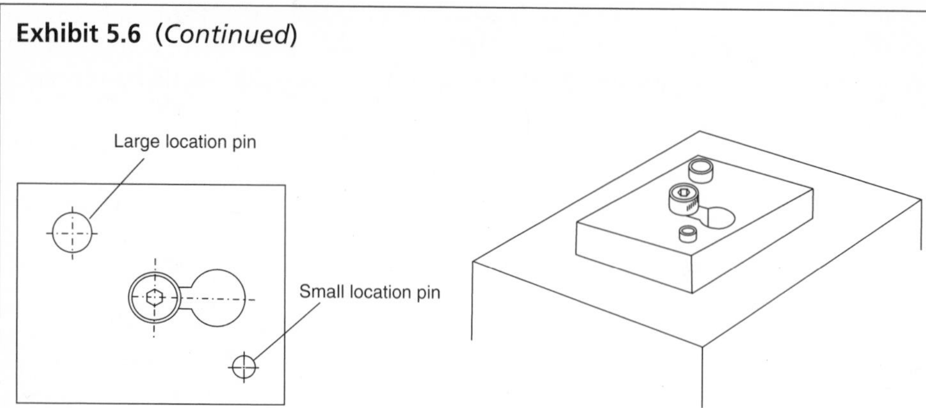

Figure 5.7(d) Possible method to locate and constrain a plate on a flat surface: 4 (keyhole slot no longer serves as a quick-release feature)

Commensurate with constraining the component in the way that is required, a generally desirable objective is for a minimum number of well chosen constraint/location features to be employed. For our purposes rapid changeover (removal and replacement) is an additional consideration – which, as shown in Figure 5.7, can be in conflict with this objective. This can apply particularly if it is also attempted to integrate 'foolproof' assembly features (enabling assembly to occur in only one possible way).

Figure 5.7 indicates the care that needs to be given, even for a very simple mechanical component. The figure applies to situations where there is to be no freedom of movement. The same attention is needed when a changeover component is to be afforded one or more degrees of freedom.

Another issue might also be taken into account: build-up of tolerances. Frequently, components will be manufactured separately with the intention of later combining them to form an assembly. A build-up of tolerances can occur as these separate components are combined. Achieving high precision can greatly assist in reducing or eliminating adjustment during changeover. In such cases it may be preferable to machine critical overall dimensions upon the completed assembly. If well designed, components can then be dismantled and reassembled in the knowledge that, assuming freedom from contamination and damage, this overall dimension will always be precisely achieved.

The issue of precision

By 'precision' we mean repeatably achieving a position or setting for the different parameters affecting the outcome of a changeover.[70] As demonstrated in Exhibit 5.5, high precision, in the context of a changeover, can reduce or eliminate the need for adjustment.

One way to achieve high changeover precision is to manufacture and secure identified components within tight dimensional and geometrical tolerances. Alternatively, rather than utilizing high-precision components, the operative can set a position manually, while also employing a monitoring device capable of indicating that the desired position has been achieved. While the use of position-monitoring devices typically has the effect of making fine adjustment easier to achieve, it does not eliminate the need for adjustment to take place.

In respect of changeover performance, the tolerances applying to the product under manufacture can also be important. As common sense dictates, the changeover required for a close-tolerance product is liable to be more complex and/or more time consuming than that for an equivalent, but low-tolerance product.[71] Where applicable, therefore, and representing a further product design consideration, changeover time savings can be possible wherever lower tolerance products can be accepted.

High-precision location of change parts alone will not eliminate the need to conduct adjustments during changeover. The issue of standardization, as described below, is also important if adjustment is to be reduced or eliminated. Standardization, when applied in terms of standard features, can dispense with the need to conduct some 'macro' adjustments to parameter settings.

The issue of standardization

Many writers assessing changeover improvement address standardization.[72–74] This is usually done in the context either of standardizing the equipment that is used[75] or, particularly, of standardizing features of that equipment.[76] Standardization can also be related to the product itself: as with the equipment that is operating upon it, standard product features can sometimes be used to enhance changeover performance.[77]

As we shall later investigate, standardization of equipment features or of product features can have the effect of eliminating blocks of changeover tasks, including undertaking adjustments.[78] Standard features can be introduced, or non-standard features withdrawn. In some cases it is also possible to achieve standardization by precisely locating change parts into other assemblies which themselves incorporate standard features – and then changing this larger assembly (or module) as a whole.[79] A good example of this technique occurs with the automated exchange of metal-cutting tools in CNC machining centres. Typically for computer-assisted machining, a range of tools, for example milling cutters, will be preset within a standard tool holder. It is this tool holder, as a complete assembly including the tool itself, that is exchanged. Again as we shall later explain (see Chapter 9), there is scope to apply this same technique to other change parts in other changeover environments.

The issue of repeatability

The issues of precision and standardization might both otherwise be thought of in terms of repeatability. In this case precision can be thought of as micro-level repeatability of the process hardware or product, whereas standardization relates to macro-level repeatability. In a changeover context a primary purpose of pursuing repeatability will be to eliminate or minimise adjustment. In other respects as well repeatability might be used to eliminate or minimise changeover effort.

Repeatability has long been crucial to enable mass manufacturing to occur. Its role remains crucial in current situations of 'flexible' (frequent, small batch) volume manufacturing. For the philosophy of mass manufacture the need for repeatability arose particularly at a micro-level where, without it, manufacturing processes become severely restricted: process parameters need to be fixed, thereby enabling the volume manufacture or assembly of successive, nominally identical products.[80] This same requirement for micro-level repeatability still exists for lean manufacture. Perhaps especially for frequent, small batch situations it is still necessary to monitor and control parameter variability – for which the use of SPC techniques might be considered[81,82] – so that reestablishing prior machine settings can swiftly enable repeat manufacture to commence.

At a macro-level repeatability is also highly important. Standard features can often either dramatically reduce or ease the tasks that comprise a changeover. A commonly reported example is standardising (repeating) die set shut heights.

5.5 A company undertaking design for changeover

As discussed previously, the machine tool industry demonstrates how an excellent changeover capability can be designed into new equipment. As noted in chapter 1, the need for better changeover performance is also becoming more apparent across a wide range of industries. Below we have isolated a manufacturer in a wholly different market. This manufacturer again clearly demonstrates how considered use of design can yield significantly better changeovers.

Zepf Inc.

Zepf Inc. emphasizes the role that design plays in the development of what is termed the 'Mark II Change Part System' for container handling. The selected excerpts reproduced below as Exhibit 5.7 are from the company's advertising material.[83,84]

Exhibit 5.7 OEM design for changeover (excerpts)

'Zepf Inc. is dedicated to offering complete engineering, manufacturing, and support services using cutting edge technology for container handling systems. We have named our combined upgraded technology the "MARK II" Change Part System.'

'Efficiency of changeovers is further enhanced by our ability to provide a multitude of colour options for correlating a complete line of different machinery to a different container. No tool changeover is accomplished by using our quick change no time star wheels and guides.'

Figure 5.8 A photograph of Zepf 'Mark II Change Part System' components

'MARK II' Philosophy

- The changeover must not require tools to install or remove components.
- The changeover is performed by any production staff.
- The changeover requires significantly less time than 'conventional' systems.
- Uptime on the line is increased when compared to 'conventional' systems.

Attributes of 'MARK II'

- No tuning or tweaking.
- No tools.
- No timing (after initial installation).
- Reduced training of line changeover staff.
- Rapid changeover between bottle sizes.
- Similar changeover procedures between machines, lines and production facilities.
- Lighter parts (Maximum 15 lbs).
- Rapid identification through colour coding.

'MARK II' Features

- Engraving options – language, line #, etc.
- Puzzle interlocks on multi-segments.

Exhibit 5.7 (*Continued*)

- Stainless steel components and hardware.
- Permanent core installation.
- Single position locating pins.
- Consistent changeover methodology.
- Lightweight.

'MARK II' Benefits

- Easy identification of change part sets.
- Faster changeovers = increased production.
- Easy to train workforce for changeovers.

Further insight into design for changeover

The use of design by Zepf Inc. assists further in understanding ways that better changeovers can be achieved. The descriptions reinforce the preceding discussion, additionally highlighting new issues including:

- Lightweighting
- Low skill requirement ('Can be achieved by any production staff'; 'Reduced training of line changeover staff')
- Being able easily to identify change parts (by colour and engraving)
- 'Out-of-the-box' operation (no adjustment)
- Consistent changeover procedures
- 'Foolproofing' ('Puzzle interlocks on multi-segments'; 'Single position pins')
- Eliminating the need to use tools
- Rapid release mechanisms
- Rapid location

Other important ways that design may be employed, for example to allow for single-person working or to provide improved access, will be described later. Reference to this company's work and an assessment of machine tool developments both indicate ways that design might be applied.

5.6 The organization–design spectrum

One purpose of the organization–design spectrum is to show that frequently any given improvement will incorporate elements of both design and organizational change. This is an important concept as it dictates that improvement by design typically cannot be isolated from improvement by organizational means. The organization-design spectrum is considered now with a view to later developing improvement tools that are valid irrespective of where emphasis for improvement (by design or organization) lies.

Evaluating design-led improvement alongside organization-led improvement

That design improvement and organizational improvement are very often combined can be illustrated by two example situations:

> **Example 1**. Take the situation of attaching a gauge to obtain on-line position data (to allow adjustment to occur more swiftly). That the gauge was added represents a change to the mechanical specification of the equipment, and therefore represents a design improvement. Yet the gauge by itself is of no value unless it is known what predetermined gauge readings are to be aimed for. This latter aspect represents the need for a revised changeover procedure. Taken together, the solution requires both organizational and design changes to occur if it is to be of use.
>
> **Example 2**. A decision is taken to bring all changeover tools adjacent to where they are to be used. Two possible options are: placing the tools on a trolley or mounting them on a rack on the changeover machine. The trolley idea would represent purely organizational improvement only if an existing trolley were to be used in an unmodified state. Any changes that were to be made to the trolley to accommodate the tools and to present them well (making them accessible and easy to identify) would represent non-substantive secondary design changes to aid the current changeover. On the other hand, mounting the tools in one or more racks on the machine would represent a non-substantive design change to the existing machine. In both cases the existing procedures would be slightly modified (see the discussion of changeover tasks, below).

These examples indicate the likely merit of an integrated approach to changeover improvement, understanding and jointly involving aspects of both organization and design. The notion of mutually combining elements of organization and design is now investigated wherein the concept of an organization–design spectrum is set out in Figure 5.9. This figure shows that solutions to changeover problems may range in their content from those that employ 100 per cent design through to those that employ 100 per cent organizational improvement. Marked is the likely emphasis on organizational changes when employing a typical consultant-led methodology for retrospective

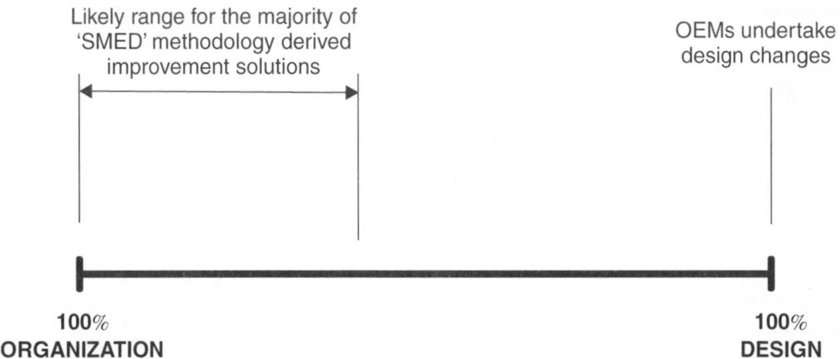

Figure 5.9 The organization–design spectrum

improvement. Similarly, it is indicated that an OEM has little direct influence on the exact changeover procedures that a user adopts, instead being able to bring about improvement to their equipment's changeover performance only by means of design.

Mapping attributes of changeover on the organization–design spectrum

At a simple level the organization–design spectrum firmly identifies that there is a role for design in changeover improvement, even if, by adopting solutions towards the organization end of the spectrum, design is frequently of a non-substantive nature. At a more detailed level it is possible to map various attributes of a changeover to the organization–design spectrum. This allows the likely relative benefits of approaching changeover improvement with an emphasis on either design or organization to be judged. A number of attributes of a changeover have so far been highlighted. These include:

- The cost of improvement
- The effort required to implement the improvement
- The time needed to achieve the improvement
- The ease with which the improvement may be sustained
- The level of improvement that might be achieved
- The skill requirement to conduct the revised changeover

These attributes are mapped to the organization–design spectrum in Figure 5.10.

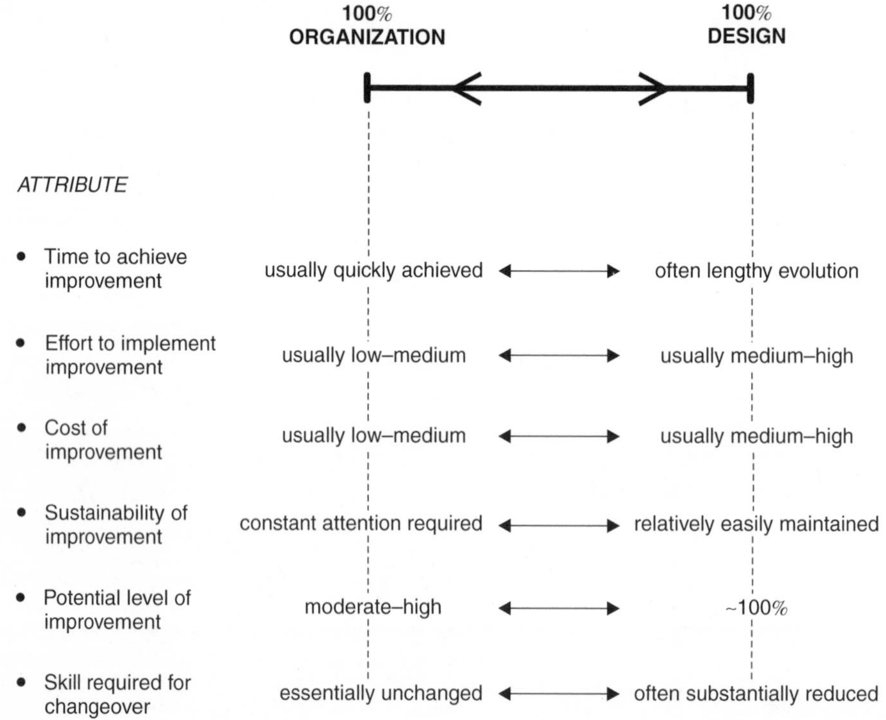

Figure 5.10 Mapping attributes of changeover on to the organization–design spectrum

Figure 5.10 shows that an insistence on largely avoiding design solutions means that potentially very valuable benefits of emphasizing design may be lost.

Good design allows for simpler, better changeover tasks to be conducted, rather than conducting existing tasks better. Because design changes allow better things to be done, a changeover that is both faster and of higher quality can be expected.

5.7 The 'Reduction-In' strategy

The 'Reduction-In' strategy is an improvement tool that we have developed. For an identified changeover task the 'Reduction-In' strategy seeks to identify an 'excess' in one or more of four possible categories. An appropriate solution can then be sought that reduces or eliminates this 'excess':[85]

- (*Reduction-in*) **On-line activity**
- (*Reduction-in*) **Adjustment**
- (*Reduction-in*) **Variety**
- (*Reduction-in*) **Effort**

This chapter now elaborates upon the derivation and application of the 'Reduction-In' strategy. The primary focus is currently upon improvement by design. Later (in Chapter 7) our approach will be extended such that a complete range of potential improvement options (represented by the full organization–design spectrum) will be encompassed. Design rules derived from the 'Reduction-In' strategy will be presented at the end of the chapter.

Iterative design improvement: one reason why the 'Reduction-In' strategy is needed

While our discussion so far has highlighted different ways in which design might be applied, there is little structure in what we have presented. What has been described can be of assistance, but as yet no method is apparent to isolate an appropriate design solution (nor indeed an organizational solution) to overcome identified changeover problems. Neither, as yet, have we described a method that permits changeover difficulties to be identified in the first place. Without structure to the improvement effort, solutions can only be arrived at based on previous experience and expertise.

Figure 5.11 shows steps of a structured, iterative improvement approach. It indicates the role of the 'Reduction-In' strategy in this overall improvement process.

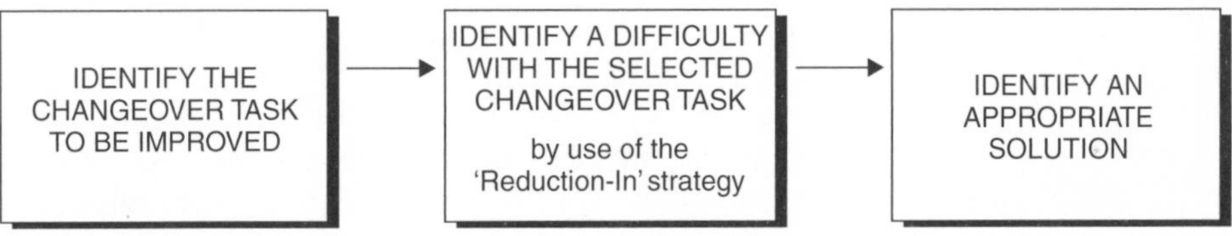

Figure 5.11 An iterative approach to retrospective changeover improvement

Further consideration of the 'Reduction-In' strategy: potential improvement mechanisms

To understand further the contribution that the 'Reduction-In' strategy can make it is useful to recognize that reduction in elapsed changeover time, T_c (see Figure 1.2), can be possible by two distinct mechanisms.[86] One mechanism is to minimize T_c without altering the current changeover tasks. The rationale here traditionally (see below) is to ensure that as much as possible is achieved before the line is halted. Although T_c is shortened by this mechanism, the overall duration of individual tasks remains unaltered. The second mechanism is to reduce the duration of individual changeover tasks, if possible to the extent of elimination. These two mechanisms are shown in Figure 5.12.

When considering Figure 5.12 it is useful to think once again about the definition of a changeover that we promote (see Chapter 1). This definition includes tasks that occur during the run-up stage. It is only once this stage has been completed that the changeover is complete. In addition, a typical changeover, as is already widely acknowledged, will include preparatory activity prior to actually halting production. These too are changeover tasks. As part of a comprehensive initiative, all changeover tasks should be subject to possible improvement.

The contrast between our interpretation of a changeover with that of a 'good piece to good piece' definition (in which the run-up phase is not included) is important. The nature and likely significance of the run-up phase should be understood. The run-up phase will largely comprise adjustment tasks.[87] The prominence of the run-up phase in many of the situations that we have studied – sometimes being up to ten times the duration of the set-up phase – indicates the attention that should be accorded to adjustment when improvement is sought.[88] Further, improvement that is sought purely by rearranging when existing tasks start will essentially have no impact on the adjustments that occur.

We argue that activity during the run-up phase must not be excluded from appropriate, direct attention. A strong emphasis on rearranging tasks into external time can be misplaced if the run-up phase is particularly extensive – because many tasks during this phase cannot easily be rearranged in this way. Furthermore, the oft-held notion of 'internal time' becomes flawed, as does any possible over-reliance upon techniques that are currently employed to minimize the 'internal time' contribution.[89]

Further consideration of the 'Reduction-In' strategy: prioritizing improvement options

The 'Reduction-In' categories do not necessarily assume any order or priority. Improvement options that each category in turn may embrace are free to be selected on merit. In other words, particular improvement options are not intrinsically accorded either dominance or precedence. For example, there is no imperative initially to seek changeover time reduction by mechanism 1 (Figure 5.12). This may be contrasted with prescriptive practice described by Shingo's three sequential 'SMED' concepts, in which Shingo apparently separated the two improvement mechanisms

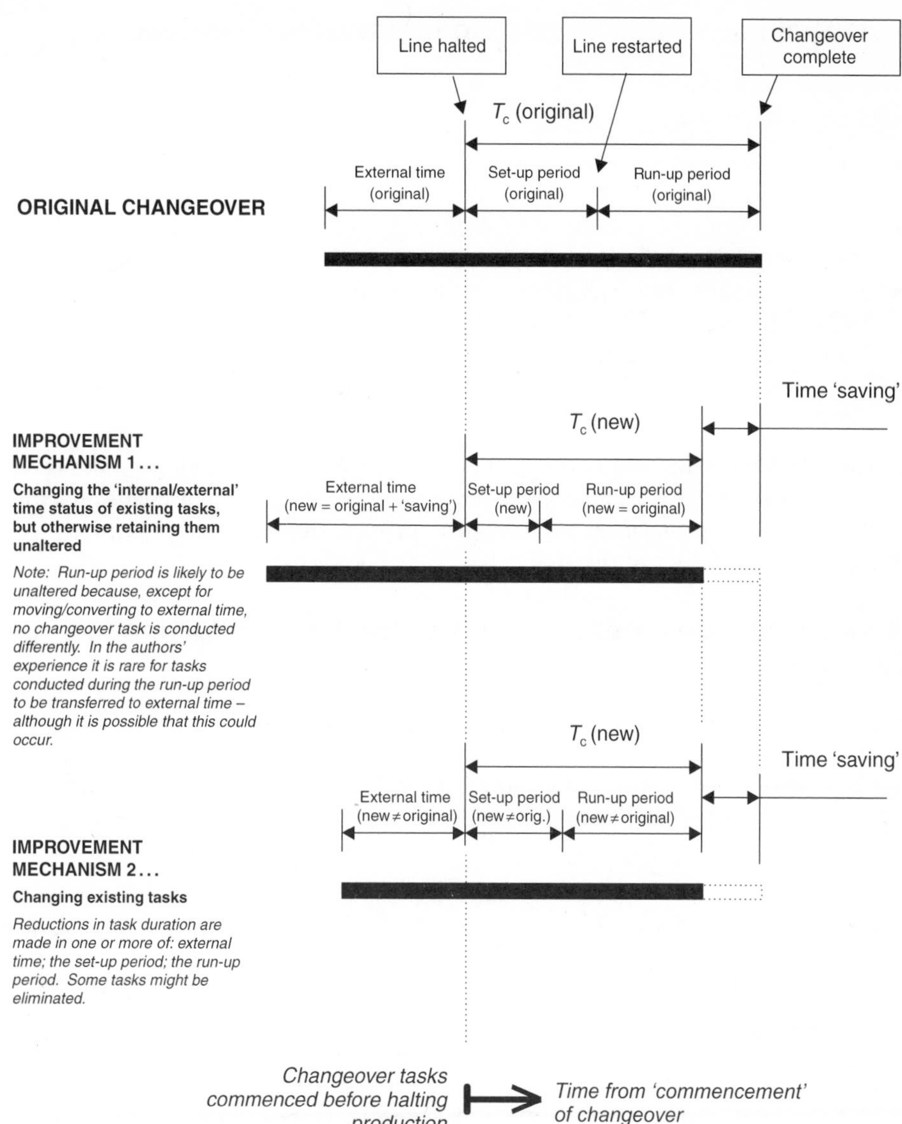

Figure 5.12 Reducing T_c by two alternative mechanisms. (Consider also that 'good piece to good piece' changeover definitions do not necessarily include adjustment tasks that occur during the run-up period)

identified by Figure 5.12.[90] Much of the 'SMED' methodology devotes attention to the former mechanism, wherein existing changeover tasks are moved ('separated') or converted into external time.[91,92] It is only when the final stage of the 'SMED' methodology is reached that attention is turned to reducing the duration of changeover tasks.

Potential limitations of existing improvement practice

The foregoing discussion alludes to some significant reservations in respect of existing improvement practice:[93]

- concern over the prescriptive nature of primary current improvement methodologies
- the prominence of gaining improvement by translating changeover tasks into external time
- problems of adequately assessing (and improving) a changeover's run-up period – or, in other words, failure to acknowledge either that a run-up period frequently occurs, or to acknowledge its significance
- poor recognition of the contribution that design modifications can make, and corresponding failure to structure improvement effort accordingly

The 'Reduction-In' strategy is a key tool to overcome these limitations. The objective of reducing or eliminating changeover task 'excesses' represents an entirely new approach to changeover improvement. It is an approach that is expounded in much of the remainder of this book.

An introductionary explanation of the four 'Reduction-In' categories

A brief explanation of the 'Reduction-In' categories is useful at this stage. In particular, the category 'on-line activity' can require explanation to distinguish it from Shingo's two initial 'SMED' concepts ('separate' and 'convert' changeover tasks into external time). A further explanation will later be provided in Chapter 7.

On-line activity

The differences between seeking improvement by reducing 'on-line activity' and seeking either to 'separate' or 'convert' tasks into external time are important. In making a distinction, it should also be reflected that a run-up period is an integral part of a typical changeover.

Reducing on-line activity means reducing the time spent conducting tasks while the line is not in a steady production state* – without fundamentally altering the nature of those tasks. A reduction in the elapsed changeover time is achieved by altering when tasks are commenced. Seeking to reduce on-line activity certainly includes translating changeover tasks into external time (as long as those tasks are not otherwise altered), but its overall scope is wider than this.

Figures 5.13 and 5.14 are provided to introduce options encompassed by 'reducing on-line activity'. As in Figure 5.12, the interval between halting production of the prior batch and recommencing full production of the new batch is marked. This is the total elapsed changeover time, T_c, including the run-up period. Figure 5.13 shows a pre-improvement situation, individual changeover tasks with respectively being conducted by three operatives. As is also shown, the overall changeover includes an automated task.

---

* While the line is halted or undergoing run-up.

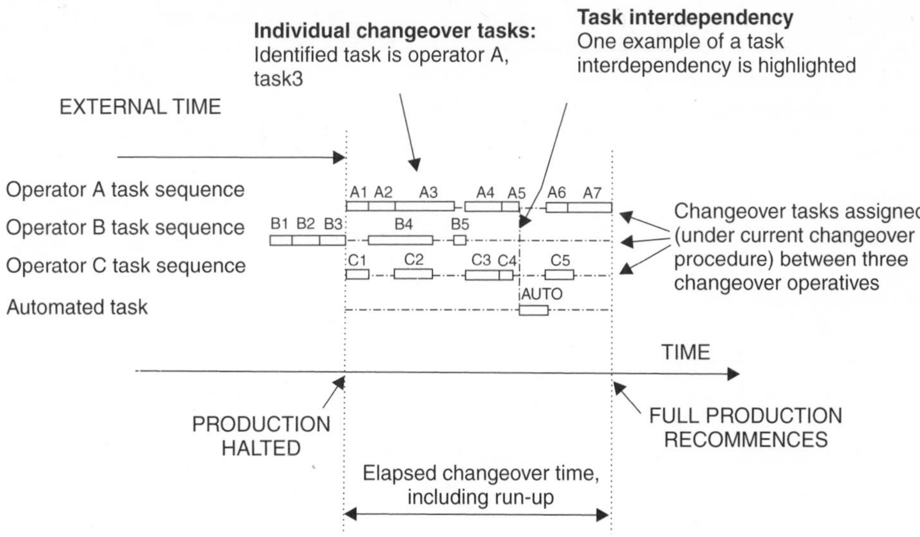

Figure 5.13 Activity during a three-operative changeover procedure

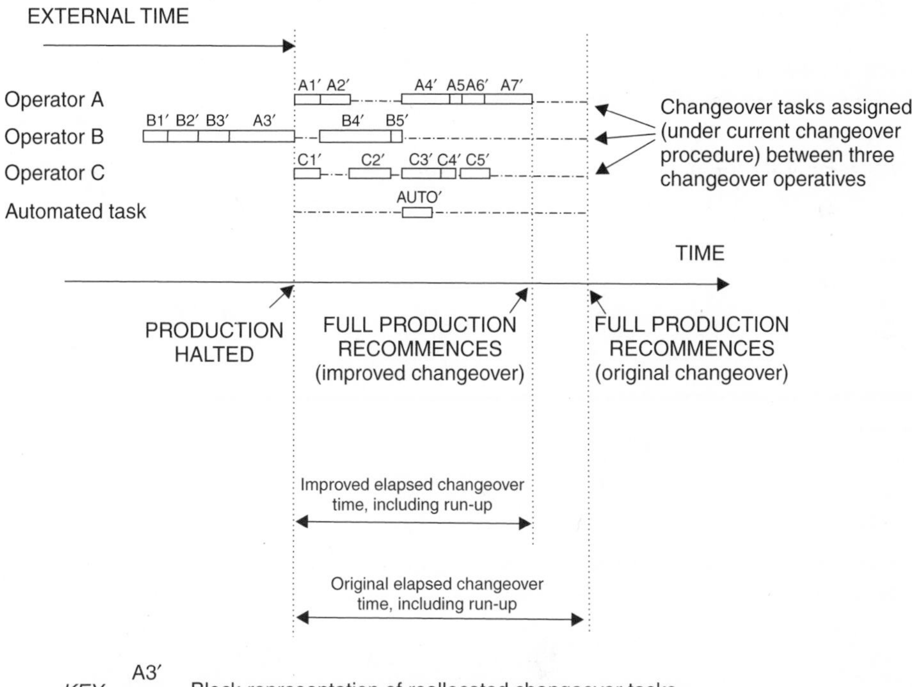

Figure 5.14 Possible task reallocation after minor task interdependency alleviation

Many tasks during a typical changeover can be freely commenced at a different time. An example might be arranging for tools to be sited adjacent to their point of use. Other tasks, however, cannot be commenced at a different time at will: task interdependencies (see Chapter 7 for a full explanation) restrict this from occurring. In other words, some tasks cannot start until other tasks have been completed. Alleviating or breaking task interdependencies can enable individual task start times to be altered – thus, as shown by Figure 5.14, enabling the total elapsed changeover time to be reduced. As is indicated, tasks need not always be newly commenced in external time for a faster changeover to come about. Significant reduction in the total elapsed changeover time can arise even if only comparatively few task interdependencies are alleviated or broken.

Figure 5.14 shows that after improvement task A3 is being conducted in external time. It is also being conducted by an alternative changeover operative. Although not explicitly shown by Figures 5.13 and 5.14, increased parallel working can also, typically, lead to reduced on-line activity. As will later be explained, the ability to increase parallel working can be highly dependent upon the ability either to increase or to reallocate resources (Chapter 7).

Finally, it is useful at this stage to appreciate the potential role of design to reduce 'on-line' activity. As has been previously described, design typically has the effect of imposing change on existing work practices. In most cases, therefore, design will change the duration of existing tasks, thus not qualifying it to reduce on-line activity. The exception to this is the use of design to alleviate or break task interdependencies – to enable tasks to be conducted at different times relative to other tasks within the overall changeover procedure. Once again, these issues will be explained further later (Chapters 7 and 9).

The primary distinguishing feature of reducing 'on-line activity' – that of not altering how existing tasks are conducted – is not apparent in any of the remaining three 'Reduction-In' categories. Thus, the option of reducing 'on-line activity' can be associated with mechanism 1 of Figure 5.12. Seeking to reduce 'adjustment', 'variety' or 'effort' are each associated with mechanism 2.

Adjustment

Seeking a reduction in adjustment, the second category of the 'Reduction-In' strategy, applies at any stage during the changeover. Thus it can apply to tasks conducted before the line is stopped, that occur while the line is halted or that occur during the changeover's run-up phase. It is sought here to alter the existing tasks, rather than, as above, altering when they commence. The scope for improvement has been indicated in some of the previous discussion in this chapter.

Variety

Significant additional work during changeover can be imposed by variety (non-standardization) – in terms of physical features of change parts and products and in terms of variable changeover procedures. Where it is appropriate to do so, curtailing this variety can often significantly aid changeover performance.

Effort

Many opportunities typically are available to reduce the effort required to conduct changeover tasks (to simplify or to deskill existing tasks). *Effort*, more so than for the other 'Reduction-In' categories, might not be seen as representing a stand-alone category. Care should be taken that this is understood when it is used, ensuring that opportunities described as 'reducing effort' are not more correctly referred to as 'reducing adjustment' or 'reducing variety'. Nevertheless, it is a classification that brings together many opportunities that do not directly reside in the other categories of the 'Reduction-In' strategy. Its scope and its usefulness will become apparent throughout the remainder of this book.

5.8 Applying the 'Reduction-In' strategy

Notes and examples in respect of applying the 'Reduction-In' strategy are now presented. The discussion anticipates that the 'Reduction-In' strategy will be used to identify an 'excess' of the existing changeover, with the intention of then seeking a solution to overcome this identified excess (Figure 5.11).

Changeover task definition

We advocate flexibility in defining a changeover task, so as not to constrain an improvement practitioner's perception of where or how improvement may be sought. A changeover task can be defined in different ways, by breaking down activity into ever more specific individual operations. For example, removing a die set from a press could be thought of as a changeover task. This definition of a task can be broken down further. A task necessary to remove the die set might be to remove six bolts securing it to the lower die shoe. Taking changeover task definition to an even more specific level, removing these six bolts might comprise separate tasks of: walking to the tool box – opening the tool box – selecting a 19 mm socket – selecting an extension drive – mounting the socket to the extension drive – etc.

We suggest that each of these task definitions should be able to apply. The improvement practitioner will be assisted if he or she has the flexibility to perceive changeover tasks at these different levels, as is appropriate.

Task elimination

Where possible, reduction in adjustment, variety or effort can be sought to the extent of task elimination. Not constraining how a changeover task is defined can be important in allowing elimination opportunities to be identified. Take, for example, die set removal, as described above. When changeover tasks are conceived at a highly specific level it might be more difficult to identify major elimination opportunities. For instance, eliminating the need to mount the socket to the extension drive (a very minor improvement) could be achieved simply by leaving a dedicated hand tool to remove these bolts permanently assembled. When broken down to this level, this opportunity might be relatively easy to identify. Similarly, mounting the assembled hand tool adjacent to the press will eliminate other tasks in this detailed task sequence. The elimination of this series of tasks altogether, though, will probably be easier to identify if die set removal itself were to be viewed as a separate task. Die set removal can be made easier to achieve (a reduction in effort) by using, for example, hydraulic clamps in place of the six securing bolts.

Iterative improvement: the possibility of eliminating earlier improvements

A consequence of an iterative approach to changeover improvement is that improvements that have previously been made can subsequently themselves be eliminated. For example, improvements such as mounting a dedicated extension bar/socket adjacent to the die set would be rendered unnecessary by any subsequent introduction of a hydraulic clamping system. Some of the previous exhibits in this chapter have also amply demonstrated this phenomenon.

The above elimination of the need to use an extension bar/socket is an example where little previous improvement work will have been rendered superfluous. In other cases the situation might not be so favourable, where significant previous work (and time and expense) will be lost. Although always a possibility, such instances of unnecessary work, delay and expense can be at least minimized by assessing – and setting – a target changeover improvement time in advance. Doing so will guide those responsible in seeking solutions where, as described below, alternative routes to an identified improvement target can be possible. The solutions that are sought likewise might be influenced by the anticipated duration of the improvement programme and the financial resource that is allocated to it, which again should both be resolved at the outset, and set as targets (Chapter 3).

Figure 5.15 shows that an initiative might be seeking an improvement in changeover performance from 10 hours to 10 minutes. This level of improvement is unlikely

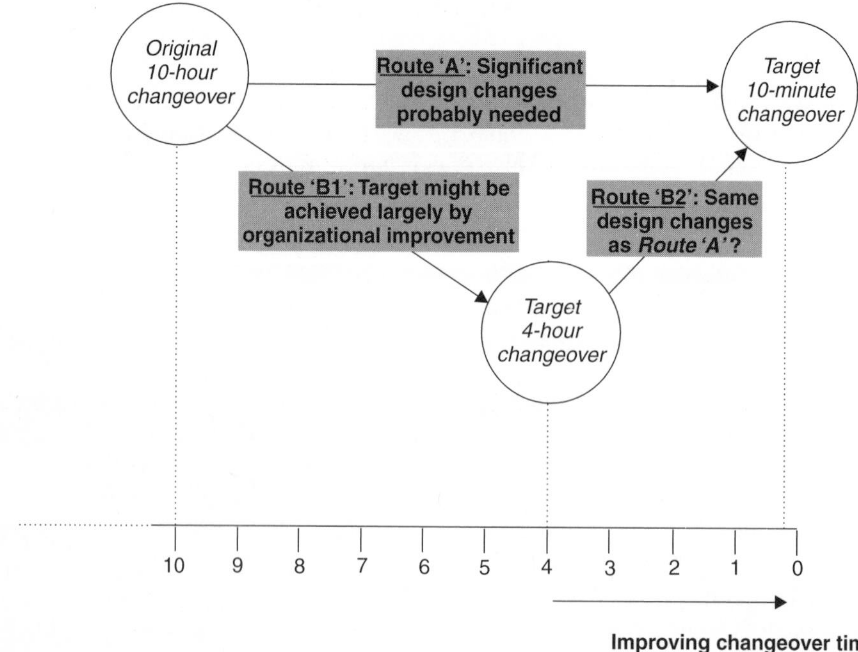

Figure 5.15 The importance of defining targets in advance, to ensure efficient improvement effort

to be achieved without significant design modification taking place. Substantive design changes might not be necessary if a 4-hour target were set instead. However, the same design modifications (that is, those undertaken to bring the changeover time from 10 hours to 10 minutes) might subsequently be needed to take this time from 4 hours to 10 minutes. Thus, if improvement had been undertaken without a clear target, it is quite possible that changes made to bring the changeover time to 4 hours would later be rendered superfluous by further changes to take the time to 10 minutes. Working in this way, at best, would represent inefficient improvement practice.[94]

Examples of changeover improvement

The iterative improvement process of Figure 5.11 seeks appropriate solutions to identified changeover problems. In other words, improvement techniques applicable to the current situation are being sought.

The great majority of potential techniques are already known. Two challenges that an improvement practitioner faces are first, determining what techniques might be applied and second, understanding how to evaluate these competing techniques in the context of the changeover currently under scrutiny. Notwithstanding how this is approached, there is also little doubt, in our opinion, that a greater knowledge of both devices and prior examples of changeover improvement will greatly facilitate the practitioner's efforts.[95] To this end we propose publishing a further book that provides examples of changeover improvement across a wide range of industries. Readers are invited to contribute to this work (see Preface). The topic of improvement examples will be addressed in greater detail in Chapter 7.

5.9 Design rules for changeover

The design of any product or manufacturing equipment (itself a 'product') will seek to optimize many often-conflicting requirements. Function, cost, durability, reliability, aesthetics, safety, ergonomics ... the list is considerable.[96,97] The overall design process has been expounded by academic authors,[98–100] whose work indicates its complexity.

Our purpose here is to describe good design practice for changeover – either for retrospective improvement or for original equipment. We propose design rules for changeover improvement that are based on the 'Reduction-In' strategy. Doing so allows potential solutions to identified changeover problems to be selected with greater ease; the design rules being a part of the iterative improvement process as shown in Figure 5.16.*

---

* The iterative process is described at this point with an emphasis on design. An overall approach that is applicable also to organization-biased improvements (towards the left-hand side of the organization–design spectrum) will be presented in Chapter 7. The identification of a changeover task for improvement, the first stage in this iterative process, will be discussed in the next chapter.

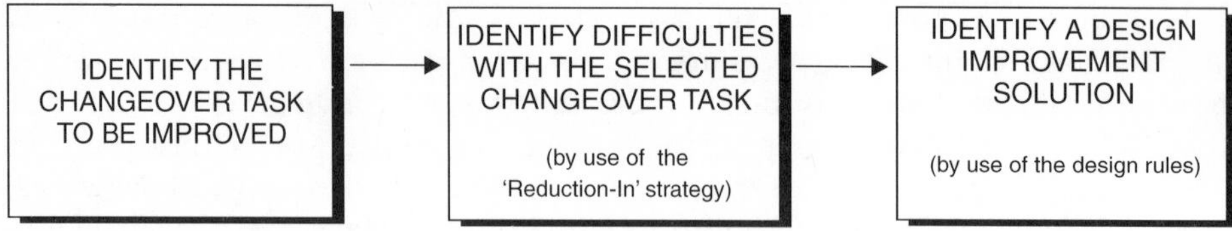

Figure 5.16 The steps involved in selecting a design solution

Design for changeover rules can be set against other 'design for' rules or procedures. These prominently include design for assembly[101,102] and design for manufacture.[103–105] Wider-ranging topics can also be included in the 'design for' nomenclature,[106] where even design for scheduling[107] has been discussed. Design for function, cost, safety and quality, among other objectives, are typically taught to students of design as an integral part of the design process. This approach to design subsequently becomes reinforced in professional working practice. Not all 'design for' objectives are mutually compatible, where some elements of design for changeover might conflict with other design objectives.

Although our proposed design rules for changeover can be used with the 'Reduction-In' strategy, as described by Figure 5.16, their 'stand-alone' use is not precluded. This might be more appropriate in any case for OEMs.

Design change to enhance maintenance

We do not consider the topic of industrial maintenance in detail, for which many authoritative texts are available.[108,109] Nevertheless, as we have previously described (Chapters 1 and 4), making maintenance faster and easier to conduct can have much in common with making changeover faster and easier to conduct. The design rules that we set out in this section might be equally applicable to reduce equipment downtime for maintenance purposes.[110] There is evidence that the design rules for changeover that we have developed are similar, in some respects at least, to what has previously been advocated for maintenance purposes.[111]

Design rules for changeover

The design rules are set out in Figure 5.17. The figure also presents some potential improvement options for each design rule.

Like the 'Reduction-In' strategy, the design rules have been developed as a working tool. They have been derived from the 'Reduction-In' strategy, with the advantage that a path can be traced in the iterative process described in Figure 5.16 (that, as yet, is described only for design).

The design rules are intended to be of use either in conjunction with the 'Reduction-In' strategy or in their own right. The design rules have been tested in diverse situations during our research. Some examples of their application are presented at the end of this chapter.

'Reduction-In' category	Design rule	A few examples of possible improvement options
On-line Activity	1 Engineer change as to when tasks can be conducted *(to enable better task allocation between existing resources, including conducting tasks in external time)*	• Provide parallel equipment, including equipment to enable presetting • Eliminate barriers to parallel working (for either manual or automatic tasks) • Provide for tasks to be conducted in an alternative sequence
Adjustment	2(a) Control the precision of change parts and make provision of features on either or both the product or the process equipment, to enable repeatable location 2(b) Employ on-line monitoring equipment 2(c) Seek 'universality' *(to eliminate the need to conduct adjustment)* 2(d) Make mechanical systems more tolerant of variation between nominally identical batches *(to help obviate the need for 'fine tuning' adjustments)*	• Add precision location features to change parts • Employ location pins • Provide for the use of setting pieces • Seek damage resistance • Linear position monitoring • Monitor die shut forces • Employ 'universal' transportation mediums – for example, pin chains or pockets
Variety	3(a) Where possible, standardize change part and product features 3(b) Foolproof information exchange and change part location	• Standard tool shut height • Standard location positions • Employ computer-aided information transfer
Effort	4(a) Add discrete mechanical devices to aid existing tasks 4(b) Modify existing components to aid existing tasks 4(c) Consider mechanization of existing tasks, including the use of automated, remotely activated devices 4(d) Improve access to the change parts 4(e) Aspire to single person execution of tasks 4(f) Aspire to eliminate the use of tools 4(g) Aspire to eliminate existing task or existing blocks of tasks	• Add tool runners • Add quick-release clamps • Introduce key-hole slots • Lightweight components • Use a power winder • Use hydraulic clamping • Improve guard removal • Split parts that otherwise have to be removed entirely • Revise modular assembly

Figure 5.17 Design rules for changeover improvement

Design change within the wider business system

We have argued that the hardware physically undergoing changeover and the procedures that are employed during changeover should be thought of as part of a wider internal business system. This wider business system might also be subject to change – to being restructured – to improve, exploit or sustain the company's changeover capability. Opportunities might be available, for example, to alter the scheduling system that is employed, to improve feedback by revising changeover problem reporting or to ensure better information exchange between production and marketing functions. As highlighted elsewhere, verification of changeover item quality can also be highly significant.

Although each of these systems can be 'redesigned', the design rules that we have presented are only appropriate to primary engineering design changes to the equipment and the products undergoing changeover.

Examples of applying the design rules in conjunction with the 'Reduction-In' strategy

Exhibit 5.8 considers a situation where a silk-screen print operation was handicapped by poor changeover of the screens. Exhibit 5.9 describes difficulties when changing over a gear set.

For each exhibit the application of the 'Reduction-In' strategy and the associated application of the design rules is described. In each case a structured approach to improvement is pursued. Even so, the identification of a final solution is still helped by the practitioner's experience – in knowing previous examples and knowing the fine detail of the machine being worked upon.

The work that was conducted during these case studies may be compared with what might have occurred had low-cost organizational improvement alone been sought.

For Exhibit 5.8 the cost of the design changes was not high. The reported savings in changeover time eventually exceeded 90 per cent, again indicating (see also Exhibit 5.2 and Chapter 4) that the relationship between the cost and the level of improvement that is achieved is very difficult to predict. Exhibit 5.9, although more costly to instigate, was not prohibitively expensive. As with almost all substantive design changes, the solution presented in Exhibit 5.9 obliges a significantly revised, simpler changeover procedure to be used. Reversion to the old procedure cannot happen unless these design changes are undone.

Exhibit 5.8 Silk-screen changeover

Loading five polyester silk screens into a sequential UV bottle printing machine was time consuming. A notable difficulty existed in setting these screens into position (adjusting them) so that the screen's consecutive colours were correctly overlaid on the product. The overall changeover of a set of five screens took approximately, on average, one hour to complete.

The 'Reduction-In' strategy and design rules were applied to this problem (which was just one part of the overall machine changeover):

Changeover task: Exchange five screens.

Existing 'Reduction-In' excess: The primary problem was the time taken to *adjust* the screens to get each of the five print images correctly positioned on the

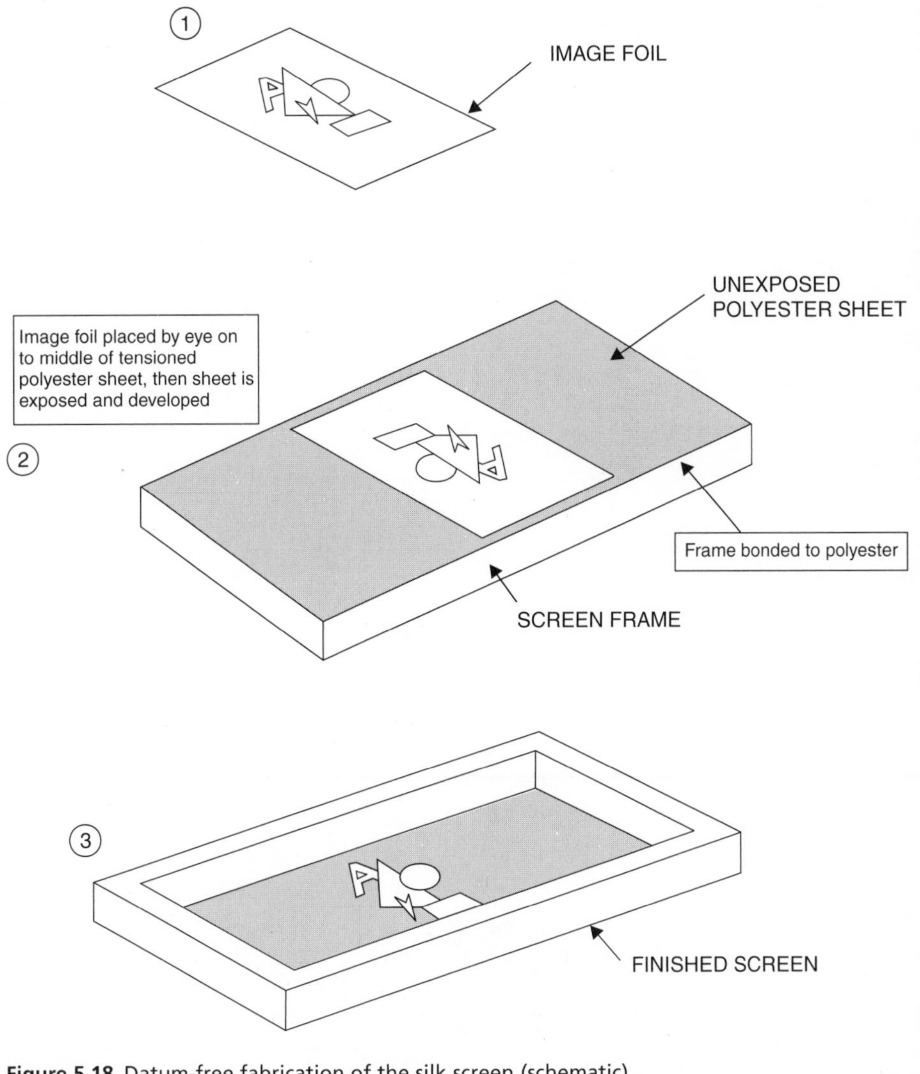

Figure 5.18 Datum-free fabrication of the silk screen (schematic)

Exhibit 5.8 (*Continued*)

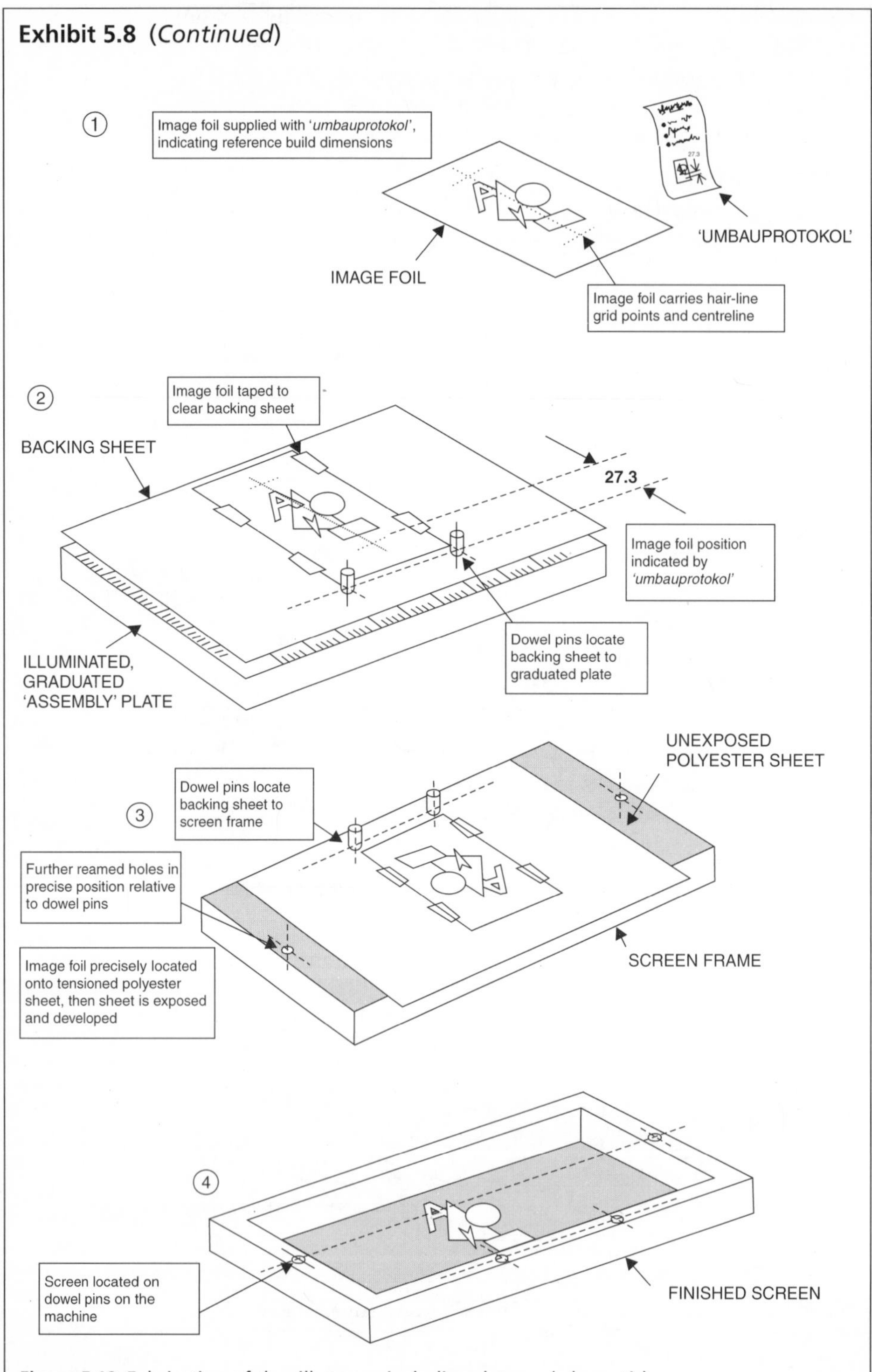
1
Image foil supplied with 'umbauprotokol', indicating reference build dimensions
'UMBAUPROTOKOL'
IMAGE FOIL
Image foil carries hair-line grid points and centreline
2
Image foil taped to clear backing sheet
BACKING SHEET
27.3
Image foil position indicated by 'umbauprotokol'
Dowel pins locate backing sheet to graduated plate
ILLUMINATED, GRADUATED 'ASSEMBLY' PLATE
UNEXPOSED POLYESTER SHEET
Dowel pins locate backing sheet to screen frame
3
Further reamed holes in precise position relative to dowel pins
Image foil precisely located onto tensioned polyester sheet, then sheet is exposed and developed
SCREEN FRAME
4
Screen located on dowel pins on the machine
FINISHED SCREEN

Figure 5.19 Fabrication of the silk screen including datums (schematic)

bottle. This took approximately 50 minutes of the 60 minutes total time for this changeover.

Proposed improvement: Attempt to eliminate the need to adjust the screen images, by precision location in the print machine.

Design rule employed: 'Control the precision of change parts and make provision of features on either or both the product or the process equipment, to enable repeatable location'.

Two elements of this improvement work were identified:

- Positioning the image precisely in the screen
- Positioning the screen precisely in the machine

The way that these problems were overcome is described in Figures 5.18 and 5.19 respectively, showing the situations before and after improvement.

The improvement involved both design and organizational changes:

- Pinning some components of the print machine together to reduce the scope for adjustment that was previously available – but was never required.
- Defining a location datum to set the image foil to the screen during exposure. Key data were incorporated on the 'umbauprotokol' (bottle image data sheet). A feint central datum line was added to the image foil.
- Writing a standard procedure to gain repeatable, non-variable work practice when manufacturing the screens.
- Defining datum positions on the screen frame relative to the image foil datums. These positions are reamed holes.
- Standardizing on just three screen sizes (from approximately 20).
- Standardizing screen polyester tensions.
- Defining a datum position on the machine to locate the screen frame, and hence the screen (which is two dowels protruding by 5 mm).

Quick-release toggle clamps were also fitted to secure the screens.

The importance of datum positions

An early improvement suggestion was to set the screens into the machine using newly affixed scales (to achieve repeat positions). This work failed to take into account the non-repeatable position of the screen image in the screen. Thus, upon using these marked screen frames for a new print design, there was found to be no time improvement for the screen changing task (in fact, because the problems that were experienced were not anticipated, the changeover took longer than before to complete).

Exhibit 5.9 Gear assembly changeover

A simple gear set required to be changed. The original machine design in this area appeared to have been made with an emphasis on functionality and cost, wherein changeover performance had been compromised. The team assessing the gear set

Exhibit 5.9 (*Continued*)

changeover believed that a redesign was possible that would speed up and simplify the existing procedure.

Attention was focused on the degree of dismantling that was required – of heavy and awkwardly shaped components. The most prominent of these components was a 'T-piece', shown in Figure 5.20 (see also Figure 5.21), into which one gear and its bearings were mounted.

Changeover task: Exchange gear assembly.
Existing 'Reduction-In' excess: Overall, significant manual *effort* was expended on an assembly that was awkward to remove and reassemble.
Proposed improvement: Try to detach the bearing block without the removal and replacement of the large T-piece.
Design rule employed: 'Modify existing components to aid existing tasks'.

Figure 5.21 schematically shows the situation after improvement. A key feature of the previous design was the bearing block's location in a central slot in the T-piece. This being so, the bearing block could not be removed until the original T-piece was broken down from the machine. This problem was overcome by moving the vertical arm of the T-piece to one side and providing this arm with a slot through which just the gear block securing screws passed. Detachment and reassembly of a new gear/bearing block assembly then became much easier to achieve. The changeover time for a single gear set fell from approximately 12 minutes to 5 minutes.

Figure 5.20 Original gear set assembly

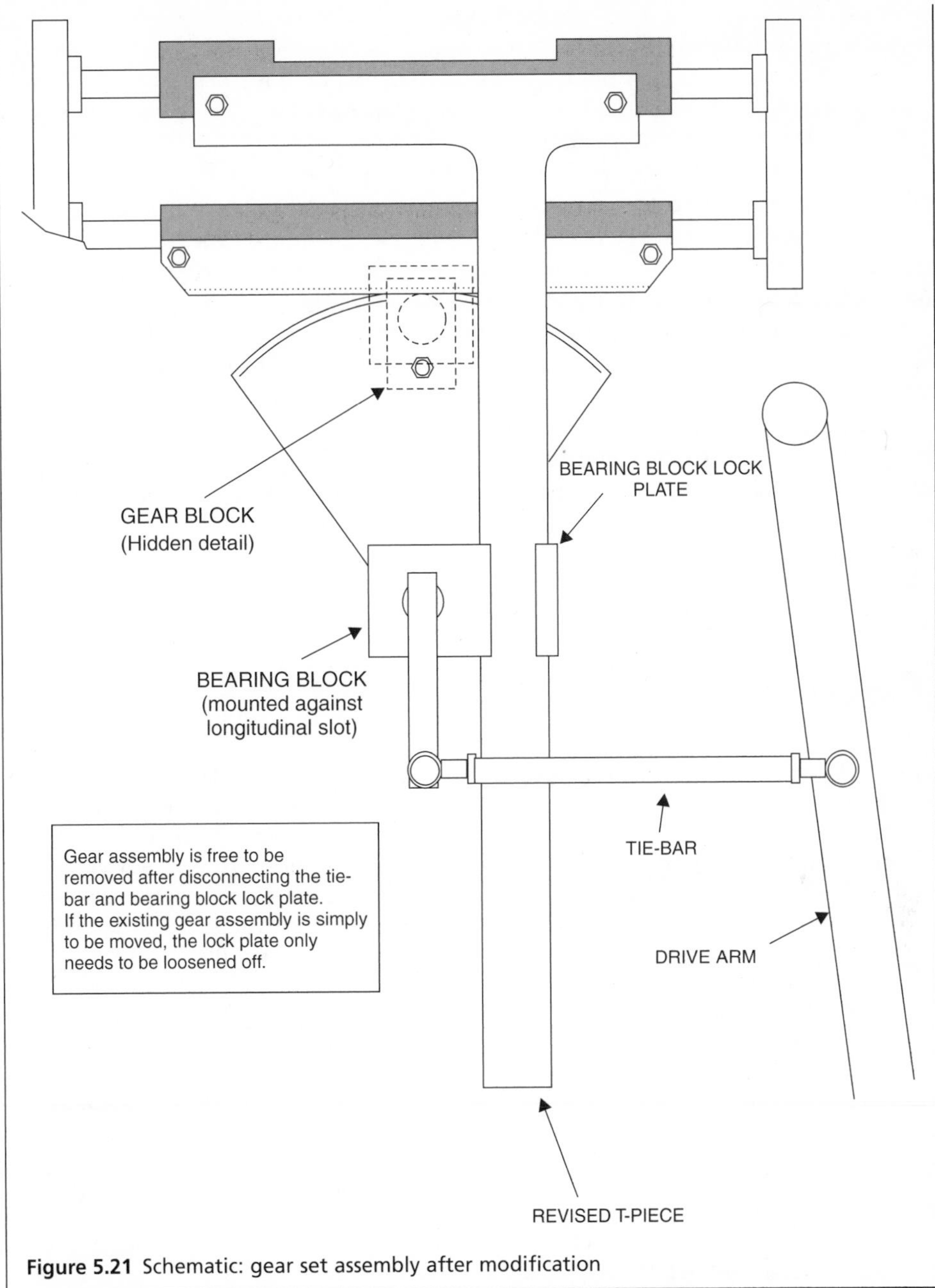
BEARING BLOCK LOCK
PLATE
GEAR BLOCK
(Hidden detail)
BEARING BLOCK
(mounted against
longitudinal slot)
TIE-BAR
Gear assembly is free to be removed after disconnecting the tie-bar and bearing block lock plate.
If the existing gear assembly is simply to be moved, the lock plate only needs to be loosened off.
DRIVE ARM
REVISED T-PIECE

Figure 5.21 Schematic: gear set assembly after modification

5.10 Conclusions

The 'Reduction-In' strategy has been introduced and design rules for changeover have been presented. The 'Reduction-In' strategy has been described so far with an emphasis on design. In later chapters a more balanced approach that considers organization-led improvements alongside design-led improvements will be described.

Notwithstanding the desirability of associating the design rules with the 'Reduction-In' strategy, attention has also been given to presenting the design rules as a meaningful 'stand-alone' description of how design might be applied. The design rules can be applied both for original design and for retrospective design. In either case, possible design improvement options will be investigated further in Chapter 9.

To understand better the contribution that design improvement can make, an alternative perspective of a changeover has been introduced that augments the time-based definition that has previously been presented. This perspective helps to clarify alternative improvement options beyond translating tasks into external time, seeking more readily to reduce their duration, simplify them or to eliminate them altogether.

The impact that design changes can make has been discussed. In some cases relatively minor design changes can yield considerable improvement. We believe that design can – indeed must – be considered within a *kaizen* retrospective improvement exercise.

References

1 Rawlinson, M. and Wells, P. (1996). Taylorism, lean production and the automobile industry. In Stewart, P. (ed.), *Beyond Modern times*, pp. 189–204, Frank Cass, London. That design can alter how a changeover is conducted has been noted by Rawlinson and Wells, who also assert that a design thrust should be considered as part of the 'SMED' methodology: 'The "SMED" system was based on two principles: first, technical changes to the dies and presses; and second, organisational changes to the labour processes involved in die changing.' Rawlinson and Wells continue: 'The technical changes to a large extent enabled the changes in the organizational process.' Examples will be provided in the current chapter to support this assertion.

2 Brooke, L. (1991). Japan's stamping master. *Automotive Industries*, November 1991. A review of Japan's Hirotec Corporation is presented. The company's core business has long been automotive component stamping. Struggling to improve its in-house die change times below 30 minutes, it is reported that the company undertook to design equipment for its own use. Brooke reports: 'So the company's engineers and workers got together and produced their own automated die change system, slashing die set [changeovers] to the sub-2 minute level.' It is further reported that these design improvements have also dramatically reduced the number of changeover team personnel (who conduct the changeovers). Partly due to these and other developments, the company states that it is expanding from being an automotive OE (original equipment) supplier into a supplier of production equipment and control systems.

Hirotec is just one example. Other companies can also be cited. Similarly, we have located companies whose role instead is to act as consultants; improving press system changeover by design by use of selected equipment from suppliers such as Hirotec.

3 Jenkins, R. (1995). Quick die changing. *Conf. on Leading Edge Strategies for the Metalforming Industry, Metalform '95*, Chicago, 12–15 March, Ch. 75, pp. 435–450. Jenkins presents an overview of how design can be employed to assist die set changeovers.

4 Laven, P. E. (1989). In Smith, D. A. (1991). *Quick Die Change*. SME, Dearborn, USA. Specific devices have been previously identified in Chapter 2. Of devices and equipment available for the metal stamping industry, Laven observes: 'The technology that can help accomplish quick die change is immense.'

5 Shingo, S. (1985). A *Revolution in Manufacturing: The SMED system*. Productivity Press Inc., Massachusetts, USA. A statement is made by Shingo (in a case study example in the latter part of the book): 'Although software improvements eliminate waste, inconsistencies and irregularities in operations, these improvements are just the first step in achieving the benefits of 'SMED'. Hardware improvements are also necessary.' Although not explicit, 'software improvements' are taken to mean organizational changes (moving and converting internal tasks into external time). It is suggested that these should occur first, as indeed is promoted by the 'SMED' methodology.

We do not agree that it is appropriate in every situation that design changes should follow organizational improvement.

6 Feng, C.-X., Kusiak, A. and Huang, C.-C. (1997). Scheduling models for set-up reduction. *Journal of Manufacturing Science and Engineering – Trans. of the ASME*, **119**, No. 4A, 571–579. Feng *et al.* are unusually forthright about the importance of design in achieving improved changeover performance. Three basic ways to achieve improved performance are set out:

- Design of products
- Design and optimizations of processes and operations
- Design of manufacturing systems

Although the latter category, of design of manufacturing systems, is restricted only by reference to cellular manufacturing, we agree with these conclusions.

7 Hay, E. J. (1989). Driving down downtime. *Manufacturing Engineering*, **103**, No. 3, 41–44. Hay has extended Shingo's raw 'SMED' methodology to a degree (see also reference 8), promoting improvement in areas beyond moving and converting internal tasks into external time. One area is 'clamping', where it is difficult to conceive how improvement can come about without change to the equipment, and in turn how this can happen without recourse to at least low-level design.

Although Hay pays considerable attention to organizational aspects, and indeed advocates a 'no cost/low cost' approach, there is evidence in Hay's writing of an understanding that design can be important. First, Hay goes some way to acknowledging a role for design when describing team composition. Hay writes: 'Shop floor ownership is vital to the success of a set-up reduction project. These people ultimately decide which ideas are good and whether they can be made to work.' He then continues: 'Supporting these shop floor people, though, are the technical people who, depending on the specific equipment, might come from process engineering, tool design, tool room, maintenance, or some combination of these functions. With shop floor and support personnel working together, all set-up reduction ideas can be implemented following sound engineering and tool design principles.'

Second, Hay explicitly describes design changes that might be carried out: 'For instance, on a press, the time to get a die onto the bed of a press is often much shorter than the time to get the die squared, centred or in some other exact position ready for clamping. The die could probably be redesigned so that the first motion of getting the die onto the bed is exact enough for proper positioning before clamping.'

8 Hay, E. J. (1987). Any machine setup time can be reduced by 75%. *Industrial Engineering*, **19**, No. 8, 62–67.

9 McIntosh, R. I., Culley, S. J., Gest, G., Mileham, A. R. and Owen, G. W. (1996). An assessment of the role of design in the improvement of changeover performance. *Int. J. of Operations and Production Management*, **16**, No. 9, 5–22.

10 Schonberger, R. J. (1982). *Japanese Manufacturing Techniques – nine hidden lessons in simplicity*. Free Press, New York, USA. Refer also to Chapter 1 for further discussion of changeover and modern manufacturing practice.
11 Karlsson, C. and Åhlstrom, P. (1996). Assessing changes towards lean production. *Int. J. of Operations and Production Management*, **16**, No. 2, 24–41.
12 Dawson, P. (1994). Total quality management. In Storey, J. (ed.), *New Wave Manufacturing Strategies*. Paul Chapman, London.
13 Imai, M. (1986). *Kaizen – the key to Japan's competitive success*. McGraw-Hill, New York.
14 Suzaki, K. (1987). *The New Manufacturing Challenge: Techniques for continuous improvement*. Free Press, New York. As others similarly emphasize: 'After all, who knows better about the work areas than the operators?'
15 Naylor, J. (1996). *Operations Management*. Pitman, London.
16 Naylor, J. (1996). *Op. cit*. ref. 15.
17 Soeda, T. (1991). Every day, and in every way, I'm getting better and better. *Manufacturing Engineer*, **70**, No. 9, 41. *Kaizen* improvement is contrasted with innovation (elsewhere described as 'fundamental shift' or 'step change' improvement): 'Innovation is another type of improvement but it usually means a sudden drastic change, often carried out with huge investment by a small number of engineers and managers. The introduction of CNC manufacturing cells can be regarded as innovation while devising small jigs or fool proof mechanisms are typical examples of *kaizen*.'
18 Jenkins, R. (1995). *Op. cit*. ref. 3.
19 Sturgess, R. H. (1995). Design of a vernier punch and die set for single minute bending set-ups. *J. of Mechanical Design*, **117**, No. 3, 396–401.
20 Rebsamen, W. (1997). Dash to the finish. *American Printer*, **219**, No. 2, 60–62.
21 Doyon, P. (1999). Changeover time minimized. *Connector Specifier*, **15**, No. 6, 14–15.
22 Smith, D. A. (1991). *Quick Die Change*. SME, Dearborn, USA.
23 Noaker, P. M. (1991). Pressed to reduce set up? *Manufacturing Engineering*, **107**, No. 3, 45–49. See also ref. 108, Chapter 3.
24 Silverton, A. C. (1985). Recent developments in the production of motor car body panels. *Proc. 2nd International Conf. on Robots in the Automobile Industry*. IFS Publications, Bedford, UK, pp. 51–69.
25 Frontzek, H. (1997). Redesigned press for cold forming fasteners and precision parts from wire. *Wire Industry*, **64**, No. 760, 243–244.
26 Nunn, R. E. and Eichhorn, K. H. (1985). Quick mold change – manufacturing flexibility for injection molding. *Proc. Injection Molding – the next five years*. Pittsburgh, USA, pp. 67–81.
27 Rozema, H. and Travaglini, V. (1995). Quick mold change systems for high volume stack molds. *Proc. Annual Tech. Conf., ANTEC 1995*, Soc. of Plastics Engineers, Vol. 1, pp. 1011–1015.
28 Luggen, W. W. (1991). *Flexible Manufacturing Cells and Systems*. Prentice Hall, Englewood Cliffs, NJ.
29 Brown, C. R. (1992). Easing the impact of just-in-time inventory management on machine set-up with quick change tools. *American Machinist*, February, 54–57.
30 Poling, D. (1987). The key to single-digit-changeover. *Modern Machine Shop*, **60**, No. 6, 78–86.
31 Williamson, D. T. N. (1968). A better way of making things. *Science Journal*, June, 53–59. Mass manufacturing techniques are described, tracing their development back beyond Ford's early twentieth-century assembly line developments to the manufacture of ship pulley blocks during the early 1800s at Portsmouth. It is reported that some of these early machines were still in use during the 1960s!

The main purpose of Williamson's paper is to describe the Molins 'System 24', which is widely acknowledged as a landmark machine, frequently being cited as representing the birth of modern flexible manufacturing systems.

32 Williamson, D. T. N. (1968). *Op. cit.* ref. 31. The batch processing environment that the Molins 'System 24' was designed to improve (in the late 1960s) is described: 'Batch manufacture is almost universally accomplished today by breaking down the work to be done on each part into perhaps 5–30 operations. Each of these may involve either passing the part from one machine or process to another or changing the 'set-up' or position of the workpiece on a machine table. Actual cutting time, expressed as a percentage of the total time the workpiece occupies any one machine table, averages 15–20%. Where the component has to be passed from machine to machine the situation gets much worse. It is seldom possible to manage a large manufacturing shop in such a way that components spend less than a day between operations and, in many companies, the time between operations is nearer one week. Even when only a few operations are involved, the queuing problems associated with machine loading give a total component manufacturing cycle between three and six months, the average being 100 days. Only about 1 per cent of this time is occupied in the actual cutting of metal.' FMSs can alleviate these problems by reducing handling and changeovers, conducting multiple machining operations (that previously would have been done on many machines) in a single set-up.

33 Shimokawa, K. (1994). *The Japanese Automobile Industry: A business history.* Athlone Press, London. Shimokawa reports that, for the Japanese automotive industry at least, a balance in the advantages of using FMSs is moving towards its responsive production capability (see also the previous reference).

34 Luggen, W. W. (1991). *Flexible Manufacturing Cells and Systems*, Prentice Hall, Englewood Cliffs, NJ. Luggen writes: 'As the first FMS systems (sic.) were installed in Europe, they followed Williamson's concept, and users quickly discovered that the principles would be ideal for the manufacture of low volume, high variety products.' Implicit in this statement is that FMSs can achieve high levels of changeover performance.

35 Anon. (1983). Block tool system – key for conventional and unmanned machining. Society of Manufacturing Engineers, Technical Paper MR83-232.

36 Anon. (1986). Revolution in changeover times. *Machine and Tool Blue Book*, Vol. 81, No. 7, pp. 53–56. Of automation and CNC control it is written: 'Presswork has always been associated with high-volume production, for two reasons. First, press tools are high cost items, thus substantial production quantities are needed before the capital outlay can be justified. Secondly, it has not been possible – until recently – to quickly change a press and its tooling, so long runs have traditionally been necessary before changeover costs could be justified. Although some reductions in tool costs have been possible as a result of advanced technology in the tool and die making industry, the tools remain high cost items. Therefore, reducing minimum economic batch sizes cannot be achieved by using cheaper tools. The problem, however, has been tackled from two other angles: tool changing has been automated; and the machines themselves have adopted programmable logic controllers (PLC) or computer numerical control (CNC) in order to slash set-up times.'

37 Becker, H. (1984). Automatisierung durch Werkzeugwechsel und Periphere Massnahmen. *Werkstatt Betrieb*, **117**, No. 2, 109–112. Automation of tool changing is discussed in punching and press environments in response to reported pressure within these industries to reduce batch sizes. Advantages that are cited include reducing the setting times, extending the non-supervised working period and improving product quality (by enhanced repeatability of settings).

38 Luggen, W. W. (1991). *Op. cit.* ref. 34. Luggen reports that there was an 'explosion' of activity to develop flexible manufacturing operations in the early 1970s, following pioneering

work as outlined above. The following literature – from an immense pool of research – gives some indication of this work. Luggen, or other similar books, can provide a useful overview.

39 Gokler, M. I. and Koc, M. B. (1997). Design of an automatic tool changer with disc magazine for a CNC horizontal machining centre. *Int. J. of Machine Tools and Manufacture*, **37**, No. 3, 277–286. Shows that sub-6 second tool changes have become possible in the 1990s.

40 McMurtry, D. R. (1980). Developments in in-cycle gauging probes and automatic tool setting probes for use on machining centres and lathes. Metrology Conf. NELEX 80 Int., National Engineering Laboratory, East Kilbride, Scotland.

41 Anon. (1997). AGVs provide manufacturing flexibility at Chesebrough-Pond. *Modern Materials Handling*, **52**, No. 2, 51–53.

42 Colbert, J. L., Menassa, R. and De Vires, W. R. (1990). A modular fixture for prismatic parts in an FMS. *Proc. 14th North American Manufacturing Research Conf.*, May, 597–602.

43 Leutgöb, A. (1986). Automatic tool change for flexible folding systems. *Strips Sheets Tubes*, **3**, No. 3, 43–45.

44 Curless, R. A. (1986). Computer control aids speed and accuracy to hydraulic tool changer. *Hydraulics and Pneumatics*, **39**, No. 8, 77–78. Tool storage and handling are described for horizontal machining centres.

45 Mason, F. (1992). A turn to set-up time reduction. *American Machinist*, February, 53–57. Presents an overview of quick-change tool holding systems (for both speed and precision). An example is presented of design changes reducing average changeover time – for this aspect of the changeover – from 7 minutes to 1 minute.

46 Kornienko, A. A. and Levina, Z. M. (1979). Improving automatic tool changer mechanisms through study of their dynamics. *Machinery Tooling*, **50**, No. 3, 14–17.

47 Willmott, P. (1991). Optimising plant availability. In *Managing in the 90's*. DTI, London. Willmott describes the increase in computer-aided machining in the UK through the 1980s: 'There are well over 50,000 CNC machines installed in the UK and at least 4,000 robots. Ten years ago these figures were in the low thousands.'

48 Suzaki, K. (1987). *Op. cit.* ref. 14. Suzaki adds: 'But, as with learning skills such as baseball, tennis, or swimming, this approach requires tenacity. It also requires an inquisitive mind.' An 'inquisitive mind' can also be assisted for changeover improvement by use of appropriate tools, such as the 'SMED' methodology or other tools as described in this book.

49 McIntosh, R. I., Culley, S. J., Mileham, A. R. and Owen, G. W. (2000). A critical evaluation of Shingo's 'SMED' methodology. *Int. J. Production Research*, **38**, No. 11, 2377–2395.

50 Shingo, S. (1985). *Op. cit.* ref. 5.

51 Zunker, G. (1995). Fifty percent reduction in changeover without capital expenditures. *PMA Technical Symposium Proc. for the Metal Forming Industry*, pp. 465–476. Zunker presents a case study of a changeover of a large press. Improvement from 90 minutes to 45 minutes, the target time, is shown graphically as the initiative progressed. Zunker emphasizes the low-cost, organization improvement aspect of this case study. The improvement graph that he presents – when it is sought to change the way that people work – mimics the pattern of improvement that we have experienced in our own research: improvement falling to a plateau as remaining options for better work practices become exhausted. Certainly Zunker's graph can be argued to be influenced by this initiative's targets, but it is logical that organizational gains alone (without any changes being made to the hardware) will inevitably be ever harder fought for.

52 Anon. (2000). www.jp.com.au/Made.html (site accessed 15 May 2000), JP Pistons, Australia. It is shown that technology used in modern mass-market automotive piston manu-

facturing is not suited to small batch manufacture: 'The manufacturing process of pistons has changed considerably since the inception of the internal combustion motor. Modern piston manufacturing is fully automated with little or no human intervention. This is not the case with JP Pistons. JP Pistons strength lies in our ability to manufacture very low numbers of pistons at a time (for example, 10–20). This ability means that we do not, and cannot, compete with the massive automated facilities of the manufacturers of mass produced pistons for modern vehicles. It also means that they cannot do what we do either.'

It is asserted that the company has the flexibility to be a successful niche manufacturer – an ability that companies satisfying large-volume demand cannot match.

53 Zunker, G. (1995). *Op. cit.* ref. 51. Zunker emphasizes that low-cost organizational improvement should be conducted first, from which he argues a 50 per cent reduction in changeover time can be expected – hardware changes are to be sought later. Zunker writes: 'First, simple method changes are installed. The first 50 percent reduction in job change can be quite easy with a minimum amount of expenditures. These reductions are the result of input from the production people on the shop floor and therefore require the least amount of time to implement. The last 40 to 50 percent reduction in setup time requires major press and tool revisions. This could result in major expenditures and long lead time for implementation.'

In this book we argue that this approach is too simplistic: while there are many situations when organizational improvement should indeed be conducted first, this is not always so. Also, we assert that it is dangerous to anticipate set levels of changeover time improvement by these different mechanisms for all different changeover situations. Zunker's assessment also fails to take account of minor design changes, of the type that we argue commonly occur, nor of impact of such minor design changes.

54 Kolb, T. and Schlechter, V. (1998). New concepts for packaging machines with flexible programmable single drive systems. *IECON Proc.* (*Industrial Electronics Conf.*), Vol. 4, pp. 2301–2305. The usefulness of programmable drive systems in respect of changeover (in place of 'hard-coupled' drive systems) is highlighted.

55 Demetrakakes, P. (1997). Quick change artists. *Food Processing*, **58**, November, 80–83. We have encountered the use of servos on a number of occasions during our research. Demetrakakes discusses the use of servo motors to assist changeovers.

56 Shingo, S. (1985). *Op. cit.* ref. 5. Mechanisation features as a changeover improvement 'technique' in the final 'streamlining' phase of the 'SMED' methodology. As discussed in detail elsewhere in this book, Shingo's 'SMED' methodology is widely used, both 'as is' and as the basis of other changeover improvement methodologies (apparent in each of the next three references).

57 Mather, H. (1992). Reducing your changeover times. IMechE seminar, Birmingham, UK.

58 David, I. (1997). The quick-change artists. *Professional Engineering*, November, 31–32. Different organizations' approach to changeover improvement are briefly investigated. Mechanization, as applied by one of these organizations, assumes a greater prominence than it does within the 'SMED' methodology.

59 Hall, R. W. (1983). *Zero Inventories*. Dow Jones-Irwin, Homewood, IL.

60 Anon. (1978). Push-button welding saves tooling change. *Welding Design Fabrication*, **51**, No. 7, 76–78. A welding machine used in the fabrication of vehicle exhaust pipes is described.

61 Manji, J. F. (1997). Robot upgrade jump-starts production, bolsters quality. *Managing Automation*, December, 44–45. The installation of an aluminium welding machine is described, upgrading the previously used equipment.

62 Grasson, T. J. (1996). Robotic tool changers boost uptime. *American Machinist*, **140**, No. 11, 2–4. An automated tool changer is described, the use of which is claimed to 'slash cycle time by expediting changeover'.

63 Edwards, A.-M. (1997). Pira spreads its gospel. *Packaging Today*, **19**, No. 7, 32–34. Edwards investigates the use of mechanization to improve changeovers in the packaging industry.

64 Mather, H. (1984). Change the system or change the environment. Fall Industrial Engineering Conference, American Institute of Industrial Engineers, Atlanta, 1984, pp. 190–194. With reference to using design to alter the existing manufacturing system, Mather more generally notes the attraction of simplification: 'The idea that complex systems can help us control the complex world of manufacturing has largely been a failure.' Mather suggests: 'A better idea is to simplify the manufacturing environment.'

65 Williamson, D. T. N. (1968). *Op. cit.* ref. 31. Williamson describes the complexity inherent in traditional batch manufacturing shops. The 'System 24' FMS development is anticipated to impose change on this environment (also read reference 32), making it inherently simpler: '[A traditional batch manufacturing shop] . . . involves a large investment in partly finished material, imposes a serious delay in fulfilling orders and requires a complex and expensive production control system to chase the work from operation to operation. It is not clear to me why such a remarkably ineffective system came to be universally selected and accepted. In a modest sized workshop the method may achieve its objective, but in today's large manufacturing units it gets completely out of hand; the cost of the inventory alone exceeds any advantage obtained from the division of labour, and good communication becomes impossible.'

66 Demetrakakes, P. (1997). *Op. cit.* ref. 55. The attraction of using no tools or a minimum of tools to conduct a changeover is sometimes reported (see also next reference). Demetrakakes strongly argues for no-tool changeovers: 'When it comes to packaging machinery changeovers, tools are the enemy of speed. They get lost. They're hard to keep handy. Most of all, they usually require mechanics – skilled, specialised, expensive personnel.'

67 Emproto, R. (1994). Label with a cause. *Beverage World*, **113**, December, 78.

68 Shingo, S. (1986). *Zero Quality Control – source inspection and the poke-yoke system*. Productivity Press Inc., Massachusetts, USA. As well as his contribution to changeover, Shingo was influential in other aspects of manufacturing, not least of which was his *poke-yoke* – otherwise referred to as 'foolproofing' or 'mistake-proofing' – concept. This work, like his changeover work, has been widely picked up since publication. Other authors' contributions can also be found in the literature.

69 Hoffman, E. G. (1996). *Setup Reduction through Effective Workholding*. Industrial Press, New York. We give no more than an introduction to this topic. The points that are addressed by Exhibits 5.5 and 5.6 are far more thoroughly covered in Hoffman's book.

70 Liang, Y., Kurihara, K., Saito, K., Murakami, H., Kumagai, K., Ohshima, M. and Tanigaki, M. (1998). Profile control of plastic sheet in an industrial polymer process. *Polymer Engineering and Science*, **38**, No. 10, 1740–1750. A description is given of previous parameter settings being reapplied for subsequent changeovers. Ease of changeover and material saving during changeover are both presented as benefits of the system that is used.

71 Stevens, D. P. (1994). A stochastic approach for analysing product tolerances. *Quality Engineering*, **6**, No. 3, 439–449. It is noted that the time to set up machines for specific manufacturing operations was three to four times the standard time for the set-up, for which overspecified tolerances for the parts that were being machined were considered to be largely responsible: 'Tolerances that were unnecessarily restrictive yielded higher scrap rates than were necessary. In addition, tighter tolerances required more precise machining with more precision tools, which resulted in longer set-ups. These problems had not been apparent in the past when large volumes of aircraft were being produced and machines used the same set-up for extended periods of time.'

72 David, I. (1997). *Op. cit.* ref. 58.

73 Sekine, K. and Arai, K. (1992). *Kaizen for Quick Changeover*. Productivity Press, Cambridge, Massachusetts.

74 Noaker, P. M. (1992). Stamp it JIT. *Manufacturing Engineering*, **108**, No. 4, 45–49. Referring to press tools, Noaker reports: 'For instance, eliminating straps and using a common tie-down height can often reduce setup time on a straight-side press from 45 minutes to 20 minutes.'

75 Vicens, B. (1989). In Smith, D. A. (1991). *Quick Die Change*. SME, Dearborn, USA. Standardization is described easing the planning effort, because with standard presses (in this example) the planners do not have to arrange for production in a nominated press.

76 Shingo, S. (1985). *Op. cit.* ref. 5.

77 Sepheri, M. (1987). Manufacturing revitalisation at Harley-Davidson Motor Co. *Industrial Engineer*, August, 87–93. Sepheri describes modification to a crankshaft design at Harley-Davidson, which eliminated some previously required tool set-up procedures.

78 Zunker, G. (1995). *Op. cit.* ref. 51. Zunker's improvement methodology is:

1 Identify elements
2 Breakdown internal/external elements
3 Reduce internal elements
4 Standardize to eliminate adjustments: (a) tooling revisions, (b) press modifications
5 Repeat process until goal is achieved

79 Demetrakakes, P. (1997). *Op. cit.* ref. 55. Demetrakakes argues how using modules can assist changeover performance.

80 Womack, J. P., Jones, D. T. and Roos, D. (1990). *The Machine that Changed the World*. Rawson Associates, New York.

81 Wetherill, G. B. and Brown, D. W. (1991). *Statistical Process Control*. Chapman and Hall, London. Many books are available on this topic.

82 Vaughan, T. S. (1994). An alternative framework for short-run SPC. *Production and Inventory Management Journal*, **35**, No. 3, 48–52.

83 Zepf Inc. http://www.hayssen.com/zepf/chgparts.htm, Zepf Inc., Florida, site accessed 20 May 2000.

84 Zepf Inc. http://www.hayssen.com/zepf/mark2cps.htm, site accessed 20 May 2000.

85 Gest, G. B., Culley, S. J., McIntosh, R. I., Mileham, A. R. and Owen, G. W. (1994). Classification methodologies for set-up reduction techniques within industry. *4th Int. Factory 2000 Conf., IEE Conf. Publication No. 398*, pp. 486–490, October, University of York. The four 'Reduction-In' categories were originally proposed and tested in respect of retrospectively describing how a changeover *has been* improved. We focus here upon proactive use of the tool – as a way to determine how a changeover *might be* improved.

86 McIntosh, R. I., Culley, S. J., Mileham, A. R. and Owen, G. W. Reinterpreting Shingo's 'SMED' methodology (paper in preparation). This case is argued in a paper that we are currently preparing. The argument is an extension of that commenced in the next reference.

87 McIntosh, R. I., Culley, S. J., Mileham, A. R. and Owen, G. W. (2000). *Op. cit.* ref. 49. This issue is investigated in more detail in our paper.

88 Shingo, S. (1985). *Op. cit.* ref. 5. The 'SMED' methodology addresses adjustment, which is cited as an improvement technique under its final 'streamlining' stage. Other writers, as we do, have accorded adjustment greater prominence. Shingo apparently employs a 'good piece to good piece' changeover model, failing to distinguish the significance of adjustment tasks in many changeover situations.

89 McIntosh, R. I., Culley, S. J., Mileham, A. R. and Owen, G. W. (2000). *Op. cit.* ref. 49. The inadequacy of the term 'internal time' is highlighted. Similarly, this issue is addressed in other sections of the current book – for example, see Appendix 1.

90 McIntosh, R. I., Culley, S. J., Mileham, A. R. and Owen, G. W. (2000). *Op. cit.* ref. 49.
91 Shingo, S. (1985). *Op. cit.* ref. 5.
92 McIntosh, R. I., Culley, S. J., Mileham, A. R. and Owen, G. W. (2000). *Op. cit.* ref. 49.
93 McIntosh, R. I., Culley, S. J., Mileham, A. R. and Owen, G. W. (2000). *Op. cit.* ref. 49. These reservations are explained in more detail. Further reservations might also be nominated.
94 Mileham, A. R., Culley, S. J., McIntosh, R. I. and Owen, G. W. (1999). Rapid changeover – a pre-requisite for responsive manufacture. *Int. J. of Operations and Production Management*, **19**, No. 8, 785–796.
95 Shingo, S. (1985). *Op. cit.* ref. 5. Shingo also firmly makes this point, writing of the 'SMED' methodology (specifically, of the 'concepts' thereof): 'Although it is of primary importance for you to understand the fundamental concepts involved, there is no doubt that knowledge of actual case examples is very effective as well.' A substantial second section of Shingo's book is dedicated to examples of changeover improvement.
96 Andreason, M. M. and Hein, L. (1987). *Integrated Product Development*. Springer-Verlag, Kempston.
97 Dieter, G. (1983). *Engineering Design – a materials and processing approach*. McGraw-Hill, New York.
98 Suh, N. P. (1990). *The Principles of Design*. Oxford University Press, Oxford.
99 Pugh, S. (1990). *Total Design*. Addison-Wesley, Wokingham.
100 Pahl, G. and Beitz, W. (1996). In Wallace, K. (ed.), *Engineering Design*. Springer, London.
101 Boothroyd, G. (1992). *Assembly, Automation and Product Design*. Marcel Decker, New York.
102 Huekstra, R. (1989). Design for automated assembly: an axiomatic and analytical method. *SME Int. Conf. and Exposition*, Detroit, AD89-416, pp. 1–22.
103 Judson, T. W. (1976). Product design for turning and milling, SME. Dearborn, USA, Technical Paper MR76-902. Like design for assembly, the topic of design for manufacture has long received considerable academic attention.
104 Maylor, H. (1997). Concurrent new product development: an empirical assessment. *Int. J. of Operations and Production Management*, **17**, No. 12, 1196–1214. Referring to DFM/DFMA (design for manufacture and assembly), it is described that there are: '. . . organisational aspects of these techniques, including programme management, communication and flow of information beyond the engineering function'. Similar observations, as we describe, apply to changeover improvement.
105 Swift, K. G. (1989). Techniques in design for manufacture. *Proc. Instn. Mech. Engrs, Part D: J. of Automobile Engineering*, **203**, 25–28.
106 Prasad, B. (1996). *Concurrent Engineering Fundamentals*. Prentice Hall, New York. Prasad describes the importance of DF'X' (Design for 'X') or 'X-ability' in terms of concurrent engineering throughout his book. The scope of issues that can be considered in a 'design for' context are outlined in his table 5.1 on page 238.
107 Kusiak, A. and He, D. W. (1998). Design for agility: a scheduling perspective. *Robotics and Computer Integrated Manufacturing*, **14**, No. 5–6, 415–427. Kusiak and He write: 'This paper attempts to simplify scheduling of manufacturing systems through appropriate design of products and manufacturing systems. An attempt has been made to generate rules that allow to design products and systems for easy scheduling.'
108 Kelly, A. (1997). *Maintenance Strategy*. Butterworth-Heinemann, Oxford. This and the accompanying text (see next reference) are two such comprehensive texts on the topic of industrial maintenance.
109 Kelly, A. (1997). *Maintenance Organisation and Systems*. Butterworth-Heinemann, Oxford.

110 McIntosh, R. I., Culley, S. J., Mileham, A. R. and Owen, G. W. Changeover improvement: a maintenance perspective. *Int. J. of Production Economics* (accepted for publication, November 2000). A more complete explanation is given.

111 Blanchard, B. S. and Lowery, E. E. (1969). *Maintainability*. McGraw-Hill, New York. This book, although published more than three decades ago, shows significant overlap with the current changeover improvement discussion. It describes maintainability as a characteristic of equipment design. More specifically, maintainability is described as a design characteristic to allow maintenance activity to be accomplished in the least time, at the lowest cost and with the minimum expenditure of support services (manpower; spare parts; tools and test equipment; contractor services). This is to be accomplished without adversely affecting equipment performance or safety.

The individual design techniques that are presented compare, in many cases, with known changeover improvement techniques. These techniques are intended to assist in the rapid dismantling and subsequent rapid reassembly of production equipment. That this is done for maintenance rather than changeover purposes, conceptually, makes little difference.

6

Shopfloor issues: Changeover auditing

From this point on this book focuses upon retrospective improvement practice on the shopfloor. An approach is described in which either design-led or organization-led improvement options are selected on merit. The discussion in these successive chapters addresses issues in the preparatory and implementation phases of the overall methodology for changeover improvement (Chapter 3).

Once strategic considerations have been dealt with it will be appropriate, assuming an internal retrospective improvement route is being pursued, for direct shopfloor involvement in the initiative. Equipment operators are notable among those who should participate. Production support and engineering staff might also become involved. With their understanding of the production environment and machinery these personnel, working as a team, will be the best equipped to take on tasks in this phase of the initiative.

This chapter describes changeover auditing, which occurs during the preparatory phase of the overall methodology. It should be read alongside subsequent chapters in this book, which deal further with shopfloor activity in the preparatory and implementation phases.

Pro-formas that we describe later within the chapter are available as Appendix 3 for use if required. Alternatively, they may be copied from the web address given in the appendix. These pro-formas may be used when recording current changeover activities. They may also be used for later group discussion of what has been recorded.

6.1 Improvement team composition

A reminder is appropriate of factors that can contribute to personnel selection, to take the initiative forward. These will include:

- Leadership skills (including the generation of, and adherence to, work plans)
- Equipment knowledge/experience
- Cross-shift coverage, as necessary

- An ability to generate and communicate improvement ideas
- Engineering experience/fabrication skills
- Documentation skills
- Interpersonal compatibility
- Planning knowledge

It should be recognized that beyond these factors, the team's ability to make progress can be severely compromised by lack of empowerment. Similarly, progress can be compromised by a lack of resources (including time and finance). As Chapter 3 has shown, there will also be some issues that a typical shopfloor team simply will be unable to progress. For these issues the company should ensure that alternative improvement channels are instigated.

6.2 Changeover analysis

Existing changeover practice should be understood before any improvement is contemplated. Different analysis techniques can be applied to do this. Our strong preference, if it can be done, is to make use of a video recorder with an inclusive timer. By this means a 'representative'* changeover can be recorded from start to finish. The option of using a video is particularly attractive in that it permits discussion of specific points to occur at length, where parts of the video can be replayed as often as necessary. The video is also attractive as it presents irrefutable evidence of what has occurred. Video evidence limits the scope for interpretation and debate on changeover events. Non-team members might also be afforded the chance to sit in on the screening of the video, to contribute and to become involved, even if indirectly, in the improvement process. Video information should be documented, thus assisting the analysis and discussion, and setting out a record of existing practice (alternatively forms could be filled in directly, using a pencil and stopwatch, when witnessing a changeover). These forms can then be presented for group discussion.

Whatever method is used to capture the current changeover, this should be done in some detail. One drawback of a video camera is that it can capture only what is going on in front of it. Where parallel activities occur, or where there are activities in widely displaced locations, it can be difficult to capture events fully. Similarly, by zooming in to record certain events in detail, other changeover activity might be missed. Conversely if a wide-angle, remote video point is chosen, a sufficiently detailed analysis of events might not be possible.

---

* Selecting a 'representative' changeover can be difficult, but is not necessarily critical. Two issues are to the fore. One: the type of changeover that is being conducted. Two: different work practices of different personnel in conducting what are ostensibly similar changeovers (perhaps, for example, involving personnel from different shifts). In each case the company could possibly decide to record a cross-section of changeovers. More useful, typically, is to ensure everyone's involvement in the analysis of a single changeover. Often having more than one changeover to analyse will complicate matters, while adding little. Any single changeover that is chosen for analysis is likely to contribute more than sufficient material to enable good initial progress to be made.

Record changeover events, even if they are already believed to be fully understood

The detailed recording of current practice should occur even when the team already feel that they understand all that goes on, and have also, they believe, a fair appreciation of the times that are involved. Table 6.1 shows discrepancies that existed when one estimate of a frequently conducted changeover was made. The danger of not knowing correct changeover times (which are highly important as well when scheduling batch production) is likely to be greater when times have not been reviewed for a long period, or where an improvement exercise is in progress.[1]

Care needs to be taken in other ways not to prejudge changeover activity. It might be erroneously believed, for example, that no scope for improvement of certain tasks is available. Even when they exist, the improvement team should also treat any historical records with appropriate caution. Historical records, or indeed fixed changeover procedures, may well not accord with current practice.

Changeover auditing in context

As well as an assessment of the immediate equipment undergoing changeover, a more general study of existing changeover practice might also be applicable. For example, as we have previously argued, the existing changeover could be influenced by what occurs in the wider context of job scheduling.

This chapter concentrates solely upon changeover activity on the shopfloor. It also only describes auditing during changeover's set-up and run-up phases: for simplicity, auditing of changeover tasks prior to halting production has been omitted. These

Table 6.1 Metal slitter changeover – estimated and actual times[2]

Action	*Estimated time (minutes)*	*Actual time (minutes)*
Establish nature of job and locate plate	–	6
Locate fork-lift truck	15	2
Unload and reload hopper feeder	5	8
Fill in record sheet and set counter	–	4
Fetch and set measuring platen	–	3
Locate hydraulic pump	5	2
Remove stacker (incl. disconnect services)	5	5
Remove scrap diverter bar	5	5
Remove first cutter set	15	5
Set cutters 2 to 6 inclusive	100	54
Reposition front finger stops	5	4
Clean cutters of oil and dirt	10	10
Replace finger bar and run a plate through	–	4
Refit scrap diverter bar	25	8
Initial stacker reset	20	10
Refit, position and conduct fine setting	15	9
Square up front finger stops	10	7*
Cut and check sample blanks	10	7*
Change scrap bins	5	4
Total	**250**	**157**

* These two tasks in practice are conducted in unison

'external' changeover tasks ideally should also be audited, if only, perhaps, at a later stage in the initiative, once some initial gains in the elapsed changeover time have already been made. This tactic of delaying attention to external changeover tasks can often be justified, at least in part, on the basis that they are likely to be modified in any case by changes to the changeover during earlier stages of the initiative (during early 'preparatory phase' and 'implementation phase' cycles of the overall methodology – see Figure 3.16). The general approach that this chapter describes can be adopted if and when external tasks are also audited.

6.3 Data recording

The data recording task, in principle, is essentially a simple one. In practice what needs to be done can be rather more complicated.

The need to record events in good detail has been described above. Individual changeover tasks will need to be defined and the times that they take to complete will need to be determined. What is used at each step of the changeover should also be recorded. What is used can involve, for example, setting data or hand tools.

What is recorded should not stop here. All these readily apparent and easily recorded details together will assist an improvement practitioner, but alone they are still insufficient to impart a full understanding of the changeover. To be more fully aware of what is happening the practitioner must also understand what individual changeover tasks are achieving and how precisely they are being conducted. The practitioner must understand *why*, as observed, each changeover task is being done. This can be rather more difficult. With an understanding of the changeover to this greater extent the practitioner should emerge in a stronger position; more capable of proposing significant, long-term improvements.

To attain this level of understanding the information that is derived from the observed changeover will thus need to be rather more comprehensive than simply noting the tasks involved, the items that are used and the times that the tasks take to complete. The word 'derived' is chosen because these latter aspects can be addressed once the original changeover recording (by a video or by other means) has been completed. Rather than being determined solely by an individual, these details are in any case more appropriately described and agreed upon by the team as a whole in open discussion, including those personnel who originally conducted the changeover. We describe in the next section what this additional derived information might comprise. We also explain some of the ways that subsequently it can be of use.

Structured changeover recording and analysis, as part of an overall initiative

Augmenting basic changeover data by subsequent discussion (above) is still information gathering within the overall methodology step of '*conduct a changeover audit*'. It is still, as well, a step in the overall improvement process, and thus should be undertaken as an integral part of this process. Ideally care needs to be taken that those participating at this stage structure their auditing in a way that prepares for the next step of '*develop an operational strategy*' – wherein specific improvement opportunities are sought. Data recording should be undertaken in a way that is compatible with this next step. If, for example, the team were applying the 'SMED' methodology, and particularly if they were concentrating upon 'moving' and 'converting' tasks into external time, information concerning the current changeover should be sought and recorded accordingly. The improvement methodology that we propose (which is fully detailed in the next chapter) aspires to be more broadly based than a typically applied 'SMED' methodology. It seeks a wide range of improvement opportunities, some of which previously might have been comparatively difficult to identify. As we amplify below, changeover auditing needs to be comprehensive to allow a full range of varied improvement options to be determined.

6.4 Deriving a detailed understanding of the changeover

The documentation we describe for use during changeover auditing is intended to be taken forward to the subsequent '*develop an operational strategy*' step of the overall methodology, wherein specific improvement options are sought and compared. Beyond helping to identify improvement opportunities, changeover documentation is important additionally in allowing improvement to be measured. It can also help in prioritizing where improvement is to be sought.

Below we describe selected aspects of the current changeover that it might be useful to review in detail:

1. What changeover tasks are conducted
2. What times taken are taken for these tasks
3. What skills are needed to conduct these tasks
4. What sequence the tasks occur in
5. What problems can occur
6. What interruptions occur
7. What change parts or components are used
8. What tools are used
9. What information is required
10. Which personnel are involved at each stage of the changeover, and when does more than one person need to become involved for specific tasks
11. Where adjustments are made and why
12. How change parts are moved and stored
13. How change parts and tools are identified
14. Where changeover tasks are hindered by poor access, or where specific changeover tasks are conducted simply to gain better access elsewhere
15. Where cleaning needs to be undertaken
16. Where locating operations occur
17. Where securing operations occur
18. Where inspection or trial operations occur

Some notes concerning these different issues are given below. In most cases the comments that are made are deliberately brief because identification and discussion of specific improvement options, which these aspects can help identify, will be dealt with in later chapters.

1 What changeover tasks are conducted: perception at different levels

Chapter 5 showed that seeking improvement opportunities can be hindered if a preconceived idea is held of what comprises a changeover task. The same problem of task definition can also be apparent when analysing the changeover.

Appendix 3 presents a blank changeover audit sheet and an example of using this sheet is presented later in this chapter. The audit sheet will become the improvement group's record, and will be crucial in helping to identify a range of potential improvement options. It will be seen that two task description levels are provided. This allows different assessments of what constitutes a changeover task to be made, much in the manner of the die set example of the previous chapter. Entries will be open to individual preference, but by making the entries in the context of a group discussion, all parties will be aware of what particular entries describe and why these divisions of changeover task have been selected.

2 The times taken for the changeover tasks

Reviewing the times taken for changeover tasks should not be difficult, if all who participate agree in their definition (including when individual tasks start and when they are completed). Again, use of video evidence would typically provide the best way for this agreement to occur. Some difficulty might be encountered if individual tasks are conducted at intervals, where the task is left part-completed as the operator moves on, temporarily, to do something else (in which case the 'task' is anyway perhaps best entered as a series of distinct, separate tasks). Another possible difficulty is that task times from the original analysis are not representative where, for different possible reasons, an operator has been working in an atypical way (an example is provided later in Figure 6.3, where one operator extended a task simply because he was awaiting his assistant to return, after having been called away from the changeover).

3 Required skills

Some tasks will require that those who undertake them are adequately skilled in a particular respect. 'Skill' might involve the likes of physical strength, legal authority (for example, for setting press systems) or manual dexterity. More usually 'skill' will represent a heightened ability to conduct tasks on the basis of training and/or experience. Care should be taken to distinguish 'skill' from demarcation: where individuals are restricted from participating in certain activities not on the basis of their ability.

An important reason for noting skill requirements is that options later may be considered, as appropriate, to eliminate the need for these skills by changing the nature of the task in question. Alternatively, if the task is to remain essentially unchanged, training of further personnel might be contemplated.

4 The sequence in which changeover tasks are completed

The sequence in which tasks are conducted can be highly significant. There are two important aspects of task sequencing. First, in many cases, a good proportion of changeover tasks are significantly independent of one another and task resequencing might allow a better distribution of the tasks between the personnel who are available to complete them, thus shortening the elapsed changeover time. Second, situations of task interdependencies often also arise, whereby one task cannot be started until a previous task has been completed. The breaking of such interdependencies, where possible, can have a significant impact on changeover performance.

The sequence in which changeover tasks are conducted is automatically recorded by noting what tasks are done at what times, for each person who conducts the changeover.

5 Problems (arising on an infrequent basis)

Any particular 'snapshot' video recording can be misleading because, perhaps, different personnel will work differently or, probably more importantly, a specific type of changeover will be witnessed (see Chapter 3). A further difficulty is that particular problems almost certainly will have occurred during previous changeovers on an infrequent basis. Some of these problems will be significant, but will not be directly captured, and thus, possibly, not flagged up for attention.

Two possible options to assess this latter issue can be suggested. The first and most comprehensive option is to record as many changeovers as possible (as suggested by good work study practice), extracting from these records all instances of infrequently occurring problems. This, though, presents its own difficulties, notably that the sheer manpower effort of gathering all these data is likely to be prohibitive. The second option is for the team to sit together and separately make note of particular problems that they recall have occurred. A difficulty of this second option can be recalling what problems have arisen, and what their impact was.

In the past we have used a hybrid of these two options. The usefulness of permanent logging of line speed has been noted earlier in the book, which determines when changeovers are occurring and how long they take. Such a logging system was employed. At the same time we left sheets on-line for the operators themselves to note when specific, out-of-the-ordinary, problems were arising. We asked that the nature and the approximate duration of such problems be documented. Correlation between unexpectedly long changeovers and the logging of comments was good.

While it is extremely important to assess problems that can occur, such problems need to be distinguishable from routine poor practice. Each require to be assessed separately. The section 'Reference changeovers' in the next chapter describes how this may be done.

Finally, account should be taken of possible tasks that arise only when particular types of changeover are conducted. Are these tasks significant and worthy of improvement attention? How will other improvements that are made affect these tasks – or be affected by them?

6 Interruptions that occur

A separate issue is that of interruptions to the changeover. The category of 'interruption' is not meant to include instances such as retrieving missing tools, which can otherwise be recorded as tasks (albeit unwanted ones) during the changeover. Rather it is meant to encompass occasions when those conducting the changeover are distracted from, or are prevented from, executing 'real' changeover tasks. This can be, for example, for breaks of the operator's own volition. Alternatively, interruption might occur through being called from the changeover to undertake another job, or when awaiting changeover items to be delivered, or when awaiting required manpower (for example, a skilled setter) to become available.

The point of recording interruptions separately is simply that they are distractions from proceeding with the changeover proper. With a will to do so they could be eliminated immediately.

7 Changeover components

All components that need to be changed, moved or adjusted should be known. Components can be documented within the changeover task descriptions, thus recording when attention to these components is required. For example, a task description entry might be: 'use 13 mm spanner to remove the 6 × M8 top shroud securing bolts'.

8 Tools

Likewise, almost every changeover will require that particular tools are used. These may range from hand tools through to specialist equipment such as designed-for-purpose power winders, or setting devices. Tools that are needed to conduct the changeover should be recorded. Some tools might be indispensable. Others will merely be helpful.

9 Information required to conduct a changeover

As with the topic of tools, the team should also record information needed to conduct the changeover, such as equipment setting data. In some cases the non-availability of such information will hinder the changeover. In other cases the changeover simply will not be able to proceed without it.

Necessary information is likely to include data relating to product quality, allowing those who conduct the changeover to verify that adjustments have been made correctly and that the new product is being manufactured to an acceptable standard.

10 Personnel/support functions

Assessing personnel involved in a changeover has been discussed in part above. Aspects to do with personnel can include issues of 'skill' and training, and the number of people available to conduct the changeover.

It is useful to record what each person separately does during the changeover. This information can then be charted (see the section 'Reference changeovers' in the next chapter), which can be particularly beneficial in determining who is available at different stages during the changeover. A better changeover task sequence can often be determined, that enables any periods of inactivity instead to be utilized.

Similarly, it is useful to note when a particular changeover task requires more than one person to be available. This can impose a significant constraint on improvement (notably on resequencing and task allocation). As such, overcoming the need for multi-operator tasks as part of the changeover can be highly beneficial.

Beyond those principally involved in the changeover, details of additional personnel who contribute to it should also be recorded. Often of particular note can be personnel who are responsible for either delivering or removing items to enable the changeover to be completed. For example, in our research the role of fork-lift truck drivers has often been crucial. Similarly, it is sometimes necessary for other staff during a changeover to become involved to 'lend a hand'. In this case their presence might be essential because they have specialist skills, or because a particular task simply requires that additional people are present. In this latter case we have experienced a need for extra people to manhandle items where, without them, completing the task would have been extremely difficult (for example, see Exhibit 5.1). In another example we have witnessed the need for one additional person to operate a press 'inching' facility, while the other personnel, who otherwise completed the whole of the changeover, undertook alignment (adjustment) tasks. In each case there is considerable potential for possible delay, while awaiting these personnel.

11 Determining where adjustments are made and why

Adjustment activity in part is related to lack of precision, when initially locating changeover components. Such activity is a major feature of many changeovers and during a changeover it is usefully identified. As noted elsewhere in this book, options exist to reduce the extent of adjustment that occurs.

12 Changeover item handling and storage

How change parts are moved and stored, including where they are stored, should be understood. As well as change parts, this same understanding should apply also to the tools and information that are required to conduct the changeover. Storage might relate to temporary storage during the changeover.

13 Identification

The way that change parts, tools and information are identified might also be noted.

14 Access

Some changeover tasks involve partial dismantling/refitting of parts or assemblies – that otherwise would not need to be disturbed – only for the purpose of gaining access. Changeover tasks that fall into this category might usefully be identified. Access difficulties might still be noted, even when no secondary dismantling needs to take place.

15 Cleaning

Cleaning is often necessary to allow precision location (see below) to occur. Cleaning of change parts themselves, or of other machine assemblies, might be required to prevent swarf or other contamination from compromising the changeover.

16 Locating

How changeover items are located can significantly influence both the ease (and hence the speed) with which they are mounted, and the degree of adjustment that they might subsequently require.

17 Securing

Securing change parts and assemblies can often be differentiated from locating those same change parts and assemblies. The two functions are different, even though they can often be combined in a single operation. Any securing operation that involves threaded fasteners, for example, is likely to be time-consuming, and represent a sound opportunity for improvement.

18 Inspection/trialling

Inspection and trial runs feature in many changeovers. Upon completion, further adjustments are often made. The process can then be repeated. Any improvement might make these iterations shorter and/or less frequent. A relatively minor improvement can therefore be experienced on a number of occasions, thereby making it more worth while to instigate.

6.5 Resist unqualified judgement of improvement options

As ever greater understanding of the changeover evolves, the team are likely to start to conceive possible improvements. These ideas might be influenced by team members' understanding of the equipment. Equally, ideas might be influenced by previous experience, or knowledge of changeover improvement techniques.

Improvement ideas are always to be welcomed. Formal submission of ideas, though, should be resisted at this stage. As yet, all that should be undertaken is a changeover audit. The team should defer describing improvement ideas until the auditing has been completed. Even when ideas do start to be discussed the team should be wary of adopting the first ideas that are proposed. Neither should the team be unduly swayed towards ideas that are the cheapest to implement, nor, for example, those that are concerned only with translating tasks into external time. What each idea will yield needs to be established. Equally, implementation issues will need to be evaluated. How does the proposal rank alongside alternative suggestions? How will the proposal be affected by possible future improvements? It is important that the overall methodology's steps are followed in sequence: these latter issues will be fully addressed at a later stage.

6.6 Changeover analysis in preparation for improvement idea generation

The mechanism by which the step of '*conduct a changeover audit*' is taken forward smoothly into the step of '*develop an operational strategy*' is now described. We are still concerned with auditing the changeover, but auditing is being done in a way that will allow subsequent steps in the preparatory and implementation phases of the overall methodology to be undertaken with greater ease. Our method involves the use of forms – changeover audit sheets – that are intended to guide the improvement practitioners.*

Changeover audit sheets – Section 1: acquiring basic changeover information

The earlier discussion on information gathering concerned both the logging of specific changeover tasks and, in wider terms, seeking further understanding of current practice. As is detailed below, it is useful for these two aspects to be recorded separately on one sheet. We term this sheet a 'changeover audit sheet'.

These two aspects can be recorded at different times. First, when recording changeover tasks, we suggest that simple logging columns be employed. These columns comprise Section 1 of the audit sheet. Section 1 columns can be completed 'live' when recording the changeover. If a video with an inclusive timer is employed, however, Section 1 would normally be completed later, hence involving other team members within a group session. Appendix 3 presents a blank changeover audit sheet, comprising a front page and a subsequent continuation page(s). Figure 6.1 shows a sample completion of the top sheet. Figure 6.2 shows a sample entry for the basic recording of changeover tasks (completion of Section 1 of the changeover audit sheet).

In terms of the basic entries in Section 1, as shown in Figure 6.2, the chart has the following elements:

- First, a task description is written down (for which two columns are allocated)
- Second, the time upon task completion is recorded
- Third, who conducts the task is noted
- Fourth, space is available for unusual circumstances or details to be noted

The first three of these elements are largely self-explanatory and have already been discussed. The fourth element, of noting unusual circumstances or details, should not be required except on comparatively rare occasions. The purpose here primarily is to highlight that abnormal conditions have been witnessed. These abnormalities may subsequently be taken into account upon analysis. An example of such an entry is provided in Figure 6.2.

Changeover audit sheets – Section 2: further data to help decide improvement options

Completion of Section 2 of the audit sheet should involve rather more thought. It is intended that these Section 2 entries should be made only after the basic recording exercise of Section 1 has been finished. It is also anticipated that all those who participate on the improvement team should jointly consider and agree upon the entries that are made. Completion of Section 2 of the changeover audit sheet entails

* These forms can, of course, be adapted to more exactly meet the particular needs of the organization that is employing them.

CHANGEOVER AUDIT SHEET

Changeover Equipment: Guillotine **Recorded By:** Richard Mc **Document Ref.:** Guillotine C/O 4		**Date / Time:** 27 Feb. 99 – p.m. shift **Changeover Personnel:** Pete Law & Steve Bell
Changeover Start Time: 12.10	**First Piece Manufactured at:** 14:42	**Changeover Complete at:** 15:05
Changeover From/ To: 170 mm Fascia to 195 mm Fascia *(Changeover type)* (16 s.w.g. both types)		

Notes:

(i) When manually timing successive changeover tasks using a stopwatch and clipboard, complete only Section 1 entries.

(ii) Section 2 entries should be completed at a later time. These entries should be made by the improvement team together, including those who have just been recorded.

(iii) If a video recorder with an inclusive timer is being used, all analysis sheet entries may be made if desired at a later time, when replaying the video.

(iv) 'Activity Type' and '"Reduction-In" Opportunity' entries might be made using the key presented below.

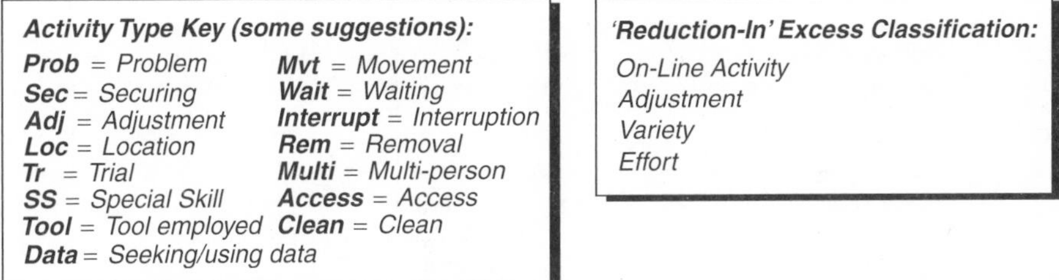

Activity Type Key (some suggestions):

Prob = *Problem*
Sec = *Securing*
Adj = *Adjustment*
Loc = *Location*
Tr = *Trial*
SS = *Special Skill*
Tool = *Tool employed*
Data = *Seeking/using data*
Mvt = *Movement*
Wait = *Waiting*
Interrupt = *Interruption*
Rem = *Removal*
Multi = *Multi-person*
Access = *Access*
Clean = *Clean*

'Reduction-In' Excess Classification:

On-Line Activity
Adjustment
Variety
Effort

Figure 6.1 Sample top sheet entries (see Appendix 3 for blank version)

more detailed interpretation of the observed changeover. The session(s) to make these entries should include those operators who have just been recorded.

The team should be clear in advance that the columns in this later section of the audit sheet are being completed for three specific purposes:

- to ensure a good understanding is gained of each individual changeover task
- to assign a 'Reduction-In' excess category, if warranted, to signify a perceived shortcoming of that existing task
- to assist in prioritizing potential improvement options

In effect, therefore, the audit sheet at this stage is serving a joint role. Strictly it can be interpreted as data gathering, but data are here being deliberately sought and considered with a view to deriving improvement options. Section 2 of the changeover audit sheet, particularly in the use of the 'Reduction-In' strategy, is thus providing

CHANGEOVER AUDIT SHEET *(cont.)*

Document Ref.: Box line – Guillotine C/O 4 **Sheet No.:** 1 of 9

Section 1					Section 2		
Task ('major')	**Task ('detail')**	**Time Complete**	**Who ?**	**Notes**	**Activity Type**	**Duration**	**'Reduction-In' Opportunity**
Removal of guards	Top guard – undo 6 x M6 securing bolts	12:17	P.L.				
	Locate socket extension bar	12:18	S.B.				
	Lift top guard and carry to temporary storage area	12:23	P.L. & S.B.				
	Temp. replace 6 x M6 bolts in threads by hand for storage	12:24	P.L.				
	Release 10 x side guard QR fittings and swing guard away	12:28	P.L.				
	Clean swarf and dirt from top guard vicinity	–	P.L. & S.B.	Pete called away to help clear paint line wreck (12:33)			
	Clean swarf and dirt from top guard vicinity (continued)	12:49	S.B.	Working slowly – awaiting P.L. return (12:46)			
	Select top tool support arm and position to support tool	12:58	P.L.				

Figure 6.2 Sample entries for Section 1 – the basic recording of changeover tasks

CHANGEOVER AUDIT SHEET *(cont.)*

Document Ref.: Box line – Guillotine C/O 4 **Sheet number.:** 1 of 9

Section 1					*Section 2*		
Task ('major')	**Task ('detail')**	**Time Complete**	**Who ?**	**Notes**	**Activity Type**	**Duration**	**'Reduction-In' Opportunity**
Removal of guards	Top guard – undo 6 x M6 securing bolts	12:17	P.L.		Rem/Tool	7 min.	Effort?/ OLA?
	Locate socket extension bar	12:18	S.B.		Wait	1 min.	OLA
	Lift top guard and carry to temporary storage area	12:23	P.L. & S.B.		Rem/Mvt/ Store	5 min.	Effort
	Temp. replace 6 x M6 bolts in threads by hand for storage	12:24	P.L.		Store	1 min.	Effort?
	Release 10 x side guard QR fittings and swing guard away	12:28	P.L.		Rem/ Wait (S.B.)	4 min.	
	Clean swarf and dirt from top guard vicinity	–	P.L. & S.B.	Pete called away to help clear paint line wreck (12:33)	Interrupt.	13 min. away	OLA
	Clean swarf and dirt from top guard vicinity (continued)	12:49	S.B.	Working slowly– awaiting P.L. return (12:46)	Clean	21 min. (atypical)	OLA
	Select top tool support arm and position to support tool	12:58	P.L.		Rem	9 min.	Variety?/ Effort?

Figure 6.3 Sample entries in Section 2 of changeover audit sheet

the link into the next stage of the overall methodology of '*develop an operational strategy*'.

The audit sheet entry perhaps requiring the most flexibility of thought – and thus perhaps the most difficult – is that of '*Activity Type*'. The purpose of the '*Activity Type*' column is to oblige the improvement team to consider exactly what the task under review is achieving. Suggested classifications for this column are summarized in Appendix 4, which also includes a brief explanation of possible ways that assessed activity might be improved upon.[3] These 'activity type' classifications generally match our description earlier in this chapter of changeover detail that is likely to be particularly worthy of recording. The intention is that inefficiencies – or 'excesses' – in the current process might be more readily spotted (Chapter 5). Excesses are summarized in the final Section 2 column of the audit sheet, in one or more of the four available 'Reduction-In' categories – 'On-Line Activity', 'Adjustment', 'Variety', 'Effort'. Figure 6.3 presents an example of a completed changeover audit sheet.

6.7 Conclusions

This chapter deals with auditing current changeover practice. This is not a trivial task. The more improvement team members are able to understand how and why current tasks are conducted, the stronger their position will be to suggest suitable improvement options.

The study of the existing physical changeover activity resides in the step of '*conduct a changeover audit*', in the preparatory phase of the overall methodology. Detailed knowledge of existing changeover practice alone, achieved by a comprehensive audit, is still insufficient to generate a full range of potential improvement ideas. The following chapters describe how the team can progress the initiative; generating potential improvement options, deciding which of these options are to be used and implementing the changes that have been agreed upon.

References

1 Stevens, D. P. (1994). A stochastic approach for analysing product tolerances. *Quality Engineering*, **6**, No. 3, 439–449. Stevens indicates the extent by which estimates of changeover performance can be adrift from actual performance. A situation at Bell Helicopter is noted: 'The time required to set up machines for specific manufacturing operations was [found to be] three to four times the standard time for the set up.'

2 Davies, P. S. (1993). Slitter changeover. Bath University Changeover Group, internal case study report, Summer 1993. Davies researched changeover activities on-site at a north London factory over a period of 3 months.

3 Gest, G. B. (1995). *The Modelling of Changeovers and the Classification of Changeover Time Reduction Techniques*. PhD thesis, University of Bath, UK. This chapter is largely based on Graham Gest's original work.

7

Shopfloor issues: Develop an operational strategy; Set local targets; Implement; Monitor

Following a comprehensive audit of the changeover, an improvement team need to identify possible improvement options. Team members then need to decide which options they are to take forward and implement. Work needs to be allocated. It also needs to be monitored that improvements have been completed, and determined what their impact is.

7.1 Overview

Figure 7.1 sets out again the individual steps in the 'preparatory' and 'implementation' phases of the overall methodology. With the exception of changeover auditing, this chapter describes all the work that an improvement team should undertake.

As in the previous chapter, documentation that might be used is described. The current chapter is usefully read in conjunction with Chapters 8 and 9, in which potential improvement techniques are described in detail.

New changeover improvement tools: matrix methodology and reference changeovers

A large part of this chapter is devoted to explaining a novel methodology to identify a wide range of potential improvement options. The methodology we describe is called the matrix methodology. Its use is advocated, alongside other activity, within the step '*develop an operational strategy*' (see Figure 7.1).

We have similarly developed a further tool for parallel use in the step '*develop an operational strategy*'. We refer to this tool as a reference changeover, which is a model of how well a changeover might be conducted were every possible organizational

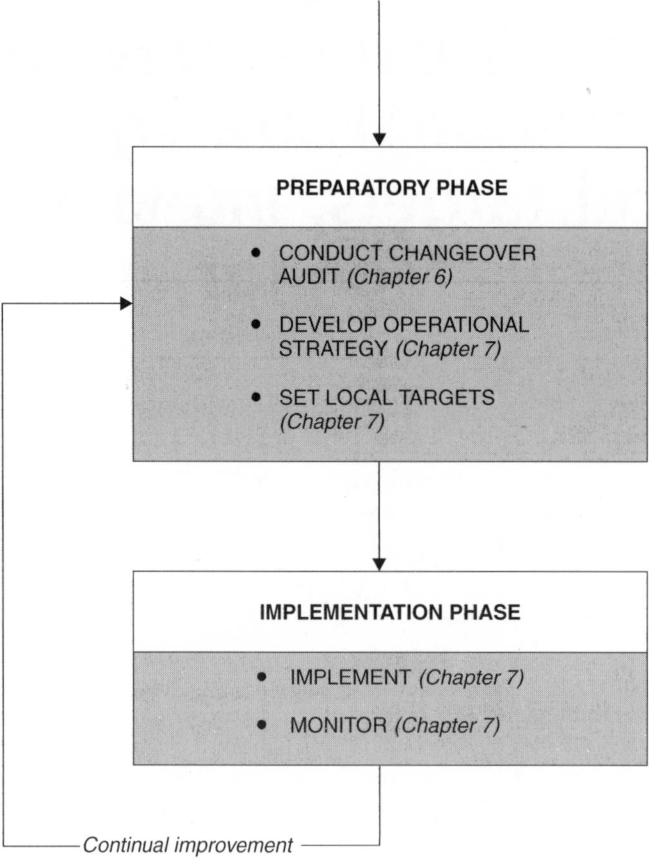
Retrospective improvement of existing changeover practice has been decided upon in the strategic phase of the overall methodology. The initiative is taken up by a shopfloor team – to execute steps in the methodology's preparatory and implementation phases.
PREPARATORY PHASE
• CONDUCT CHANGEOVER AUDIT (Chapter 6)
• DEVELOP OPERATIONAL STRATEGY (Chapter 7)
• SET LOCAL TARGETS (Chapter 7)
IMPLEMENTATION PHASE
• IMPLEMENT (Chapter 7)
• MONITOR (Chapter 7)
Continual improvement

Figure 7.1 The latter two phases of overall methodology for changeover improvement (see also Chapter 3)

improvement to be completed. Derivation and use of a reference changeover is explained in detail. A reference changeover model, like the matrix methodology, helps to identify potential improvement options. It also helps to clarify their impact.

Allocating and progressing work

Some topics, like allocating and progressing work, which constitute the latter steps described by Figure 7.1, are covered more than adequately within general manufacturing and management texts. These topics are not changeover-specific. One possible approach that team members might adopt is explained. Other approaches that a company might be more familiar with could equally be used.

Changeover improvement techniques

Improvement techniques that might be applied are summarized in relation to the matrix methodology. The techniques themselves, often, will be familiar to those who have experience in changeover improvement. In most cases they are techniques that are commonly used and have been extensively written about elsewhere. Some techniques that are described, however, will be less familiar. The matrix methodology clarifies what particular techniques can achieve.

Improvement process 'map'

The work that is about to be undertaken will not necessarily be simple. The role of the team leader in particular in steering a path for the team's efforts should not be underestimated. Part of the team leader's role will be to understand the mechanics of the overall improvement process, including the tools that might be used (for example, the matrix methodology) and the techniques that can be applied. To get the best from a group situation the team leader similarly will need to be aware of the capabilities and likely behavioural characteristics of each of the team members.

The overall improvement process, starting with a strategic assessment of changeover improvement, is arguably now entering its most crucial phase. Likening the gaining of better changeover performance to a journey of improvement can be useful, particularly when setting out for the first time. Those involved can be made aware of the desirability of having a 'route map'. For shopfloor improvement activity Chapters 6 and 7 can be viewed in this way, guiding team members in their efforts. The guide we provide – the route map – is complex. We make no apology for this: we believe the route will be unexpectedly difficult, with probable poor paths and dead-ends to negotiate. Without a comprehensive map, inefficient and possibly unsustainable improvement is likely.

Momentum

Despite our attention to conducting structured improvement, any shopfloor team should not allow itself to become too involved in a conceptual improvement exercise, constantly refining the best way to proceed. Preparing for improvement is important, but there will always be uncertainty that the best possible solutions have been identified. This should not become an issue. Progress should be sought by identifying sound improvement options. Other opportunities or refinements will be identified later. The key here is action: a team should strive for constant forward progress, rather than enduring frequent periods of apparent delay. An improvement exercise might be viewed as burning away old and inefficient practice. Fuel – incrementally executing improvements – is constantly required to keep the fire burning.

7.2 The matrix methodology: preliminary discussion

A comprehensive audit of the current changeover, recently completed, will have afforded a detailed understanding of the changeover. Use of the matrix methodology follows on from this assessment. Its purpose is simple: it assists identification of solutions to overcome the problems that the audit has highlighted.

For the next few pages, we briefly describe our logic in developing the matrix methodology,[1] which is founded on three basic tenets:

- that the improvement process is aided by a three-level hierarchical awareness of: changeover improvement concepts; changeover improvement techniques; changeover improvement examples
- that a classification can be made of existing changeover difficulties, which in turn can aid seeking appropriate solutions
- that improvement options should not be restricted; that practitioners should be free to conceive of a full range of potential improvement options featuring either design or organizational improvement at any stage of an initiative.

These individual points are briefly explained further.

Seeking improvement: a hierarchy of changeover concepts, techniques and examples

Shingo[2] makes clear that his widely used 'SMED' methodology represents more than simply a collection of diverse changeover improvement techniques. The point is made that a changeover practitioner should not rely on a knowledge of improvement techniques alone. Shingo indicates that it is highly desirable to have a structure, or methodology, that guides the practitioner in the use of possible improvement techniques. We share this view.

Like the matrix methodology that we describe here, Shingo's 'SMED' methodology is primarily targeted at workshop practitioners. In most respects, however, these two workshop methodologies are significantly different from one another.*

Shingo's work intimates three levels of awareness. At a top level Shingo presents a conceptual framework (the individual 'SMED' concepts). Next, specific improvement techniques are assigned to these individual concepts. At a final level, examples relating to individual improvement techniques are described. Figure 7.2 interprets this aspect of Shingo's work.

The matrix methodology is intended to update Shingo's original work: it has been developed to overcome limitations that we argue can apply when strictly adopting the 'SMED' methodology[3] (see p. 255). Nevertheless, this new methodology has as one of its foundations the hierarchy that Figure 7.2 illustrates.

More generally, without ascribing it specifically to the 'SMED' methodology, a hierarchy of changeover *concepts*, *techniques* and *examples* would be as shown in Figure 7.3.

A classification of existing changeover difficulties

The topic of classifying existing changeover difficulties has been discussed in Chapter 5. In this context we introduced the 'Reduction-In' strategy. Four categories were

---

* Both are also quite different – including their purpose – from the overall methodology described in Chapter 3.

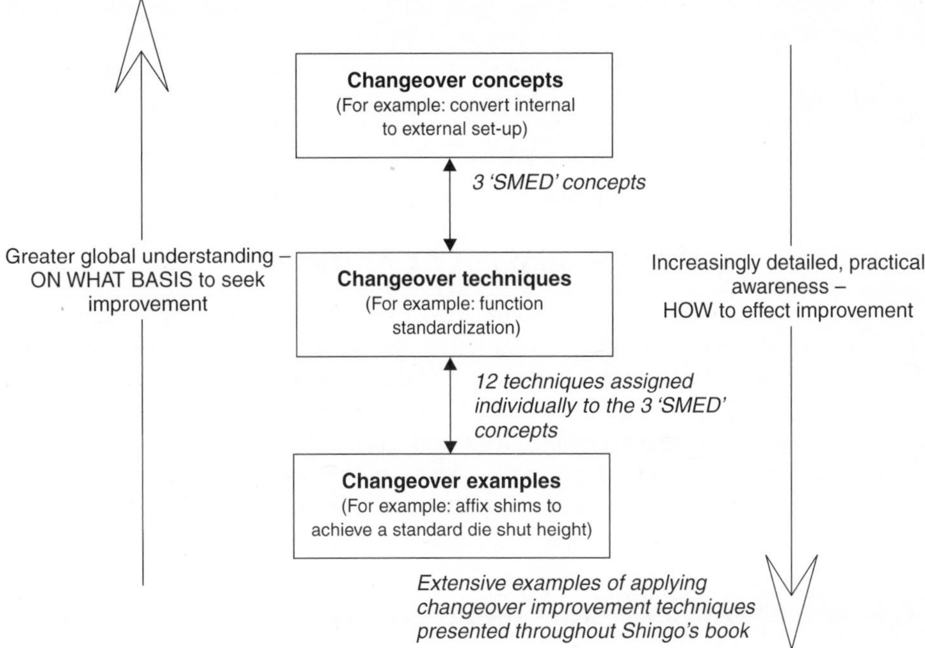

Figure 7.2 Hierarchy of changeover concepts, techniques and examples, interpreted from Shingo's work

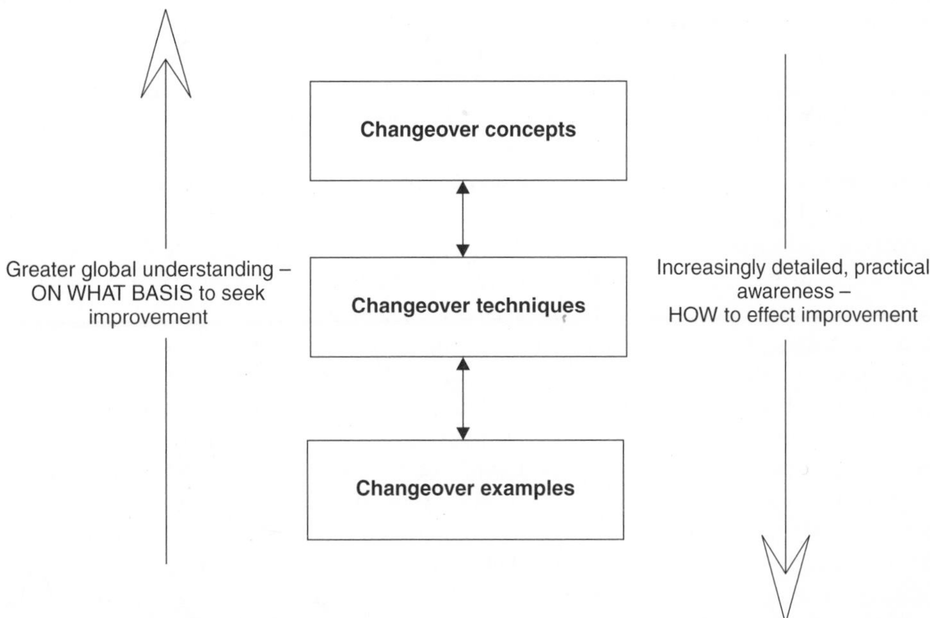

Figure 7.3 Hierarchy of changeover concepts, techniques and examples as a foundation of the matrix methodology

nominated in which potential 'excesses' of the existing changeover might occur: 'On-line activity', 'Adjustment', 'Variety' and 'Effort'.

The 'Reduction-In' strategy – seeking to reduce or eliminate identified 'excesses' of existing changeover practice – has been applied earlier, forming the basis of the design rules of Chapter 5. However, wider application of the 'Reduction-In' strategy is possible, embracing a full range of potential improvement opportunities – at any point along the organization–design spectrum (Chapter 5).

Improvement options should not be restricted

This book questions the wisdom of adopting a prescriptive approach to changeover improvement. Notably, we have questioned the dominance of seeking improvement by translating tasks into external time. Additionally it can be asked, in diverse changeover situations, whether applying both improvement concepts and improvement techniques in a predetermined, fixed sequence will always represent the most effective improvement route. A more prominent, more formal recognition of the role of design might also be beneficial (see also Chapter 5, and Case study 1 in Chapter 10).

7.3 Developing the matrix methodology

The three axes of the matrix methodology[4]

The matrix methodology is so-called because of the 3-axis matrix upon which it is based. For reasons that are now explained, this is a $2 \times 3 \times 4$ matrix, thus yielding a total of 24 positions or cells.

We described in the previous section how a hierarchy of changeover improvement concepts-techniques-examples can assist the practitioner. This hierarchy is used to define one matrix axis. Shown by Figure 7.4, the matrix thus has three levels. These levels respectively are the changeover improvement *concepts* level; the changeover improvement *techniques* level; the changeover improvement *examples* level. There are eight available positions at each of these successive levels. In this way the matrix structure indicates a minimum of eight top-level (as yet undefined) changeover improvement concepts for this new workshop methodology. Figure 7.4 also shows how equivalent matrix positions or cells correspond to one another between levels. This describes, when the matrix methodology is used, and much like the 'SMED' methodology, that particular series of changeover improvement techniques and examples each correspond to a specific changeover improvement concept.

Further similarity with the 'SMED' methodology exists in that the number of entries at each matrix level cascades down. For each single changeover improvement concept (of eight in total) there will be, say, four or five corresponding changeover improvement techniques. In turn, at the final level, a book (see Preface) is required to do justice to the innumerable potential changeover improvement examples that exist.

To assist us in making useful entries to the matrix positions (which is done in the next section in this chapter) we have chosen to employ both the 'Reduction-In' and organisation-design classifications presented in Chapter 5. These classifications, as shown by Figure 7.5, define the remaining two matrix axes, and apply at each level.

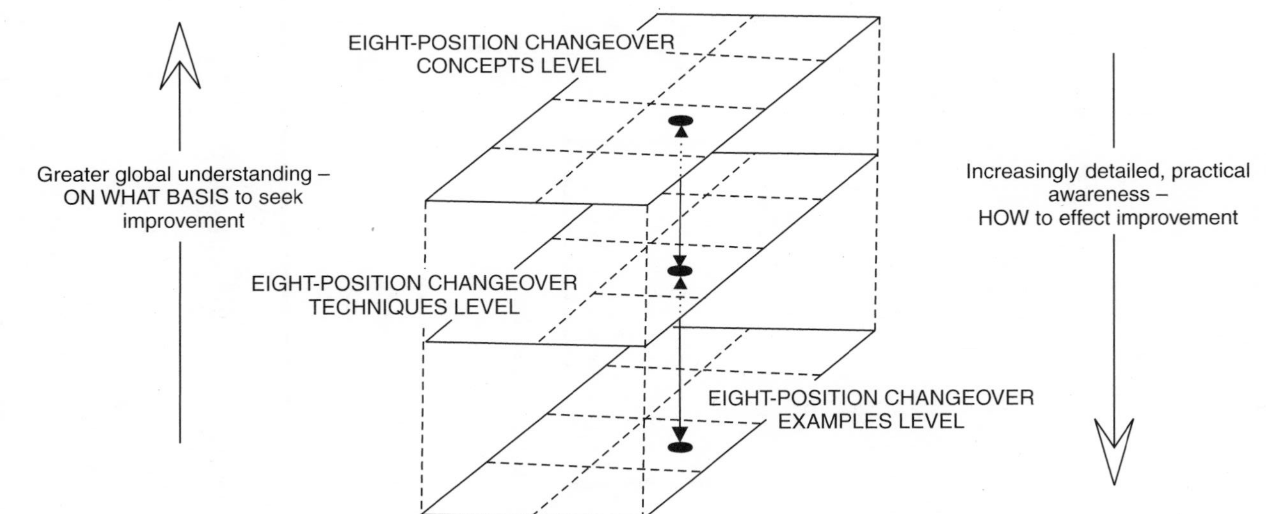

Figure 7.4 Three-level matrix format, with eight corresponding matrix positions at each level (matrix positions at successive levels correspond to one another)

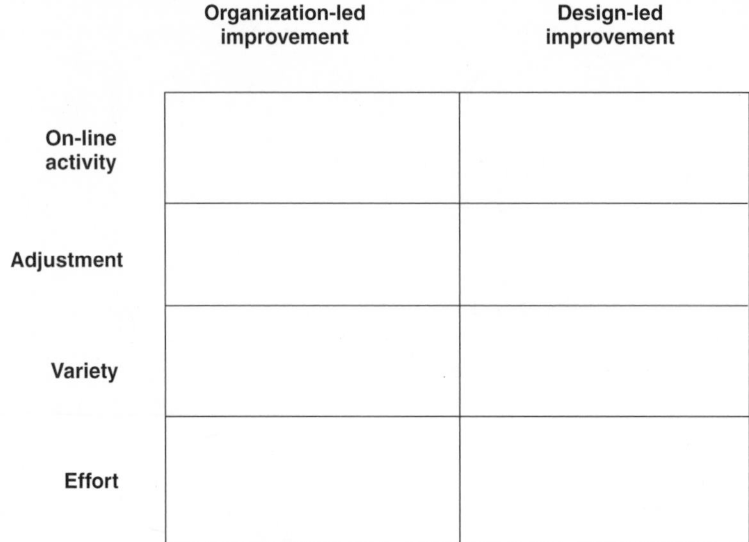

Figure 7.5 The remaining two axes of the matrix

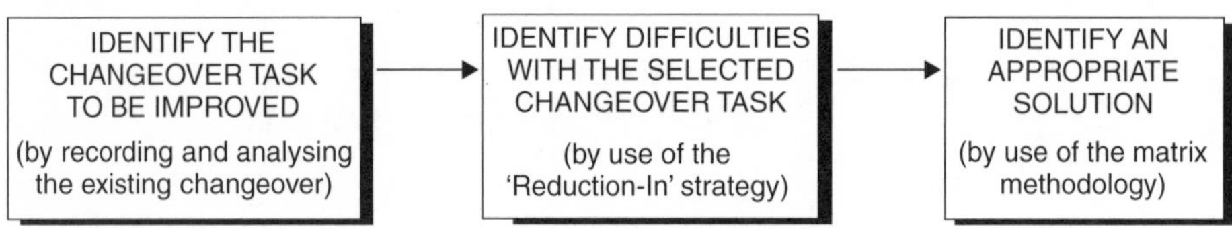

Figure 7.6 The steps involved in selecting an appropriate improvement

Iterative improvement

The matrix methodology is suited for use in an iterative, *kaizen* exercise, as shown by Figure 7.6*. The intention is that successive improvements, one by one, are sought and executed. The process is one of continual incremental improvement.

7.4 Matrix methodology: matrix concepts and techniques

Matrix concepts and corresponding matrix techniques are now explained. More detailed descriptions of these possible changeover improvement techniques will be provided in the next two chapters of this book.

The purpose of matrix methodology is to generate as many improvement ideas as possible, irrespective of their perceived cost, perceived benefit or perceived difficulty of implementation. Evaluation of the different ideas is to occur later.

* This figure can be seen to be a 'completed' version of earlier figures in Chapter 5.

The eight matrix improvement concepts

The matrix concepts are set out in full in Figure 7.7. Subtle distinctions between organization-led improvement and design-led improvement are made.

Explanation of these eight changeover improvement concepts is given below. A description of potential improvement techniques is also presented (which are later summarized in Figure 7.8). This section is set out:

1 A summary explanation of reduction in on-line activity
2 A description of improvement techniques applicable to the matrix's **Organization: on-line activity** concept.

	Organization-led improvement	Design-led improvement
On-line activity	Reduce the contribution of existing tasks to the 'set-up' and 'run-up' phases of a changeover *(by reallocating tasks and/or resources)*	Reduce the contribution of existing tasks to the 'set-up' and 'run-up' phases of a changeover *(by permitting task and/or resource reallocation)*
Adjustment	Minimize adjustment tasks *(by reducing 'trial and error' adjustment)*	Minimize adjustment tasks *(by reducing the need for adjustment)*
Variety	Standardize the way the changeover is conducted *(by standardizing changeover tasks)*	Standardize the way the changeover is conducted *(by standardizing physical conditions)*
Effort	Eliminate or simplify existing tasks *(by working in a better prepared and more efficient manner)*	Eliminate or simplify existing tasks *(by equipment or product modification)*

Figure 7.7 The matrix improvement concepts

3 A description of improvement techniques applicable to the matrix's **Design: on-line activity** concept.

4 A summary explanation of reduction in adjustment
5 A description of improvement techniques applicable to the matrix's **Organization: adjustment** concept.
6 A description of improvement techniques applicable to the matrix's **Design: adjustment** concept.

7 A summary explanation of reduction in variety
8 A description of improvement techniques applicable to the matrix's **Organization: variety** concept.
9 A description of improvement techniques applicable to the matrix's **Design: variety** concept.

10 A summary explanation of reduction in effort
11 A description of improvement techniques applicable to the matrix's **Organization: effort** concept.
12 A description of improvement techniques applicable to the matrix's **Design: effort** concept.

1 A summary explanation of reduction in on-line activity

It is necessary to distinguish exactly what reducing on-line activity can contribute. Importantly, seeking to reduce on-line activity *does not* directly equate to 'maximize external activity' or 'minimize internal time'.

A reduction in on-line activity requires that existing changeover tasks are not fundamentally altered, which is a defining characteristic of this 'Reduction-In' category. If, therefore, a task involves manually connecting compressed air to the bottom die assembly of a press tool, then after improvement the same task would remain. Improvement is sought by changing when tasks are conducted. It may be necessary to reallocate resources (including sometimes increasing resources) to enable this to occur.

Different options are available to reduce on-line activity. First, as much work as possible can be translated into external time – the period before the line is halted. Second, tasks can be conducted more in parallel. Third, changing when tasks are conducted can also allow periods of operator inactivity to be minimized (assuming that more than one operator conducts the changeover). The reader might wish to look again at Figures 5.13 and 5.14. With reference to the last of these three options, it is shown that improvement is gained by 'compacting' tasks together better.

A cautionary note is needed that conducting tasks more quickly (more efficiently) does not qualify for inclusion in this 'Reduction-In' category – because the task is not commenced at an alternative time.

Reflecting again on Shingo's 'SMED' methodology, care should be taken that the scope of reducing on-line activity is wider than 'separating' and 'converting' changeover tasks into external time. To facilitate reducing on-line activity, attention to barriers that currently limit the extent of task reallocation to a different start time may be necessary. This is the breaking or alleviation of task interdependencies: in other words overcoming barriers imposed by other tasks as to when an identified task can commence (see also Chapter 8).

2 A summary of improvement techniques applicable to the matrix's **Organization: on-line activity concept**

Reduce the contribution of existing tasks to the 'set-up' and 'run-up' phases of a changeover *(by reallocating tasks and/or resources)*

Like all organizational change, this matrix position (as opposed to its design-biased counterpart) tends to represent options for swift, low-cost improvement.

In Chapter 5 the objective of the 'Reduction-In' strategy to reduce task 'excesses' to the point of elimination was explained. In terms of on-line activity this objective is to reduce (and if possible, eliminate) tasks' contribution to the changeover's set-up and run-up phases.

Techniques that can be nominated in the corresponding improvement techniques level of the matrix include :

- **Reallocate tasks to external time** Many changeover tasks, typically, are needlessly conducted during the set-up period of a changeover. Tasks that fall into this category are often concerned with gathering all the required components and checking their condition – i.e. tasks that could be done prior to stopping the line or machine. Some tasks might also be needlessly conducted during the changeover's run-up period. Moving any tasks into external time, where possible, should immediately reduce the elapsed changeover time.
- **Reallocate tasks to be conducted more in parallel** It might not be possible for some tasks to be conducted in external time. Nevertheless, task start times might still be altered such that tasks are newly conducted in parallel, in either the set-up phase or the run-up phase of the changeover.
- **Reallocate tasks such that they are 'compacted' together better** Again for tasks that cannot be conducted in external time, reduction in the elapsed changeover time might be achieved by distributing tasks more effectively (see also Figures 5.13 and 5.14).
- **Breaking or alleviating task interdependencies (by organizational change)** Breaking or alleviating task interdependencies means overcoming, at least partially, limitations imposed by other tasks as to when a particular task can be conducted. Breaking or alleviating task interdependencies often needs to occur to allow full advantage to be taken of the options outlined above. Two notable possibilities here are:
  - **Employing parallel equipment** Parallel equipment would often be introduced specifically to enable tasks to be conducted in external time (in Shingo's terms: to be 'converted' into external time). Parallel equipment can range from complete replicated lines, through to just replicating identified line elements.
  - **Engaging additional personnel** Engaging additional personnel, in many situations but by no means all, can help to reduce the elapsed changeover time.

For example, additional personnel can permit tasks to be conducted in parallel, whereas previously this was not possible. This option, like employing additional equipment, constitutes an increase in resources.

3 A summary of improvement techniques applicable to the matrix's **Design: on-line activity** concept

Reduce the contribution of existing tasks to the 'set-up' and 'run-up' phases of a changeover (*by permitting task and/or resource reallocation*)

In respect of applying design, improvement options are comparatively limited. Options are limited primarily because, in most circumstances, equipment modification imposes change on how changeover tasks are conducted. Nonetheless, one important option can be included in this matrix position:

- **Break task interdependencies (by design change)** Design change similarly can be employed to break task interdependencies. For example, mounting additional guards might allow manual tasks to be completed in parallel to automated tasks (where previously this was not permitted for safety reasons).

4 A summary explanation of reduction in adjustment

Seeking a reduction in adjustment can apply at any stage during a changeover. It can apply to tasks that are conducted prior to halting the line, right through to tasks at the conclusion of the changeover's run-up phase. Improvement might be brought about by conducting existing tasks in a more organized manner, or by altering existing tasks.

Adjustment is highlighted not least because of its potential impact: it can often be highly significant in terms of both the elapsed changeover time and the changeover's quality. Reduction in adjustment 'excesses' is sought, if possible, to the extent of eliminating individual adjustment tasks from the changeover.

5 A summary of improvement techniques applicable to the matrix's **Organization: adjustment** concept

Minimize adjustment tasks (*by reducing 'trial and error' adjustment*)

Many changeovers at some point will include making an adjustment and then checking that adjustment. Thus equipment settings are gradually refined. A much faster approach, if it can be arranged, is to go with confidence straight to the correct (dead-stop) setting. Doing so eliminates time-consuming 'trial-and-error' adjustments. If not possible wholly to eliminate 'trial-and-error' adjustments, steps might be taken to minimize their impact.

Some techniques that might allow improvement to be made are:

- **Know of, and use, predefined settings** Rapid, repeatable, precise equipment set-up should be possible if predetermined settings are known and used. These settings will usually have been recorded from a prior set-up for the manufacture of the same product (although in some cases settings can be predetermined). Caution is still needed, however, that these predetermined settings can become worthless where variability within the equipment or product has not been taken account of. Case studies will be presented in Chapter 10.
- **Define inspection parameters to signify adjustment completion** If inspection is being used to verify that settings have been satisfactorily established, then exactly what is being inspected should be known, as should the tolerances that should apply. Thought can also be given to tools that might aid inspection.
- **Understand possible adjustment interrelationships** Some equipment is such that making one adjustment will compromise other adjustments. While it is probably preferable to design out such anomalies, it should also be possible to overcome problems of this kind by knowing how such adjustments interrelate with one another.
- **Address precision requirements (by organizational change)** High precision should be regarded as being more difficult/expensive/time consuming to attain than low precision. The tolerance bands that should apply to each possible adjustment should be known. These tolerance bands should be as wide as possible, commensurate with the finished product achieving satisfactory quality upon manufacture. There is no point in striving to make an awkward adjustment with a high degree of precision, when instead a much more lenient tolerance could apply.
- **Seek to avoid component damage** Component damage can increase either or both the need to conduct adjustments and the time that these adjustments take to complete. Component handling should take account of the need to avoid damage.

6 A summary of improvement techniques applicable to the matrix's Design: adjustment concept

Minimize adjustment tasks *(by reducing the need for adjustment)*

Design too can make notable contributions in respect of adjustment. This has been discussed in general detail in Chapter 5, where an explanation was made of a change-over as achieving the correct relative position between the components that it involved. Some specific options for using design include:

- **Automate adjustment** Particularly where access is poor, an opportunity might be taken to automate or partially automate an adjustment task. Adjustment typically will involve stages of first moving components, and then of detecting their position. Many opportunities to apply automation can arise.
- **Address the precision of location features** Precise location of components by design can eliminate or reduce the need for subsequent adjustment.
- **Consider using position and condition monitoring devices** Manual adjustment might still occur, but the position that is achieved can be constantly fed back to the operator. This might be by using a scale. In more sophisticated systems, typically where greater accuracy is needed or where access to read a scale is difficult, a digital display might be employed. Likewise, condition monitoring can sometimes be necessary, which might also be remotely conducted.
- **Make changeover items more robust (less prone to damage and wear)** Damage to changeover items – including damage apparent as wear or deterioration – can affect the ease with which a changeover is conducted. For example, previously determined settings can cease to be applicable, in which case non-optimal 'trial-and-error' adjustments are likely to be resumed. The incidence of damage to changeover items might be reduced by careful use of design – where 'damage-proofing' of components might be contemplated.
- **Upgrade item quality** In some cases there might be scope by means of secondary design (Chapter 5) to improve the quality of items that are delivered for use in the changeover.
- **Make equipment more tolerant of product variation** There can be variability between nominally identical products from one batch to the next. By definition there will be differences between products across the product range. Scope can exist to eliminate making adjustments during changeover to accommodate these differences. For example, products might be transported in oversized pockets – rather than using a transport medium that has to be tailored to suit.

7 A summary explanation of reduction in variety

Like adjustment, variety is highlighted as an area of special attention. Variability in how people work can be highly significant in the context of changeover performance, as can physical variety of changeover items.

When any task is conducted in more than one way, one of these ways most probably will be superior. In respect of organizing the changeover, 'Reduction-in variety' seeks to identify which is the best way of conducting each task, and then to ensure that this way of working is standardized. Standardization should apply to all personnel, on each of the factory's shifts and across the full complement of similar equipment within the plant (or indeed within the company as a whole). A standard changeover procedure should be derived, agreed by everyone and then used by everyone. The use of standard procedures implies that operator training will have to be undertaken.

In respect of the hardware that is used, standardization should apply to features of items that are involved in the changeover (change parts; consumables; product). Reducing or eliminating physical variety can have a corresponding impact on changeover tasks – where tasks would otherwise have to be completed because of this variety.

As well as standardization of work practice and of physical features, a further means to reduce variety between successive changeovers is to sequence changeovers to control the type of changeover that takes place.

8 A summary of improvement techniques applicable to the matrix's **Organization: variety concept**

Standardize the way the changeover is conducted (*by standardizing changeover tasks*)

Some techniques described below have not always featured strongly in previous improvement programmes. They include:

- **Identifying, agreeing and implementing standard procedures** Standard 'best-practice' changeover procedures should be defined and implemented. These procedures ideally should be quite comprehensive, including issues such as hand tools and the location and format of data that are used. A standard procedure will set out in what sequence tasks are to be conducted and who is to conduct them. Deriving a standard changeover procedure and ensuring that it is always adopted can represent an extensive body of work.
- **Prior checking that required changeover items are present** Changeover items, including data and tools, should be ready in advance in their assigned location (not in a variety of locations). If applicable, it should be checked that these items are also in their assigned format. Variation from standard (position and format) should be eliminated as this will detract from the changeover.
- **Prior checking that item and data quality standards are achieved** As well as being in their correct location and format, the quality of items should be verified before the changeover commences.
- **Batch sequencing** Task variety between successive changeovers can be reduced by controlling the type of changeover that takes place (see '*Scheduling considerations*', Chapter 3).

9 A summary of improvement techniques applicable to the matrix's **Design: variety concept**

Standardize the way the changeover is conducted (*by standardizing physical conditions*)

It should be sought to reduce or eliminate physical variability in features of changeover components. Variability can exist at either a 'macro' or a 'micro' level (Chapter 5). It is here sought to address 'macro'-level physical variability ('micro'-level variability, which can be thought of as changeover item tolerance, is considered above under the classification of '*Reduction-In adjustment*').

Variety in macro features from one changeover to the next frequently means that additional work is introduced: if more standard features were present a range of tasks would typically cease to be necessary.*

- **Act upon standard features of change parts** Features can be standardized when new components are bought or fabricated. They can also be retrospectively standardized, by machining existing components and/or by adding elements to them. An often-cited example is the permanent fixing of spacing pieces to die sets to ensure a standard shut height across all the dies that are likely to be used. If existing features cannot be standardized, is there scope to manipulate change parts (to clamp, move or position them) upon other features, or relocated features, that can be standardized?
- **Act upon standard product features** In seeking feature standardization, the product itself might also be considered. As with change parts, designing the product with features that are standard throughout its product range (if possible) might be desirable.
- **Standardize features of the machine system** There can be benefit in standardizing features of the machine system, such as the format of signs, indicators or instructions, or any one of a host of apparently minor details that can simplify the changeover procedure.
- **'Foolproof' how changeover items are assembled/located** By design, limit variety in the way that changeover items can be assembled, including attention to the way that items are located.

10 A summary explanation of reduction in effort

Like other 'Reduction-In' categories, the scope for reduced effort can be considerable. Unknowingly (at least until the team have a chance to assess their own performance) most changeovers will be conducted in a relatively haphazard and inefficient manner. Typically there will be numerous opportunities to make the work easier – and faster – to complete. Reduction in effort is not constrained by retaining existing tasks in an essentially unchanged manner: freedom exists to make improvement in any way that is seen fit.

It should be noted that in many changeover situations the complete elimination of existing changeover tasks can be particularly important.

---

* Design therefore contributes to minimizing the types of changeover that are experienced – without compromising manufacturing flexibility.

11 A summary of improvement techniques applicable to the matrix's **Organization: effort concept**

Eliminate or simplify existing tasks *(by working in a better prepared and more efficient manner)*

This concept is concerned with making it easier to conduct changeover tasks, by cutting out unnecessary work (eliminating superfluous tasks) and by refining existing work practice. It is thus being sought to conduct the changeover in a better prepared, more efficient manner.

- **Eliminating superfluous activity** If a specific task does not warrant being included in the first place it can simply be eliminated, without altering the changeover in any other way. We have asserted previously that changeover tasks encompass everything that contributes to the elapsed changeover time. It might be possible, for example, to eliminate making a particular x-plane adjustment when currently more that one adjustment in this plane occurs. In addition, and perhaps more significant, there might be opportunities to eliminate changeover 'tasks' in which time is used wholly unproductively. For example, breaks during a changeover might be stopped.
- **Addressing 'problems' during changeover** By 'problems' we refer to unexpected events or activities during changeover. An example might be incurring a torn web on a continuous print process. In this case material will have to be fed once again through the line. Some 'problems' are likely to recur during later changeovers. Mechanisms should be in place to record the problems that occur, and then to resolve their cause.
- **Conducting individual changeover tasks more efficiently** It will usually be possible to work more efficiently: focusing upon conducting each task in a prescribed 'best practice' way. As a simple example, unnecessary movement when conducting a task might be eliminated. Conducting a changeover more efficiently means that tasks will be conducted more quickly and, often, to a higher standard. Formal training might be needed to ensure that best practice is universally adopted and maintained. Working more efficiently should not be confused with working harder.
- **Employing the best tools/handling/storage aids** There is little point in striving to work efficiently while using poor or inappropriate hand tools. The use of better tools, including power tools, might be considered. Other changeover aids can be applicable, including: data; signposting; handling devices; storage devices. Many of the tools and devices that could be used will be available for purchase as proprietary items. Other devices will need to be specially fabricated. Tools/devices (especially if small) and data are often advantageously stored adjacent to their point of use.
- **Ensuring excellent identification** A part that can be readily identified is easier to find. Easy identification should apply to all items that are used in the changeover, including tools and data. A predetermined, clearly marked location in which the

item is stored should also be available (either for permanent storage or for temporary storage during the changeover). A major benefit of sound identification can be avoiding discovering mid-way through a changeover that the wrong item is being used (it does happen!).

- **Ensuring cleanliness** Changeover items that are clean are easier to identify; to handle; to locate precisely; to work upon. Cleanliness should apply as well to the work area, which usually can be substantially prepared before the line is stopped. There can sometimes be safety implications when due cleanliness is not maintained.

12 A summary of improvement techniques applicable to the matrix's Design: effort concept

Eliminate or simplify existing tasks
(by equipment or product modification)

Design, like organizational improvement, can be freely employed in any way that is deemed appropriate. Reduction-In effort equates to simplifying existing tasks – for example, by reducing physical work; reducing the concentration/skill required; reducing preparation for the task (for example, accessing and using data, tools or further personnel). In each case it is sought to complete the task more quickly (and possibly, as well, to a higher standard). Ultimately Reduction-In effort means simplification to the point of task elimination. In general terms, for example, automation can be thought of in this way, where many tasks typically will be eliminated from the perspective of the operator. Automation is not the only way that tasks might be eliminated. Especially after deriving a reference changeover model (see next section), many ways to eliminate identified tasks can be sought.

Design-led techniques to reduce or eliminate 'effort' include:

- **Add devices to aid existing tasks** There is often scope to add devices to aid existing changeover tasks including, for example, many different proprietary quick-release devices. Similarly, devices can be added to facilitate handling. For example, ball-runners or hoists might be installed.
- **Modify hardware or products to aid existing tasks** New devices are not added: only the existing hardware or the product is modified. In practice this is another wide-ranging improvement technique. Hence, knowledge of associated examples (at the examples level of the matrix) can be particularly useful. Notable opportunities can include employing 'keyholes' (Figure 5.7) or lightweighting components.
- **Improve access** Changeover tasks can be significantly inhibited by poor access. As demonstrated by earlier exhibits in Chapter 5, options can exist to make access to specific points (for example, clamp bolts) easier. Alternatively, it can be made easier to remove obstructive machine elements, thereby affording required levels of access more simply and more quickly. As noted below, improving access is also often usefully considered at the same time as mechanizing tasks – which can obviate any previous access difficulties.

- **Mechanize activity** Even when access is not a problem (above), relatively complex mechanization/partial automation might still be justified. Less complex mechanization options might also be considered: their impact can often belie their comparative simplicity and cost. Many proprietary devices, for example hydraulic clamping systems, are available.
- **Avoid the need for more than one person to be present** One-person tasks are typically inherently simpler than multi-person tasks. Some particular attributes of multi-person tasks might be identified. First, there are likely to be problems of optimally dividing and organizing the individual task elements. Second, each person is unlikely to remain fully engaged – there will be periods of waiting, even if optimal work distribution is achieved. Third, arranging for necessary personnel to be available (both to start the task, and then to remain for its duration) can be a significant problem.
- **Reduce task content** Altering the content of a task means that essentially the same things as before are done but, effectively, there are fewer of those things to be done. An example would be using fewer clamps to secure a change part.
- **Conduct automated tasks more quickly** Automated changeover tasks intrinsically cannot be improved except by recourse to design change. Sometimes the speed of the automated task may simply be 'turned up'. Often subtle design changes can be made that otherwise allow the automated task to be completed more quickly, or to a higher quality. Although changed, the automated task, from the operator's perspective, could remain apparently exactly as before.
- **Separate/combine items** It can be advantageous to separate existing changeover items into two or more pieces for the purpose of the changeover. Benefit is gained, if so, by moving only that part that needs to be moved, from what was originally a complete component. This technique often will have a knock-on effect in eliminating other tasks that previously had to take place (those necessary to allow the item to be moved as a single entity). Conversely, in different circumstances, benefit might be gained by combining previously separate items as larger entities to aid, for example, their handling and positioning.
- **Avoid the need to use hand tools** The comments in respect of single-person working (above) similarly apply in respect of the use of tools: it is potentially advantageous if hand tools or special tools do not need to be used. This can usually be accomplished, for example, when using quick-release devices or hydraulically actuated clamps.

7.5 The techniques level of the matrix methodology

Figure 7.8 summarizes techniques that correspond to the matrix's eight concepts. Knowledge of these techniques and their place within the matrix can be very useful.

7.6 Using the matrix methodology: conclusions

Although no examples have yet been provided, the way that the matrix methodology is to be used should be apparent. As shown in Figure 7.3, the methodology seeks to

	Organization-led improvement	Design-led improvement
On-line activity	• Move tasks where possible into external time • Conduct tasks more in parallel • 'Compact' tasks more together (reduce periods of operator waiting) • Break task interdependencies (org.)	• Break task interdependencies (des.)
Adjustment	• Know of, and use, predefined settings • Define inspection parameters to signify adjustment completion • Understand possible adjustment interrelationships • Address precision requirements (org.) • Damage avoidance	• Automate adjustment • Address precision requirements (des.) • Consider using position/condition monitoring equipment • Make items more robust (less prone to damage or wear) • Upgrade item quality
Variety	• Implement standard changeover procedures • Precheck that changeover items are present as required • Precheck that changeover item quality standards are achieved • Consider batch resequencing	• Act upon standard change part features • Act upon standard product features • Standardize machine system features • 'Foolproof' location
Effort	• Eliminate superfluous tasks • Address problems that arise • Efficient work practice • Employ the best tools/handling/storage aids • Ensure excellent identification throughout • Ensure cleanliness	• Add devices to aid existing tasks • Modify hardware/product to aid existing tasks • Improve access • Mechanize activity • Single-person working • Reduce task content • Conduct automated tasks more quickly • Separate/combine items • Avoid using hand tools

Figure 7.8 The matrix improvement techniques (summary)

assist the improvement practitioner at different levels. It provides either detailed guidance or more global guidance as to where improvement might be found. The practitioner is invited to move between the different matrix levels to gain insight into potential improvement opportunities (an example is presented in Chapter 10).

A summary has been provided of assigned improvement techniques. A more complete explanation of these different techniques is given in the next two chapters.

Good prior understanding of each of these techniques will assist the practitioner in his or her use of the methodology.

The improvement practitioner can be led into the matrix by knowing (following an audit of the changeover) 'excesses' that apply to each changeover task. Alternatively, if not using the 'Reduction-In' strategy, and especially if the practitioner has good changeover improvement experience, the matrix might be used instead only as a checklist of possible improvement opportunities. Even so, a comprehensive audit should still have taken place.

The purpose of the methodology is only to prompt improvement ideas. There are no rules as to where improvement ideas are to be sought first, nor where ideas are to be sought in preference. Indeed, we suggest that that improvement ideas are sought from as many diverse matrix positions as possible. Doing so encourages all potential options at least to be considered, and helps counter any bias that might otherwise be present in the team's approach. Improvement ideas will later be evaluated (see further within the current chapter), once they have been identified.

7.7 A reference changeover model

Another tool that we have developed is what we term a reference changeover model. Like the matrix methodology, it is a tool for use in the overall methodology step '*develop an operational strategy*'. Its use requires an entirely different approach to that employed when using the matrix methodology.

Reference changeovers set out to describe how well a changeover might be conducted after organizational improvement alone has been carried out. Deriving a reference changeover model is principally an exercise in identifying the slack – inferior or unnecessary work, or poor task allocation – that occurs within existing practice. A reference changeover does not describe a limit to the improvement that is possible because explicitly it does not consider improvement resulting additionally from design change.

A reference changeover is a *pictorial* model. Its accuracy – and its usefulness – is reflected in the effort that is made in deriving it. Its usefulness is not restricted to evaluating the limit of organizational improvement. As will be shown, deriving a reference changeover should provide a fresh perception of possible improvement opportunities. It also allows insight into the minimum possible contribution of design change, if any, to reach specified improvement targets.*

Further benefits of deriving a reference changeover are also available. A reference changeover can be very helpful in evaluating the probable impact of different improvement options (both those that are organization based and those that are design based). A reference changeover also has particular use in determining the benefit of breaking task interdependencies, and in approximating the likely extent of a run-up period.

---

* We do not imply that as little design input as possible should necessarily be made – although this may be appropriate. Understanding in advance the likely minimum input of design can be beneficial. Particularly, such understanding allows design's use to be better planned for, for example gaining better familiarization with design improvement techniques and examples, or perhaps engaging necessary assistance (a case study in Chapter 10 elaborates upon these issues).

A less structured approach to seeking better changeovers

Consider what has been described in recent chapters: the 'Reduction-In' strategy; rigorous changeover auditing; the matrix methodology. These are components of a highly structured iterative process, as set out earlier in Figure 7.6. This structure is intended to impart an awareness of a full range of viable improvement options, some of which might not otherwise have been considered.

The approach in deriving a reference changeover need not be as rigorous as that outlined in Figure 7.6. This is because improvement options that the reference changeover considers are far narrower in their scope. In short, the *prima facie* objectives of a reference changeover are simpler. Its derivation, accordingly, is also simpler. In practice a reference changeover model will provide wide-ranging insight into potential improvement options – beyond those considered in its derivation. This insight should complement that gained when employing the matrix methodology.

Evaluating and agreeing reference performance for each separate changeover task

A necessary first step is to agree 'optimum' performance in turn for each of the individual tasks the changeover comprises. The word 'optimum' needs to be qualified. It is not used to mean the best possible performance that can be achieved. Taking an extreme example, an operator might work unrealistically if being observed while under threat of dismissal. Rather, 'optimum' is intended to mean honest, repeatable good performance, without hindrance or distraction. In work study terms this is sometimes referred to as 100 per cent effort, or the basic time to complete each changeover task. Deciding this level of performance can be difficult. Within a team, those who conduct the changeover tasks should help to decide what constitutes optimum performance. The raw data to enable the decision to be made, typically, will be provided by an audit video of the changeover. The video will show bad

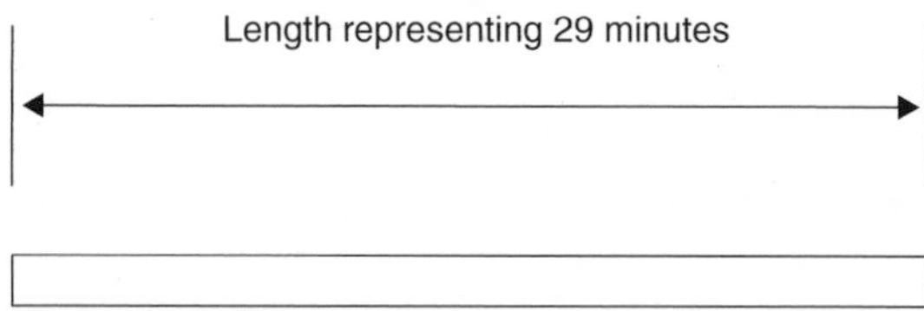

Figure 7.9(a) Block diagram of machine guard removal task, including elements of poor practice: time as recorded on video during changeover audit – 29 minutes

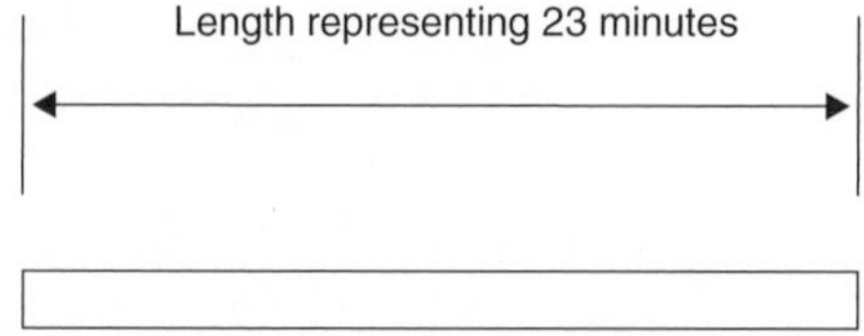

Figure 7.9(b) Block diagram of machine guard removal task: agreed 'reference performance' – 23 minutes

practice that can be deducted from the recorded task times (although allowance could also be necessary that the changeover operatives might actually be working *too fast* while they are being recorded).

Once isolated, and once an optimum or reference time for each task has been agreed, the building blocks for constructing the pictorial reference changeover model are available.

Figure 7.9 shows a possible changeover task that has been evaluated in this way. A task is represented as a simple horizontal bar whose length represents its duration. Figure 7.9(a) is the time as originally recorded, which might include elements such as waiting for hand tools to become free, unnecessary walking around the machine or placing guards in needlessly awkward and/or distant locations. This and other such inefficient practice could have led the task to be longer than the time of 23 minutes, shown in Figure 7.9(b), which represents the optimum performance time that has been jointly agreed by the team.

Other model foundations: tasks in external time and changeover item quality

The reference changeover model requires that optimum performance occurs within the limitations of the existing manufacturing system (that is, without changing existing resources in any way). A reference changeover model, if the elapsed changeover time is to be as short as possible, requires that all tasks that can be conducted in external time are conducted in external time. The model similarly requires that optimum distribution of tasks between available resources (notably personnel) also occurs. In addition, all changeover items that are to be used should be of requisite quality (fit for purpose) – compatible with conducting the changeover without hindrance or distraction.

Consider these points further. The first requirement that all possible tasks are conducted in external time means that all items used during the changeover will be immediately available. It should also be noted that the option of 'converting' tasks to external time – which we elsewhere assess as the provision of parallel equipment – is not considered because the reference changeover deliberately models what can be achieved strictly within the confines of existing resources.

There can be difficulty in establishing what constitutes requisite changeover item quality. A particular instance might be wear of changeover items, which has gradually led to degradation of changeover performance over time. It can sometimes be difficult to assess what task times should be expected.

A good way of assessing the soundness of the reference changeover model is to question whether all organization-based improvement options described by Figure 7.8 – except those that require increasing resources – have been taken into account.

Exhibit 7.1 Deriving the reference changeover model of Figure 7.12

An example is presented below from a changeover that we have studied. Team members decided how their performance could be improved – without in any way changing the equipment or increasing resources.

Exhibit 7.1 (*Continued*)

The example shows how tasks were allocated (by the improvement team) to the two operators who were to conduct a print changeover. Some restriction (as will be investigated further in Chapter 8) was imposed by different operators' respective levels of responsibility and skill/experience.*

Before showing the final reference changeover model, it is instructive to know the extent of poor practice that originally existed. This, once identified, will not be modelled. Figures 7.10 and 7.11 show selected details from two early recordings of changeovers on this equipment. The extent of poor practice that is apparent is considerable. Although only two changeovers are shown here, nine successive changeovers in fact were observed. The mean time for these nine changeovers exceeded 3 hours.

From this existing level of performance (changeover time that exceeded 3 hours†) the team derived the 45-minute reference changeover model that is shown in Figure 7.12. This model changeover applies to two people conducting the changeover. It applies exclusively to a four-plate/four-colour changeover. If these circumstances change – if the number of people conducting the changeover changes or if type of changeover changes – then this reference changeover model no longer applies. It can also be seen that some changeover tasks are automated. Quite extensive task interdependencies arose in particular around these automated tasks.

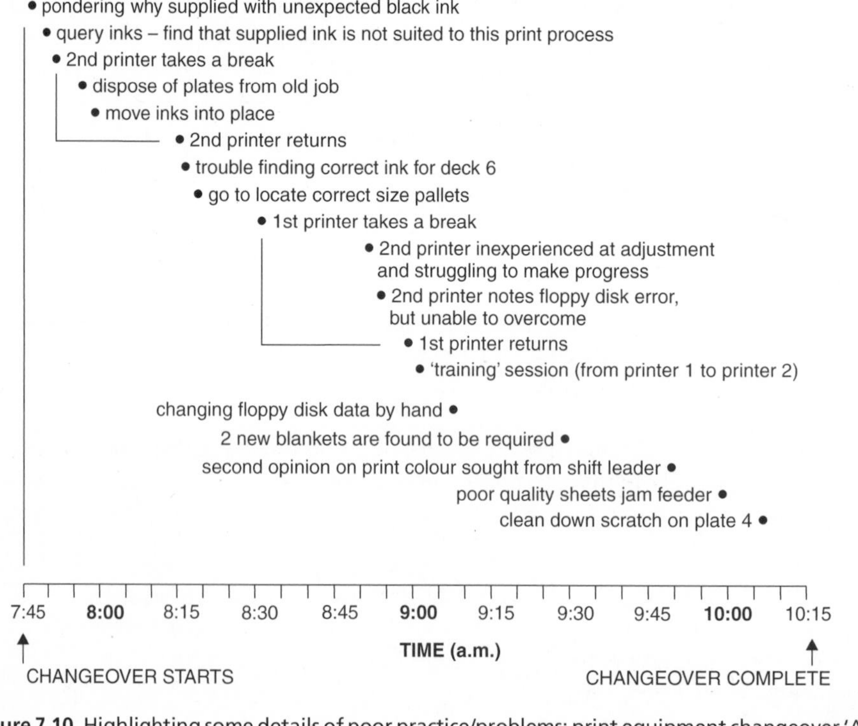

Figure 7.10 Highlighting some details of poor practice/problems: print equipment changeover 'A'

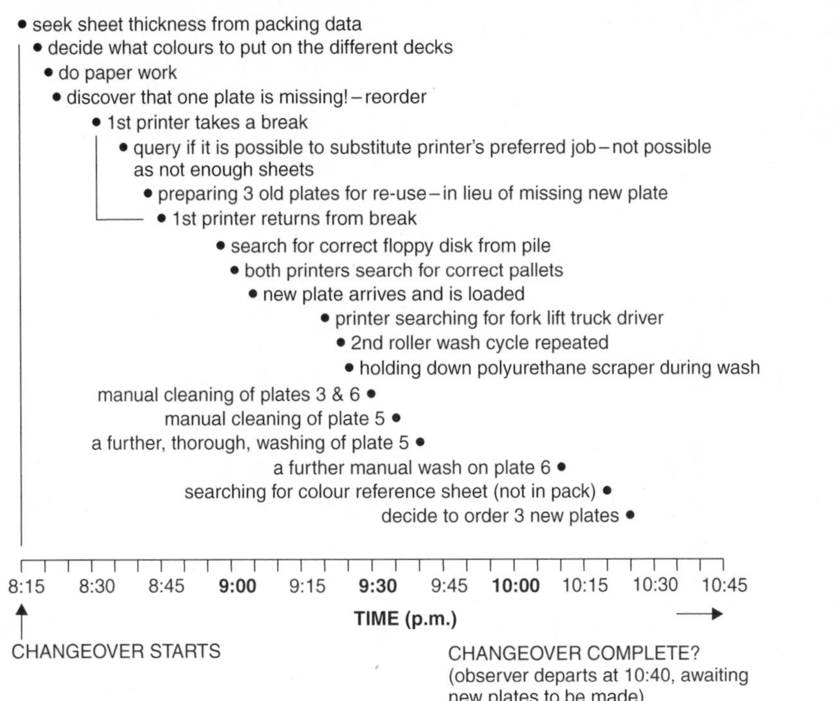
• seek sheet thickness from packing data
• decide what colours to put on the different decks
• do paper work
• discover that one plate is missing! – reorder
• 1st printer takes a break
• query if it is possible to substitute printer's preferred job – not possible as not enough sheets
• preparing 3 old plates for re-use – in lieu of missing new plate
• 1st printer returns from break
• search for correct floppy disk from pile
• both printers search for correct pallets
• new plate arrives and is loaded
• printer searching for fork lift truck driver
• 2nd roller wash cycle repeated
• holding down polyurethane scraper during wash
manual cleaning of plates 3 & 6 •
manual cleaning of plate 5 •
a further, thorough, washing of plate 5 •
a further manual wash on plate 6 •
searching for colour reference sheet (not in pack) •
decide to order 3 new plates •
8:15
8:30
8:45
9:00
9:15
9:30
9:45
10:00
10:15
10:30
10:45
TIME (p.m.)
CHANGEOVER STARTS
CHANGEOVER COMPLETE?
(observer departs at 10:40, awaiting new plates to be made)

Figure 7.11 Highlighting some details of poor practice/problems: print equipment changeover 'B'

---

* This particular print changeover example will be picked up again in Chapter 10, when it will be presented in greater detail as a case study. Some of the more subtle issues of management pressure that can compromise realistic improvement will then be investigated.

† As usually occurs, there was no assessment made of the quality of the changeover at this stage: only the time that the changeover took to complete was studied.

7.8 Making use of a reference changeover model

Using the example of Figure 7.12, some ways that a reference changeover model can be helpful are now introduced. These benefits are *in addition* to the major benefit that has already been achieved in deriving the model; of assessing poor performance, or slack within the existing changeover.

Seeking to reallocate tasks: task interdependencies

The opportunity should be taken to reassess how individual tasks are allocated, including the sequence in which they are conducted. In doing so opportunities should be sought to: conduct tasks more in external time; conduct tasks more in parallel; 'compact' tasks more together (see Figures 5.13 and 5.14). In this context a reference changeover model can give significant insight into the associated issue of resources:

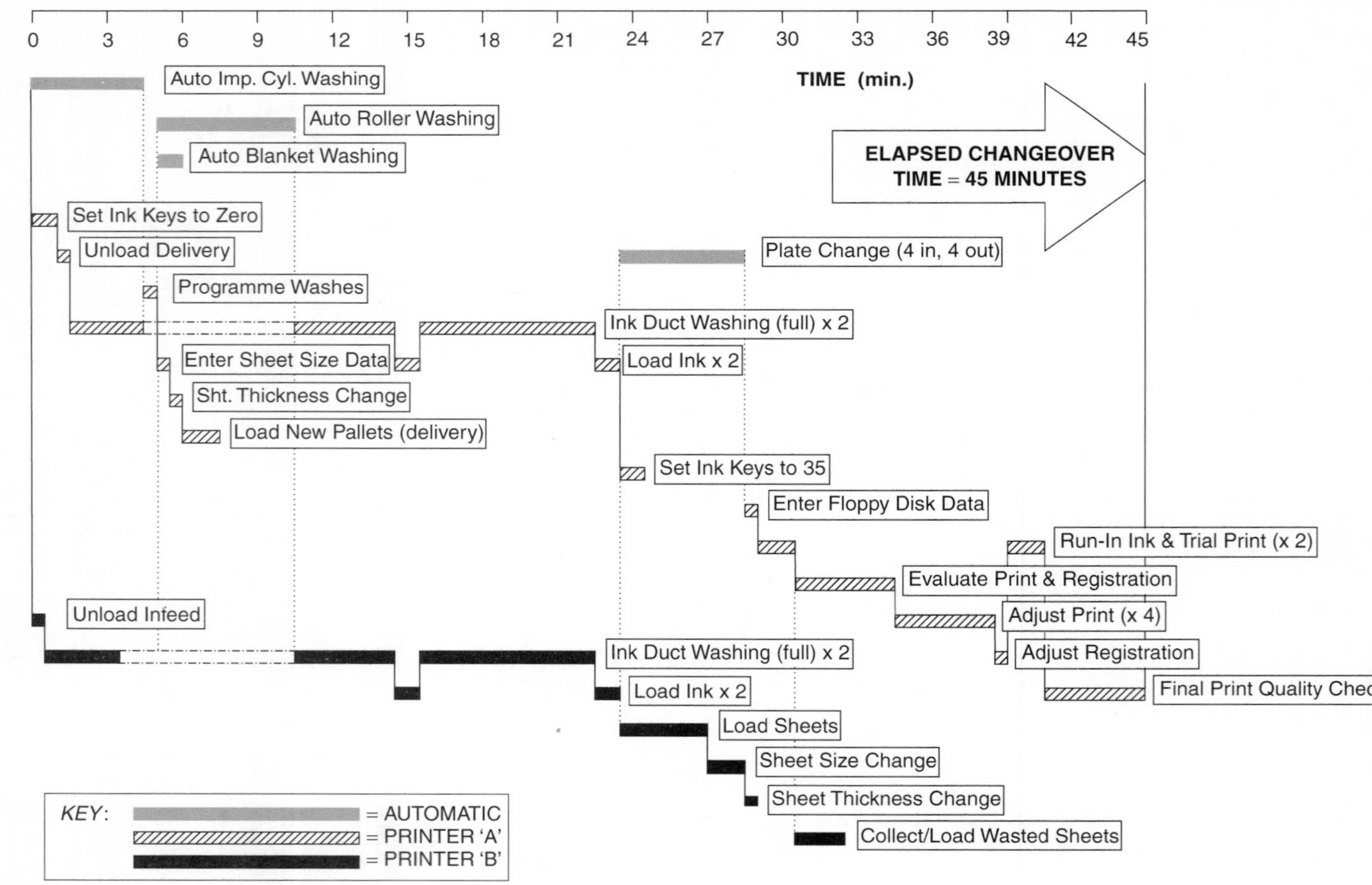

Figure 7.12 Four-colour/four-plate/two-person reference changeover

where it is also necessary to reallocate resources (to be able to conduct the individual changeover tasks) or where additional resources need to be provided (typically, greater manpower or additional/parallel equipment).

Task-reallocation possibilities will be restricted by interdependencies between tasks, which refers to limitations imposed by other tasks as to when a specific task can be conducted (this topic will be covered in more detail in the next chapter). The reference changeover model is a powerful tool to determine the extent of these restrictions, and to identify opportunities for improvement.

Where to focus improvement effort

As above for the specific issue of reallocating tasks and resources, a reference changeover model is also important in respect of all other potential improvement options (as highlighted in Figure 7.8). First, it helps team members to pick existing changeover tasks that it will be particularly beneficial to improve upon. Second, it allows the likely impact of those improvements to be evaluated more easily.

A reference changeover model can be of particular assistance in determining both the possibility and the benefit of eliminating blocks of tasks – something that hitherto has not always been easy. For example, exactly what is involved when removing a non-changeover assembly to gain access can be identified (see Exhibit 5.2). The time that this block or series of tasks contributes to the elapsed changeover time would be clearly seen. By using a reference changeover model the desirability of eliminating such a block of tasks can readily be determined.

Item quality in the context of adjustment iterations

An adjustment iteration refers to a repeated trial-and-error cycle of tasks that involve making an adjustment and then testing that adjustment. One particular opportunity to eliminate blocks of tasks (above) might be to reduce the number of adjustment iterations that occur. Often adjustment iterations, if present, will be witnessed during the changeover's run-up phase.

An instance of, say, four or five anticipated iterations could be a reflection of poor item quality (Chapter 3); item quality that can usefully be improved upon – even if a currently accepted quality standard is met. Whatever their cause, work to overcome instances of adjustment iterations will often warrant a high priority.

7.9 Deciding which improvement options to adopt

We have chosen to describe both changeover auditing (Chapter 6) and identifying potential improvement options (to this point in the current chapter) in considerable detail. This work will take a team much of the way through the *preparatory* and *implementation* phases of the overall methodology. As yet, however, no final decision as to how improvement is to occur (from the potential options that have been identified) should have been taken. Still within the overall methodology step of '*develop an operational strategy*', it now needs to be decided which of the suggested improvement options to proceed with.

Reflecting again upon the need for thoroughness for non-prescriptive improvement

Our approach demands significant preparatory work before improvements are actually made. This preparatory work might seem excessive. This opinion perhaps will be held especially by those who are adopting our approach for the first time. Certainly this work might appear to be particularly extensive when compared with a more usual prescriptive approach to changeover improvement. This, though, is the essence of our alternative approach: we are deliberately seeking a non-prescriptive improvement format. If it is going to be successful, a non-prescriptive approach requires extensive understanding in advance of both the existing changeover and of different improvement options. The best improvement option(s) can then be selected on merit. Conversely, a prescriptive approach is inherently prefocused on which improvement options to adopt (or not to adopt), irrespective of their suitability to the current changeover situation.

Issues to consider beyond the likely time saving that will be made

Before finally deciding what improvement options to adopt, and formally setting these options down as targets, it is important that all pertinent issues are considered. These issues will extend beyond the time saving that the improvement is likely to provide. Some issues are:

- **Know each team member's capabilities** It is desirable to understand the capability/skill/experience of team members (or those who are otherwise to implement any proposed changes), to ensure change can be realistically achieved as planned. In other words, it is desirable to understand the team's limitations.
- **Recognize that changeover time improvement is not all that is being sought** The impact of proposed changes should be assessed beyond their contribution to changeover time. Notable considerations will include changeover quality; sustainability; whether benefit might be derived from deskilling; whether there is benefit in reducing the number of operators needed to complete identified tasks.
- **Maintain full team involvement** Team members ideally should all be engaged in implementing improvements. There should not be, under normal circumstances, occasions when some team members are busy for, say, the next few weeks, while other members have little or no work to do. Other motivational or involvement issues might also need to be addressed, especially at the start of an initiative. Rather than try to kick off with a 'big hit', it might be prudent to undertake a series of simple initial improvements, or a single improvement that all team members can contribute towards together. Choosing an average idea that is going to be enthusiastically worked upon is also probably a better option than selecting what appears an excellent idea, but one that raises little interest from those who are to execute it. In these and other respects the team leader can have an important role in directing the progress of the initiative.
- **Assess a comparative cost-effectiveness rating for each idea** Importantly, the cost of any proposed change needs to be evaluated against its effectiveness. In this way the team will be better able to progress the initiative within its global cost and improvement targets (set as the final step of the initiative's strategic phase – see

Figure 3.16). As this is a comparative assessment, a high-cost/high-reward idea could rank equally against a low-cost/low-reward idea.

- **Consider implementation time** The time to implement the proposed improvement also needs to be assessed. Is there, for example, benefit in pursuing a major change that takes perhaps 5 weeks to implement, when many other smaller changes might otherwise have been completed in this time?
- **Consider if improvements are likely to be superseded** Are improvements being considered that later (by subsequent improvement) stand a good chance of being substantially revised, or eliminated altogether?
- **Maintain a safety awareness** Are the proposed changes in any way influencing safety?

Finally deciding what improvement options to adopt: a decision chart

Team members should collectively decide which of the proposed improvement options to pursue. A simple, visible weighting exercise is all that is required because the exercise need not be (and will not be) wholly precise: judgements are being made at all stages.

Appendix 5 presents a document that might be employed. Using a hypothetical example, Figure 7.13 shows how this document would be completed. Different options that are being considered are written down. A qualitative assessment is then made in latter columns. A 2-tick rating (✓✓), for example, strongly endorses an option's selection. Conversely, at the other extreme, a double-graded negative rating (××) indicates that significant problems are likely. Figure 7.13 also shows the use of a separate symbol (⧫), which signifies that external assistance (to the team) is needed.

Explaining the assessment chart's columns

The columns employed take account of the issues that are discussed above. First, the **changeover task** that it is being sought to improve is written down, including its current time. Individual **improvement ideas** are then entered. The **estimated time saving** for each idea should be written down.

The entry to be made in the **cost–benefit assessment** column need not be highly accurate and thus need not be considered at great length. Nevertheless the assessment should be good enough to give a meaningful comparison between different improvement ideas. The comparative assessment here is of an idea's cost against the time saving that it will make.

Other impact of implementing the idea needs to be evaluated. Entries are intended to embrace any likely benefit (or penalty) arising beyond changeover time saving. Ease of sustaining the improvement is one factor that might be considered. Another important consideration might be the idea's likely contribution to changeover quality.

The **implementation issues** column covers a number of different issues. First, it is to be used to qualify if an idea is likely to be superseded. It should also take account of motivational/involvement issues of the team members. Other issues, for example a need to stop the line to make the modification, should also be judged. Finally, the column should assess the need for external expertise to execute the idea.

Assessment: ✓✓ = Strongly positive; ✓ = Positive; × = Negative; ×× = Strongly negative

External involvement: ⌖

Changeover task	Improvement idea	Estimated time saving	Cost–benefit assessment	Other impact	Implementation issues	FINAL RANK
Remove gear cover to gain access to drive gear collet (10 minutes)	Split gear cover – to remove only its bottom LH corner	4 minutes	× (difficult machining needed=costly)	–	–	
	Use just 8 retaining bolts, instead of 20	3 minutes	✓✓	× (sealing; hot oil?)	✓✓ (very easy)	
	Use ratchet and socket in place of spanner, and mount on a special rack adjacent to gear cover on machine frame	2 minutes	✓	✓	–	
	Mount small receptacle on machine frame temporarily to store 20 M6 bolts	90 seconds	✓✓	✓ (reduces chances of loss) × (if adopted, likely to be superseded)	–	
	Substitute securing bolts with proprietary quick release fasteners	9 minutes	✓ (costly: need extensive adapting to fit)	✓✓ (no skill needed, no tools needed, no temporary storage needed)	×× (no space available to mount!)	

Remove gear cover to gain access to drive gear collet (10 minutes)	Secure with bolts acting on swinging 'C' plates	8 minutes	✓✓	✓ (bolts storage problem eliminated; no possibility of dropping bolts into machine)	✗ ⌖ (line stoppage and external fabrication)	2
Adjust drive gear position (11 minutes – typ.)	Provide and use setting piece to eliminate trial and error adjustment	9 minutes	✓✓	✓	✓ (easy; fabrication of setting pieces possible by team members)	1
	Mount scale and adjust to scale	5 minutes poor visibility: hence unlikely to be able to adjust first time with sufficient accuracy)	✓✓	✗ (need to make available and use position data) ✗ (difficult to use)	✗ (need to halt production to mount scale)	

Figure 7.13 Improvement option assessment sheet

Notes

As shown on Figure 7.13, it is often prudent, in appropriate columns, to make brief notes to remind all participants on what basis a particular entry has been made. Occasionally it might be desirable to make more than one entry in a column.

The assessment sheet's output

The final column of this sheet – which is its output – is a ranking of the proposed ideas. The ideas that are selected via this weighting exercise will then be written down and displayed in the next stage of the overall methodology: setting local targets (see below). We suggest, typically, that comparatively few ideas should be taken forward simultaneously, consistent with distributing the improvement workload across each of the team participants.

7.10 Set local targets

Local targets can be documented on two levels. First, it should be considered what information is to be made available for general consumption. Typically we suggest that this information should be largely unrestricted, so that all who work in the factory are as aware of what is going on as they choose to be. Information, including diagrams, can be set out on a display board.

Second, a document is needed to help progress the improvements that the team have decided upon. Written down should be what is to be done, who is to conduct these activities and when completion is expected. A chart that fulfils these functions is described in the next section.

7.11 Implementation phase activity

The method we describe here involves a simple, visible commitment chart. This is known as a 'PDCA' chart – 'Plan'; 'Do'; 'Check'; '(Re-)Act'. It sets out the team's improvement targets, and further provides the documentation required for this final phase of the initiative.

Self-action

Team members are required to complete the improvements – or to oversee completion – themselves. As well as completing the physical changes, there is also an onus on self-reporting of progress, including monitoring that the changes have had the impact expected of them.

Using a PDCA chart

An example of using a PDCA chart is shown in Figure 7.14 (available as a blank in Appendix 6). The document is to be used in a way where it is accessed by all the team members. All improvements to be undertaken are entered.

The four PDCA (or, better, P-D-C-A) categories are set around a circle and blanked in as they are completed:

Completion of the four PDCA stages can easily be indicated:

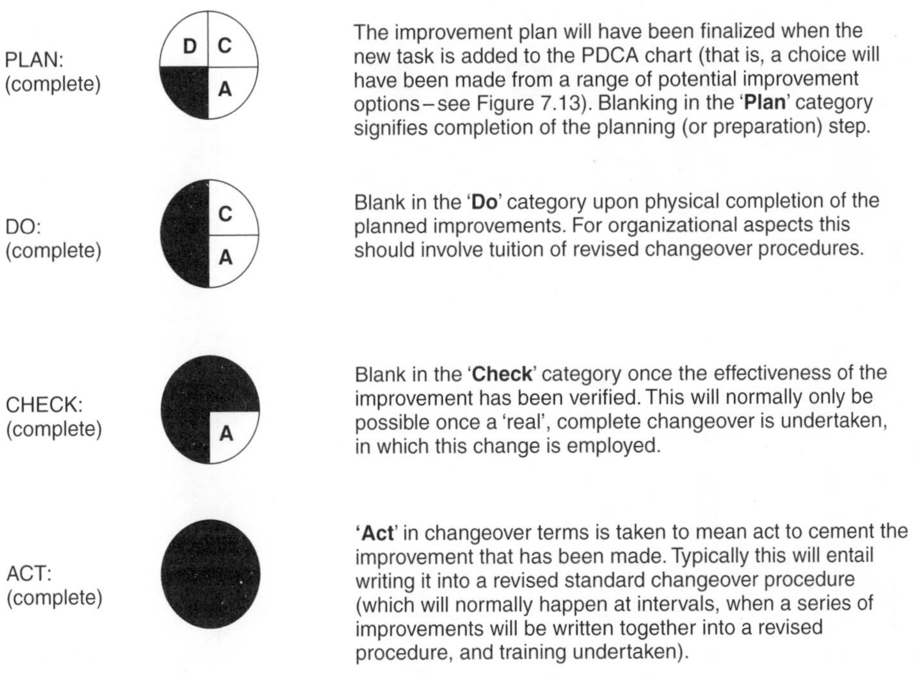

PLAN:
(complete)
D
C
A
The improvement plan will have been finalized when the new task is added to the PDCA chart (that is, a choice will have been made from a range of potential improvement options – see Figure 7.13). Blanking in the '**Plan**' category signifies completion of the planning (or preparation) step.
DO:
(complete)
C
A
Blank in the '**Do**' category upon physical completion of the planned improvements. For organizational aspects this should involve tuition of revised changeover procedures.
CHECK:
(complete)
A
Blank in the '**Check**' category once the effectiveness of the improvement has been verified. This will normally only be possible once a 'real', complete changeover is undertaken, in which this change is employed.
ACT:
(complete)
'**Act**' in changeover terms is taken to mean act to cement the improvement that has been made. Typically this will entail writing it into a revised standard changeover procedure (which will normally happen at intervals, when a series of improvements will be written together into a revised procedure, and training undertaken).

It is seen that tasks that are in-progress are easily identifiable from tasks that have been completed. Hence the state of current progress is readily witnessed.

7.12 Conclusions

This chapter describes work to be undertaken by an empowered internal changeover improvement team. The description is of a structured, visible, participative work programme. For many initiatives this work will represent the crux of the overall improvement effort.

The work that the chapter describes is not trivial, largely because we advocate undertaking improvement that has been both carefully identified and carefully evaluated. Specifically, we aim to identify solutions to overcome the changeover's audited difficulties. This approach is different to a more usual – and inherently simpler – prescriptive improvement approach.

Date: 10 February 1999

CHANGEOVER IMPROVEMENT WORK PLAN: Scroll line changeover
(PDCA Chart)

DESCRIPTION OF IMPROVEMENT BEING MADE / BENEFIT SOUGHT	RESPONSIBILITY	START	TARGET COMPLETION	STATUS	NOTES
Provide and use setting pieces to adjust guide rail position (*Benefit*: adjust with precision straight to desired position, saving an esstimated 9 minutes)	PDH	11th February	28th February	 D C A	
Replace bolts securing gear cover with 'swinging "C" plates' (*Benefit*: save estimated 8 minutes. Eliminate the need for tools and the need to remove and store bolts)	GBB	11th February	24th February	 D C A	
				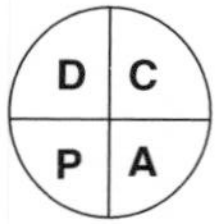 D C P A	
				 D C P A	

Figure 7.14(a) An example of a PDCA changeover chart. (In this case just two ideas are recorded. The chart can be compared with its state in mid-April (7.14(b)), which indicates the progress that is being made)

Date: ~~*10 February 1999*~~
~~*28 February 1999*~~
~~*17 March 1999*~~
15 April 1999

CHANGEOVER IMPROVEMENT WORK PLAN: Scroll line changeover
(PDCA Chart)

DESCRIPTION OF IMPROVEMENT BEING MADE / BENEFIT SOUGHT	RESPONSIBILITY	START	TARGET COMPLETION	STATUS	NOTES
Provide and use setting pieces to adjust guide rail position (*Benefit*: adjust with precision straight to desired position, saving an estimated 9 minutes)	PDH	11th February	28th February	A	To be written into revised standard procedure, mid-April
Replace bolts securing gear cover with 'swinging "C" plates' (*Benefit*: save estimated 8 minutes. Eliminate the need for tools and the need to remove and store bolts)	GBB	11th February	24th February	A	Proprietary parts ordered from XYZ company, 13th February – awaiting delivery. In revised std. procedure: April
Mount gear change data table on main gearbox frame (*Benefit*: data reference immediately adjacent to point of use. Est. saving 1 minute – and errors less likely)	ARS	28th February	11th March	C A	
Build a trolley suitable for gear change components and tools (*Benefit*: change parts prepared – and checked – in advance and brought to point of use together. Est. saving 11 minutes)	PDH / GBB	28th February	21st March	D C A	

Figure 7.14(b) Continuing the example of a PDCA changeover chart

The chapter has presented two important new tools. Their role is to help in the identification of potential improvement options. These two tools, the matrix methodology and the reference changeover, appraise the current changeover in different ways. Both, however, seek to prompt improvement ideas appropriate to the difficulties that the current changeover experiences. These two tools are usefully applied together.

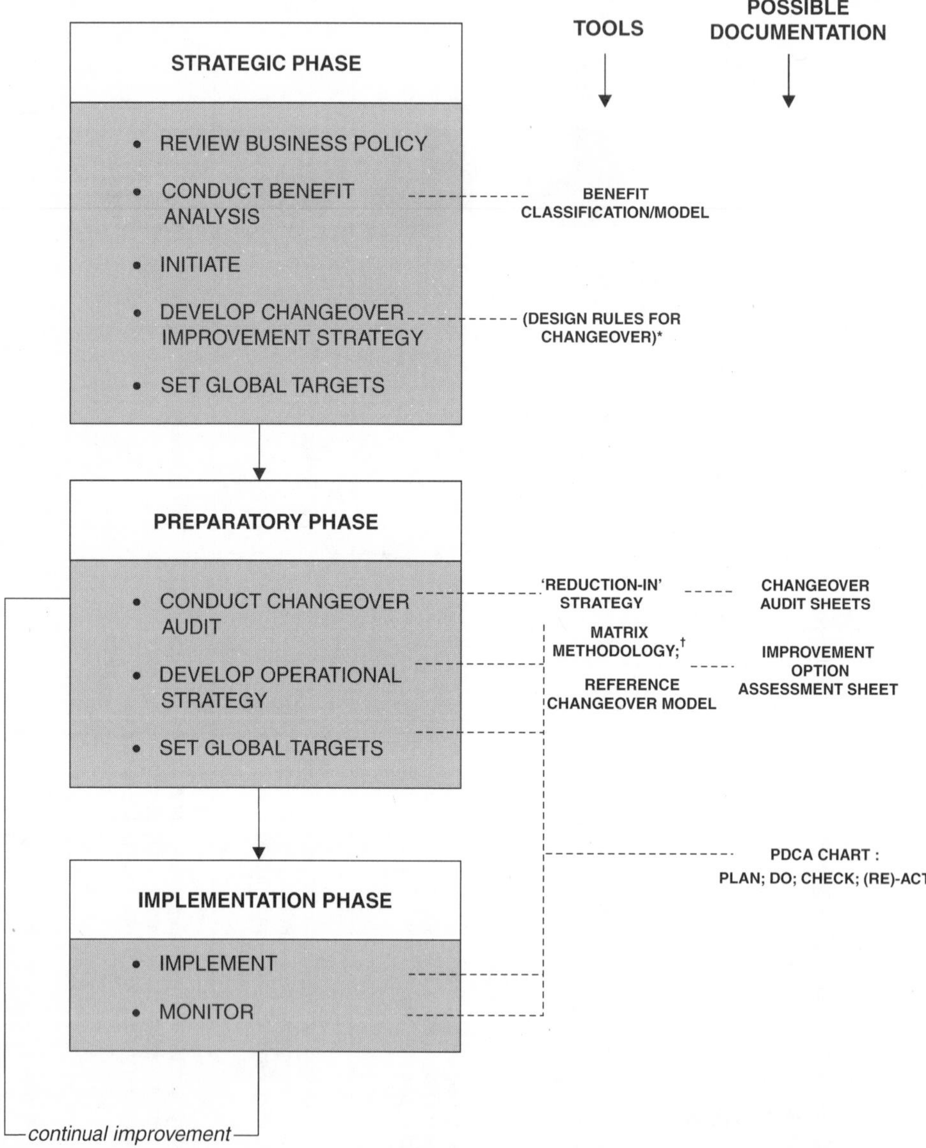

Figure 7.15 When tools and documentation are employed in relation to the overall methodology. * For use principally by OEMs (see also Chapter 3 and Figure 3.3). Decision to use OEMs taken in step '*develop operational strategy*'. †Use of the matrix methodology will be greatly assisted by understanding the varied improvement techniques explained in detail next in Chapters 8 and 9

The chapter also describes other important steps within the preparatory and implementation phases of the overall methodology. It is important to remember, however, that the work this chapter describes only applies to retrospective improvement on the shopfloor – it does not represent all that is required of a company changeover improvement initiative.

Chapters 6 and this chapter have detailed our approach for retrospective, team-based, *kaizen* improvement. We have described tools and documents that might be employed. Figure 7.15 sets out when these aids can be used. In addition, for completeness, we show aids that can be related to the earlier strategic phase of the overall methodology.

References

1 McIntosh, R. I. (1998). *The impact of design on fast tool change methodologies*. PhD thesis, University of Bath. The matrix methodology has evolved over an extended period during our research in different manufacturing environments across Europe.

2 Shingo, S. (1985). *A Revolution in Manufacturing: The SMED system*. Productivity Press Inc., Massachusetts.

3 McIntosh, R. I., Culley, S. J., Mileham, A. R. and Owen, G. W. (2000). A critical evaluation of Shingo's 'SMED' methodology. *Int. J. Production Res.*, **38**, No. 11, 2377–2395. The points that our paper makes are also argued at different stages within this book.

4 McIntosh, R. I. (1998). *Op. cit.* ref. 1. The matrix methodology that we present in this book is significantly advanced over earlier versions, first described in the McIntosh thesis.

Postscript

The matrix methodology does not necessarily unambiguously assign individual improvement techniques to different matrix positions. Neither does it assume that these techniques will always be uniquely interpreted. Similarly, the matrix concepts do not necessarily stand wholly alone from one another. This postscript briefly explains that these possibilities do not limit the methodology's usefulness. It also contrasts use of the matrix methodology with Shingo's 'SMED' methodology.

The matrix methodology in practice

First, when using the methodology, some observations might be:

- The eight positions at the concepts level of the matrix are not mutually exclusive. A specific improvement idea might be arrived at, say, via the concept '*Eliminate or simplify existing tasks, by equipment or product modification*'. Exactly the same idea might perhaps have also arisen via the concept '*Standardize the way the changeover is conducted, by standardizing changeover tasks*'.
- Likewise, it might not be possible to assign a potential improvement technique within the matrix in a wholly unambiguous manner. A particular technique arguably might be sited in more than one matrix location. Or, thought of another way, improvement techniques might overlap.

- The boundary between design-led and organization-led change will not be clear-cut. Even defining where design change occurs might be difficult. Our basic distinction is that design change occurs when alteration is made to the hardware used during a changeover. Consider the situation of stepping up the voltage to an existing power winder to make it work faster. This change is probably correctly seen as an organizational change, just as using different hand tools represents an organizational change, but an argument can be made that changed voltage represents a physical change to changeover hardware. Repair, to aid changeover, to equipment that had become damaged might equally be difficult to classify.
- Whereas categories of 'Effort' and 'On-line activity' are argued to be significantly distinct from one another, 'Adjustment' and 'Variety' can be interpreted as focused sub-sets of the 'Reduction-In effort' category. The attention given to the categories of 'Adjustment' and 'Variety' reflects their potential impact.
- It is not necessarily easy to define a changeover task (Chapter 5).

These observations do not inhibit the usefulness of the matrix methodology. Because there is no prescribed sequence either to the order or to the origin of improvement – at any point in the improvement process – such possible lack of 'exactness' is not important. What *is* important is that a full complement of meaningful improvement ideas is generated. The route by which an idea is prompted is of little consequence. This situation is wholly different from other workshop methodologies, where at any time, if strictly adhered to, only a limited range of prescriptive improvement options will be available to be selected from.

The matrix methodology, particularly at its conceptual level, promotes a hitherto unusually wide range of improvement options. Design-improvement options are deliberately featured. It should be remembered in practice that many, if not most, improvement ideas will incorporate elements of both design change and organizational change (see Chapter 5). The matrix methodology is no more than a mechanism to access improvement ideas. In this context pinning down an idea's design and organizational change content becomes a redundant exercise.

Contrasting the matrix methodology with the Design Rules for changeover

Figure 7.6, showing the iterative improvement process of which the matrix methodology is a part, is very similar to Figure 5.16. Figure 5.16 was presented in the context of the Design Rules for changeover improvement. Figure 7.6 was presented in the context of wider improvement, in which more than design change alone is being sought.

The design for changeover rules were highlighted in Chapter 5 particularly for use by remote (to the shopfloor) design agencies: a deliberate emphasis was given to undertaking design away from the shopfloor environment, by individuals who do not have direct involvement in the shopfloor programme. The design rules are also available for use by OEMs, who once again, typically, cannot easily contribute via a mechanism of incremental shopfloor improvement.

The range of fundamental options that the design rules describe are nevertheless closely analogous to the design-led options that are cited within the matrix methodology. This should be expected: their derivation has been approached in the same way.

Seeking assistance when required

Depending upon the shopfloor team's composition and influence, not all the solutions that the methodology might highlight typically can be executed without additional help. All involved in the initiative as a whole should be aware of this potential limitation. If an idea warrants implementation by external personnel, mechanisms should be in place to enable this to occur.

The Matrix Methodology and the 'SMED' Methodology

Shingo has unquestionably made a very substantial contribution to the topic of changeover improvement. Nevertheless our own research and the work of others both indicate that there is potentially scope to refine his contribution further. Our arguments in support of this case have so far not been fully set out in one place within this book. It is useful here, perhaps, simply to summarize the areas that are of concern to us. The points made below are elaborated upon elsewhere, as highlighted in the index. Further, we have written academic papers on this subject that might also be referenced [McIntosh R. I. *et al.*, 2000, A critical evaluation of Shingo's SMED methodology, *Int. J. of Production Research*, **38**, No. 11, 2377–2395; McIntosh R. I. *et al.*, Changeover improvement: reinterpreting Shingo's SMED methodology (paper under review at the time of going to press)].

In making the observations below it should also be remembered that the 'SMED' methodology is a workshop improvement tool. Many of the wider issues that can substantially impact upon the outcome of a changeover initiative – as have been detailed principally in Chapter 3 – are not necessarily within its scope.

In summary, we are concerned by :

- the prescriptive nature of the 'SMED' methodology
- the prominence of gaining improvement by translating changeover tasks into external time
- problems of adequately assessing (and improving) a changeover's run-up period – or, in other words, failure to acknowledge either that a run-up period frequently occurs, or to acknowledge its significance
- recognizing the contribution that design modifications can make, and structuring improvement effort accordingly
- the descriptive inadequacy of improvement options afforded by the 'SMED' concept of '*streamlining*'
- distinguishing between 'separating' and 'converting' changeover tasks into external time

The matrix methodology aims to overcome these difficulties, prompting consideration of a full range of potential improvement options, and conveying what these options might achieve. Like the 'SMED' methodology, it is a tool to assist retrospective improvement by workshop teams. Its use should therefore always be considered as part of a global improvement effort (see also Figure 7.15 and Chapter 3, Part 2).

8

Organization-led improvement techniques

This chapter explains possible organizational-led improvement techniques. For consistency these techniques are set out in the same order as in the previous chapter, where they were described in relation to the matrix methodology. Design improvement techniques will be considered in Chapter 9.

Organizational improvement techniques have traditionally been favoured when addressing changeover performance, principally on the basis that 'quick changeover activity does not cost a lot but quick set-up technology does'.[1] Considerable reduction in changeover time can usually be expected.[2] Then again, as we have described, strong arguments might also be made in favour of design-led improvement. A good improvement programme will be one that is able to select those techniques, whether organization-biased or design-biased, that are most appropriate to the existing changeover situation.

For the most part the sections presented below simply add substance to the summaries that have already been presented in Chapter 7. A notable exception is in describing the technique of breaking task interdependencies, which, perhaps not being immediately apparent, is explained in some detail. Conversely, the technique of task sequencing (which the shopfloor team might have relatively little influence upon) has been fully explained previously in Chapter 3.

Overlap between some of the techniques presented in these next two chapters can exist. For reasons that have previously been discussed, this does not exclude citing techniques in the way that we have chosen.

It is not intended that examples of applying these techniques should predominate here. Nevertheless a few examples are provided, to give greater insight into some of the opportunities that might be available.

8.1 Reallocate tasks to external time

This improvement technique – like others in this immediate series – is concerned with changing when tasks are conducted, rather than changing *how* tasks are conducted. The mechanism by which improvement is sought, therefore, is task reallocation. The technique aims for as many tasks as possible to be completed prior to halting the line. There is no reduction in the total work done, only when it is done.

The technique is similar to Shingo's 'separate' internal tasks into external time.[3] Caution is needed, however, that our technique permits tasks to be moved freely from either the set-up period or the run-up period. Like Shingo, we are not concerned at this point with conducting individual tasks otherwise more efficiently, nor eliminating tasks from the changeover.

Tasks that are ripe for improvement are often concerned with gathering all changeover components into place and checking their condition. There should be no need, for example, to retrieve a bolt to secure a die set while the press is idle (during changeover).[4] Instead this task should have been conducted previously, whereby the bolt is put in position ready to be used, prior to halting production.

Another potentially important option concerns timing the commencement of tasks such as pre-heating or purging (for example, for some food processes). These tasks need to be integrated as part of the overall changeover procedure (see also 'Efficient work practice' later in this chapter). There can often be scope to commence these tasks in external time, thus significantly reducing the elapsed changeover time.

Separating tasks where possible into external time can have wide application, particularly at the start of an initiative. Applying this technique can often have considerable impact.

Not all tasks can be started at will in external time. Often it will be necessary to break or alleviate task interdependencies (see below) before a particular task start time can be altered in this way. Indeed, this important topic of task interdependency can equally apply when only seeking to alter start times wholly within the interval when the line is undergoing work (in other words, altering task commencement times within the elapsed changeover time, T_c). In respect of altering when tasks can be conducted one particular issue is now introduced: the availability of equipment.

Availability of equipment

It is useful to understand the particular relevance of providing additional equipment when seeking to conduct tasks in external time[5,6] – or, indeed, when seeking to conduct tasks more in parallel. Employing parallel equipment is often necessary for this to be possible. Providing parallel equipment permits tasks, in Shingo's terminology, to be 'converted' into external time.[7]

Parallel equipment can range from small components through to large parallel assemblies. Even complete lines can sometimes be employed, that are changed over wholly in external time (while the alternative, parallel line is fully operational).[8] Such examples should not deter the practitioner from considering benefit that might be achieved by relatively modest provision of parallel equipment. Additional tools might be provided. For example, a single long extension bar might be used to undo each of twelve securing bolts. Simply by providing an additional extension bar a second person potentially could start to undo some of these bolts as a parallel operation.

The advantages gained by parallel equipment might not always be immediately apparent to a novice changeover practitioner. In this regard (as for other improvement techniques) it can be important for inexperienced team members to make themselves aware of examples of what others have done.

In seeking opportunities for additional equipment it must be considered that other issues, notably the personnel that are available to conduct the changeover, or perhaps space limitations, can limit their usefulness (see below).

8.2 Reallocate tasks to be conducted more in parallel

Reallocating tasks into external time is not the only means by which the elapsed changeover time can be reduced: task start times might also be beneficially altered within the set-up or run-up phases of the changeover. Two options here are to 'compact' tasks together better or to conduct tasks more in parallel. Figure 8.1 illustrates the improvement technique of conducting tasks more in parallel.[9]

The start time of only comparatively few tasks might be freely reallocated within the changeover's set-up and run-up phases. As noted in the previous section, other issues typically also have to be considered if opportunities to work in parallel (or to compact tasks together better) are to be increased. An important issue here is likely to be the availability of personnel.

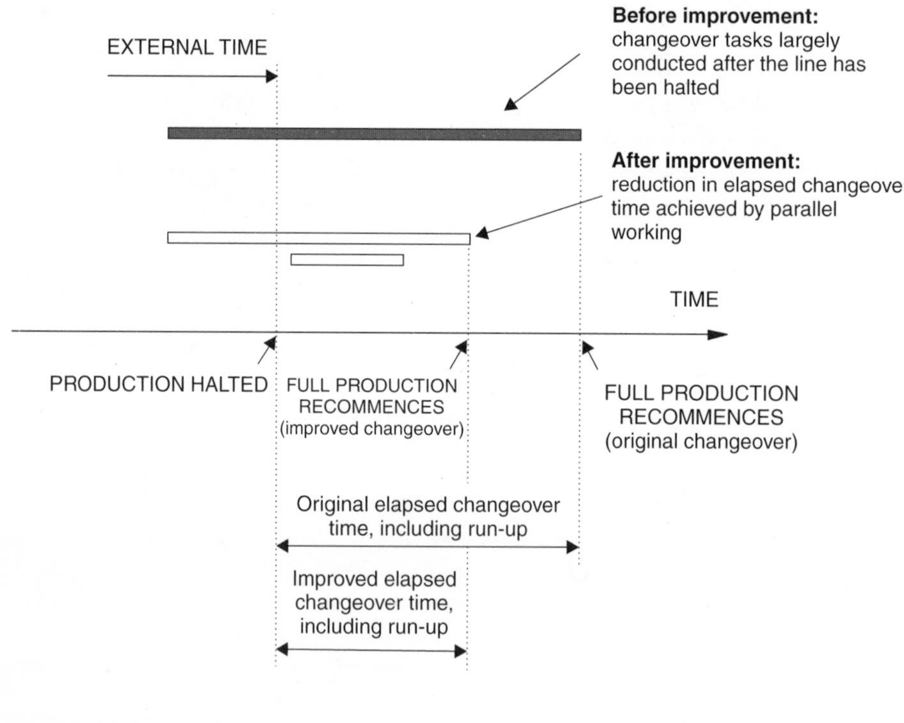

Figure 8.1 Newly conducting some changeover tasks in parallel, without changing task duration

Availability of personnel

For many changeovers (but by no means all) engaging additional personnel can allow selected changeover tasks to be completed more in parallel, and hence allow the changeover to be completed more quickly. To be of greatest use, these personnel might need to be trained so as to be conversant with their respective tasks, including awareness of their work in relation to what others are doing. In some complex changeovers the successful integration of additional personnel might be relatively difficult.

The option of engaging additional personnel can sometimes be viewed within a company as an unwarranted expense, whereby these personnel could be more profitably engaged elsewhere. As with other potential improvement techniques, it can be desirable to undertake a simple financial justification (see Chapter 4). In most cases, in our experience, the case for using more people to conduct a changeover is surprisingly strong.

8.3 Reallocate tasks so that they are 'compacted' together better

Figures 5.13 and 5.14 describe this technique. Time is saved by reducing operator 'waiting' time during the set-up and run-up phases of the changeover (where a task is not available for that particular operator to conduct). Waiting time can be reduced by changing when tasks start.

8.4 Breaking task interdependencies (by organizational change)

This potentially important technique should usually be viewed as an integral part of the three techniques described immediately above (where it was in fact described terms of availability of equipment and personnel). It is discussed separately here for clarity. In our experience it is a technique that is not always fully appreciated. Both understanding the impact of task interdependencies and identifying possible opportunities to break task interdependencies can be substantially helped by building a reference changeover model. This section presents a detailed explanation of the technique, which essentially seeks to allow the greatest possible utilization of all existing resources.

By task interdependency we refer to constraints imposed by other changeover tasks on any identified changeover task, limiting when this identified task can be conducted (at particular times in the overall changeover procedure). Before many changeover tasks can start, for example, other tasks have to have been completed first. This, though, as we describe, is not the only way in which the requirements of other changeover tasks can impose limitations.

In practice, opportunities to break task interdependencies can arise both through organizational improvement and by design change. For completeness, all opportunities are summarized now (the particular contribution that design can make will be explained in detail in Chapter 9).

A hypothetical changeover example presented below is set out as a reference changeover.[10] A real-life changeover is also revisited from Chapter 7. Like the hypothetical example, this latter example demonstrates the technique's potential impact. It too adopts the reference changeover model format.

How breaking task interdependencies can aid changeover performance

Task interdependencies restrict changing the order in which tasks are conducted (when they can commence). This in turn imposes limitations on the way that existing resources can be utilized.

In an ideal (but almost certainly unattainable) situation it would be possible to complete any changeover task entirely independently of any other changeover task. If these conditions existed it would then be possible to conduct changeover tasks in any chosen order. Given that sufficient resources are also available, notably personnel, it would also be possible, under these hypothetical circumstances, to conduct any number of changeover tasks in parallel.*

The improvement technique of breaking task interdependencies is that of overcoming – at least to an extent – restrictions that limit task reallocation. Breaking task interdependencies might allow, for example, some changeover tasks to be conducted in external time where previously this was not possible. Breaking task interdependencies can also allow significantly better task distribution between limited available resources during the set-up and run-up stages of the changeover. The same man-hours of work will need to be expended (the changeover tasks would remain essentially the same). In the latter case this same effort, by better use of resources, would be concentrated within a shorter period. Thus a shorter elapsed changeover time would be experienced.

It is useful to view breaking task interdependencies (to be able to reallocate changeover tasks) as a distinct and separate option to reallocating resources. The perspective here is that of matching changeover tasks to existing, limited resources – rather than matching resources to the changeover tasks.†

Some possible restrictions to conducting changeover tasks freely, at any chosen time

Different restrictions that might prevent freely reallocating individual changeover tasks may be assessed. These include:

- **Sequence requirements** Put simply, some tasks require other tasks to have been completed before they can commence. As an example, checking adjustments by trial production can occur. The adjustments need to have been completed first.

---

* A point is being made by citing an exaggerated situation. The situation as described is also probably wholly unrealistic because no account is being taken of likely external time activity.

† Having made this distinction, it should be appreciated that resources and tasks need to be matched together in an optimum way that minimizes changeover time. Thus, it is best to consider both reallocating tasks to resources and reallocating resources to tasks simultaneously. Resources should be as fully utilized as possible (where periods of resource idleness can extend a changeover). Or, where there is absence of resource to conduct a task, resources might be increased. Resource utilization is likely to be particularly low in external time.

- **Safety** Safety can be an issue. For example, prevented access to areas in which testing is occurring might delay other tasks in that area from taking place.
- **Interdependency brought about by limited resources (general)** Without exception changeover tasks require certain resources to be available to allow their completion. These resources might include electrical or hydraulic power sources. More significantly, changeover operatives will probably need to be available to conduct the task. Other necessary resources might include hand tools of some description, and space in which to work. For any changeover task to be completed satisfactorily, all the resources that it requires need to be available. Tasks cannot be resequenced if resources are employed in conducting other tasks.
- **Manpower: restrictions brought about by limited availability of labour** One person usually will only be able to conduct one changeover task at a time. Changeover tasks that require two or more people to be present (perhaps lifting or aligning tasks) mean that these people cannot be engaged elsewhere.
- **Manpower: restrictions brought about by limited skill availability** As we have noted previously, 'skill' might involve physical strength, legal authority or manual dexterity. More usually 'skill' will represent a heightened ability to conduct tasks on the basis of training and/or experience. Whenever changeover skill is concentrated only in the hands of certain personnel, this can have a detrimental impact on the changeover – because only specific personnel will be able to conduct particular changeover tasks.
- **Demarcation restrictions** Sometimes demarcation restricts who is permitted to conduct specific tasks. The consequences can be exactly the same as have been described immediately above in relation to 'skill'.
- **Physical (space) restrictions** Often there is only limited equipment access during changeover. It can be physically impossible for more than one person to conduct different tasks for which this limited access is jointly afforded. Thus these tasks can only be completed separately.
- **Tool restrictions** Different changeover tasks might require the use of an identical tool. These tasks cannot be completed in parallel if only one such tool is available.

Perceptions of task independence/task interdependence

The technique of breaking task interdependence has as its ultimate objective making changeover tasks genuinely independent of one another (with the objective in turn of commencing these tasks at an alternative time). This will often be impossible, but this is not to say that changeover tasks cannot usefully be made *more* independent.

The extent by which task interdependence can be alleviated is often not fully appreciated. Some of the restrictions set out above might, at first sight, seem to be difficult to overcome. Given a will to do so, however, substantial change can often be possible.

Any one of the possible restrictions suggested above might apply. Or, more than one restriction might apply in unison. Each restriction on its own limits the scope to reallocate changeover tasks. Providing additional personnel, for example, can sometimes only provide a comparatively limited reduction of changeover time – because only one restriction is being addressed.

An example of the impact of breaking task interdependencies

This first example is hypothetical. It is presented to illustrate points made in this section. The example indicates the potential of reallocating tasks while simultaneously reallocating (if necessary increasing) resources.

The example is presented in Figures 8.2 to 8.6. The changeover is successively modelled as changes are considered. One possible format for a reference changeover model is employed, whereby the two operators' tasks are linearly described. First, in Figure 8.2, the tasks that require to be conducted are set out. Their reference performance times are presented (see Chapter 7). Two people are available to conduct the changeover. Only one of these two people is a senior technician – who is highly skilled and legally qualified to set press tools. For legal and other reasons, the company has deemed that this senior technician is the only person permitted to conduct specified tasks. When it is possible to conduct only a single task at any given time tasks cannot be conducted in parallel, irrespective of two people being available), it will still be this senior technician who is at work. Only relatively unskilled jobs are conducted by the changeover assistant. The availability of the assistant does not stop the senior technician conducting unskilled tasks.

Only two individual tasks in Figure 8.2 show benefit when two people conduct them in unison. These tasks are task 'A' and task 'Q'. Both tasks are to do with manipulation of the main guard (removal and reassembly). Two people working on the guards as an organized pair significantly reduce the time that these tasks take to complete. However, two people working on a single task means that it is impossible to conduct other tasks at the same time.

For comparative purposes, Figure 8.3 shows the modelled changeover time if the senior technician alone conducted the changeover. A time of 206 minutes results, excluding run-up.

When two people conduct the changeover this time of 206 minutes is nowhere near halved. This is because significant task interdependencies exist that largely prevent free allocation of the changeover tasks between these two operators. As well as the restriction of who is permitted to conduct specific tasks, limited access (see below) is also influential, as is the need to conduct certain tasks only after other tasks have been completed.

Access problems arise because only a small space on the left-hand side of the machine is available in which to work. This limited space is used when removing the machine's side cover and when conducting internal tooling tasks through this aperture. The same space is required for an operator to stand to work upon the conveyor assembly. This space cannot be used by two people at the same time.

The two-person changeover time that is modelled, shown in Figure 8.4, is 183 minutes, excluding run-up.

Consider overcoming the interdependencies imposed by these space restrictions. One possible improvement might be to build a simple platform (a design change) that permits separate access to the conveyor assembly, thus enabling changeover tasks to be performed considerably more independently upon it. No other change takes place. With the same resources available, as indicated by Figure 8.5, a 159-minute changeover time is now possible.

Further significant restrictions to reallocating tasks are imposed by who is permitted by the company to conduct the separate changeover tasks. Legal authority to

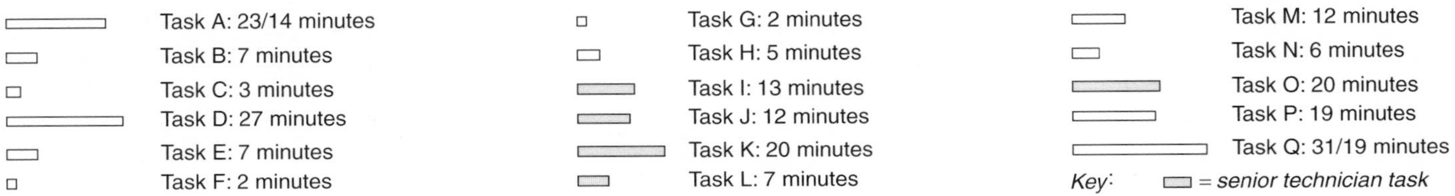

Task A: 23/14 minutes
Task B: 7 minutes
Task C: 3 minutes
Task D: 27 minutes
Task E: 7 minutes
Task F: 2 minutes
Task G: 2 minutes
Task H: 5 minutes
Task I: 13 minutes
Task J: 12 minutes
Task K: 20 minutes
Task L: 7 minutes
Task M: 12 minutes
Task N: 6 minutes
Task O: 20 minutes
Task P: 19 minutes
Task Q: 31/19 minutes

Key: ▭ = *senior technician task*

Task A: 23 minutes (one person) / 14 minutes (two people)
Remove machine guards

Task B: 7 minutes
Clean LH work area prior to dismantling commencing

Task C: 3 minutes
Remove LH cover plate

Task D: 27 minutes
Dismantle conveyor assembly

Task E: 7 minutes
Remove swarf from conveyor assembly

Task F: 2 minutes
Turn machine to TDC

Task G: 2 minutes
Loosen, and swing guide plate to one side

Task H: 5 minutes
Remove upper cutting tool

Task I: 13 minutes
Position replacement upper cutting tool on existing mounting pins
*** *task required by company to be conducted by senior technician* ***

Task J: 12 minutes
Set appropriate tool cutting stroke
*** *task required by company to be conducted by senior technician* ***

Task K: 20 minutes
Lower tool and adjust bottom shear plate to suit
*** *task required by company to be conducted by senior technician* ***

Task L: 7 minutes
By 'inching' the machine, check cutting action, and if necessary adjust cutting stroke
*** *task required by company to be conducted by senior technician* ***

Task M: 12 minutes
Position guide plate

Task N: 6 minutes
Fit new gasket and replace LH cover plate

Task O: 20 minutes
Adjust conveyor assembly throw
*** *task required by company to be conducted by senior technician* ***

Task P: 19 minutes
Reassemble conveyor assembly

Task Q: 31 minutes (one person) / 19 minutes (two people)
Reassemble guards

Figure 8.2 Changeover task descriptions, times, original sequence and restrictions (note the difficulty of conducting these tasks in external time)

Figure 8.3 One-person changeover model: 206 minutes, excluding run-up (see Figure 8.2 for task descriptions)

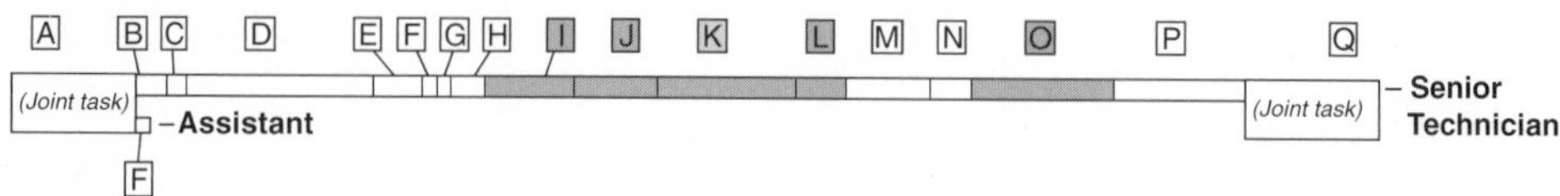

Figure 8.4 First two-person changeover model: 183 minutes, excluding run-up (see Figure 8.2 for task descriptions)

work on the press tooling is one issue. The different skill levels of the two operators are also important. A much more equal distribution of work becomes possible were these interdependencies to be relaxed. The impact of a very simple change is shown by Figure 8.6. Task 'O' – adjusting the conveyor assembly throw – does not have to be conducted by the senior technician for legal reasons. Significant benefit is gained if this task instead is conducted by the assistant. If the assistant already has the necessary skills to conduct this task, all that needs to happen for a new changeover time of 120 minutes to be possible is for the existing task demarcation, for this one task, to be removed. Alternatively, training the assistant to conduct this task also might be necessary, in which case improvement is gained by simultaneously breaking an interdependency and enhancing the manpower resource (in terms of skill) that is brought to the changeover.

Although breaking task interdependencies can be required for many tasks, there will sometimes already be considerable scope to change when a task is conducted. Task 'F' in particular – turning the machine to top dead centre (TDC) – has been substantially reallocated because of its comparatively limited interdependency. Interdependencies still apply to task 'F', but not to an extent that prevent significant reallocation of the task.

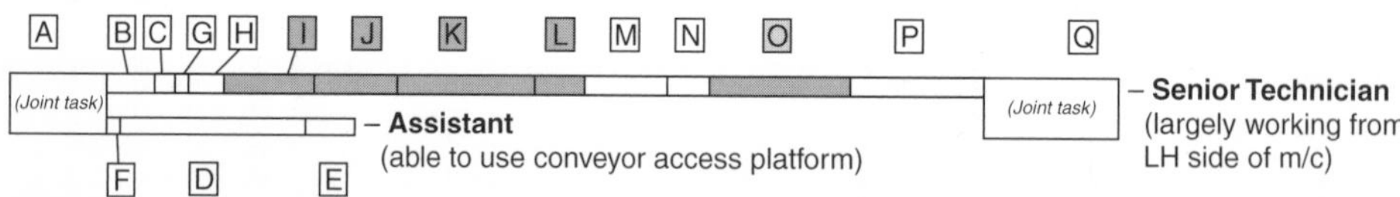

Figure 8.5 Second two-person changeover model, with separate access platform having been built for the conveyor assembly: 159 minutes, excluding run-up (see Figure 8.2 for task descriptions)

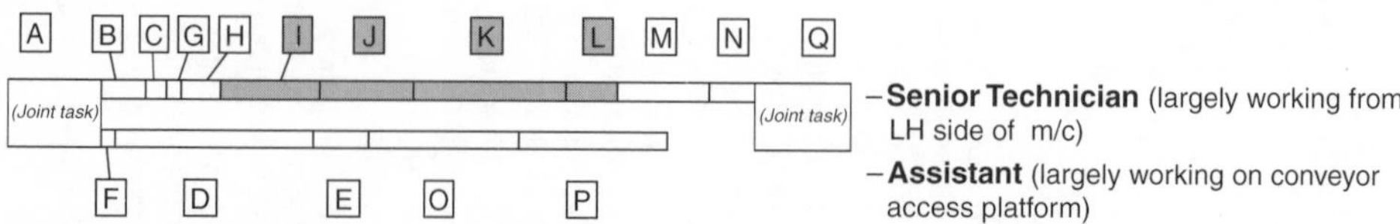

Figure 8.6 Third two-person changeover model, with separate conveyor access platform and task 'O' derestricted: 120 minutes, excluding run-up (see Figure 8.2 for task descriptions)

The improvement apparent in Figures 8.4 to 8.6 is achieved by alleviating (not necessarily fully breaking) interdependencies. This, as well, is only necessary for a few tasks: not every task needs to have its interdependencies broken. For each of these few tasks, despite alleviating a prominent interdependency, some other interdependencies still remain (for example, TDC cannot be set for this machine until after the main guard has been removed). The objective has been to distribute work fully between the two operators available to conduct the changeover. There is no benefit to be gained in breaking task interdependencies further, beyond a point that allows this to occur.

A further example of the impact of breaking task interdependencies

Figure 7.12 described a reference changeover model. Figure 8.7 considers the simple improvement for this changeover of continuing other tasks while the automatic wash cycles are executed. A time saving of 6 minutes on the elapsed changeover time would be made (according to the model) if this task interdependency was broken. This automatic task was originally being conducted separately for reasons of safety, thus a way of ensuring safe working while the wash cycles are taking place needs to be found.

To what extent should task interdependencies be broken?

As discussed in the hypothetical first example, thought should be given as to the extent that it is appropriate to break task interdependencies. Major considerations should include the equal allocation of work and the translation of tasks into external time. It should also always be assessed whether better improvement techniques exist that should be undertaken instead (that is, in preference to breaking task interdependencies and reallocating tasks).

Seeking equal allocation of work could apparently mean, for the first example, distributing one operator's workload of 206 minutes into two workloads for two operators of 103 minutes each. This is not strictly true, because having two people available to conduct certain tasks (for example, removing the main guards) changes the time that these tasks take to complete. Also, this approach ignores any advantage of reallocating tasks into external time. A better approach, typically, is to construct a reference changeover model, and then simply seek improvement opportunities from this model. In doing so all opportunities to reallocate tasks can be more easily assessed. The model can then be revised. For example, it will be readily apparent when a shared workload has been achieved – no matter what the duration of this shared work is.

8.5 Know of, and use, predefined settings

There is typically the opportunity to repeat previously used machine settings whenever a product is being remanufactured. There should be no need to re-establish manufacturing parameters from scratch. Even when a first batch of a new product

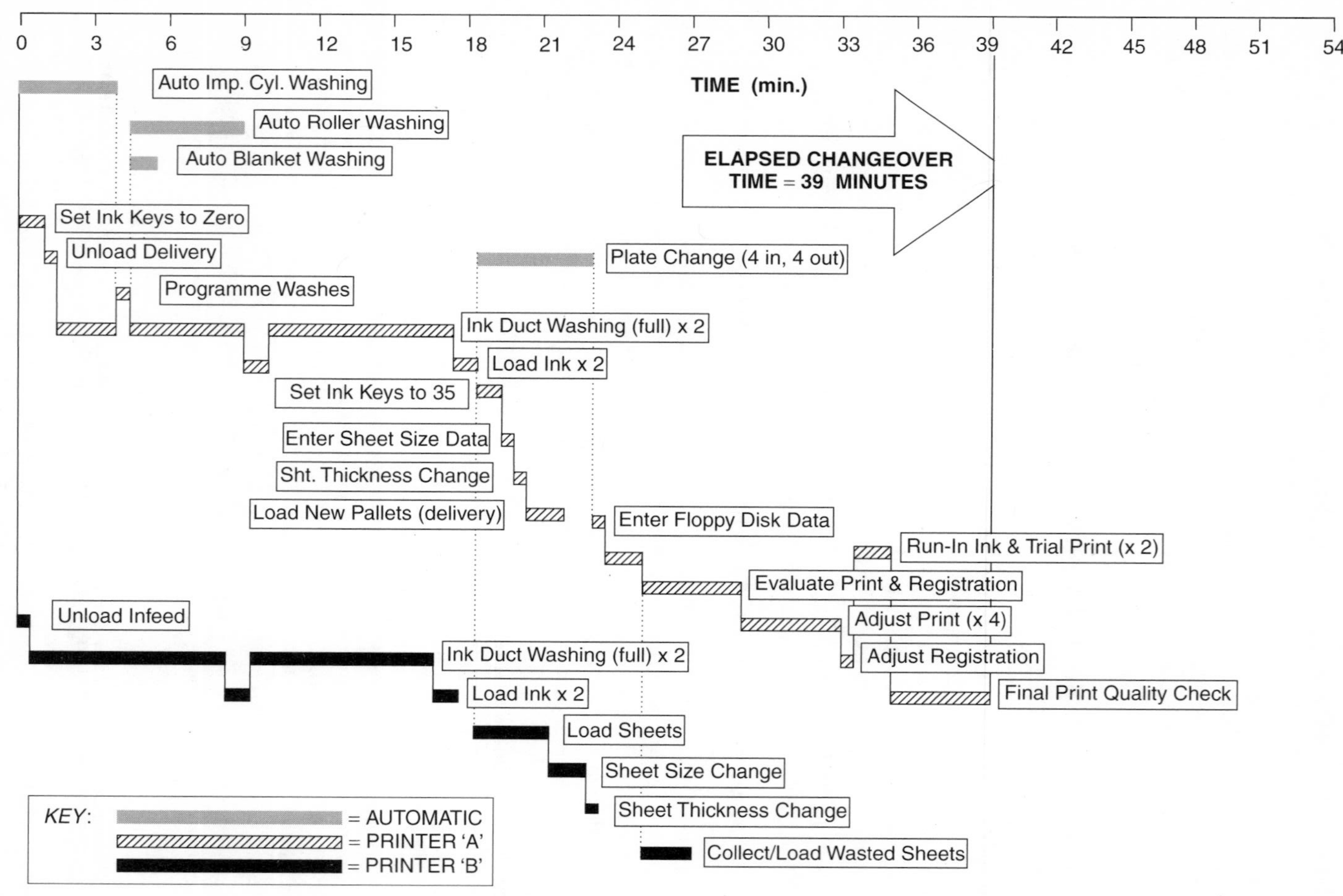

Figure 8.7 Revised four-colour/four-plate/two-person changeover, with ink duct washing now independent of auto roller washing

is being made, there might still be scope to predetermine what settings should be used. Time taken in conducting 'trial-and-error' adjustments as part of the changeover can be avoided by going straight to the required setting(s) wherever possible. At the same time, how settings are to be repeated needs to be considered. For example, the use of special setting pieces might be contemplated.

The use of predetermined settings can be rendered useless by incomplete understanding of the equipment – where all possible adjustments, and the way that adjustments interrelate (see below), have not been taken into account. Caution is therefore needed that predetermined settings fully encompass all the parameters necessary for manufacture of the new product. Case studies will be presented in Chapter 10.

8.6 Define inspection parameters to signify adjustment completion

Almost certainly, it will be necessary to check that the product meets the specification that is required of it. Inspection may be necessary at different stages during the manufacturing process. Exactly what is being inspected should be known, as should the tolerances that apply. It also has to be understood what machine parameters need to be changed (adjusted) to bring identified product parameters within specification. With complex equipment this is sometimes not at all easy.

Tools should be considered to aid making these inspections as swiftly as possible and as precisely as is necessary. Sometimes simple gauges, for example simple graduated rules, can be employed. In other cases more specialist gauges might be used, including those that are designed specially for the task in question. In this respect the use of 'go/no go' gauges can be particularly advantageous.

As noted, product inspection need not only occur as the product emerges at the end of a machine or production line. Interim inspection stages can also be necessary. Again it should be known where adjustment needs to be made, and by how much, to achieve any desired change in the parameter that is being inspected. Note also that the improvement practitioner can have the option of *increasing* the number of inspections that are made – to allow better control of the changeover (and by doing so, ultimately saving time overall).

8.7 Understand possible adjustment interrelationships

As alluded to above, there can be a degree of interrelationship between adjustments. For example, more than one adjustment might be available that moves an assembly or component in an x-plane direction. Taking great care to make one adjustment well, when a subsequent adjustment then causes the position to be lost, is poor changeover practice. Interrelationships between adjustments should be understood (again, this is not always easy). An overall adjustment procedure should be developed accordingly. Often some – sometimes many – possible adjustments can simply be pinned or otherwise fixed in place, thereby eliminating them from the changeover.

8.8 Address precision requirements (by organizational change)

This technique requires that precision requirements are matched to the changeover task in hand, commensurate with achieving a final product of satisfactory quality. The principal option here is to alter the precision of manual adjustments (that is, alter the tolerance that applies). If lower precision can be permitted, adjustments typically can be made more quickly. For the benefit of the changeover as a whole, for selected adjustments, it can sometimes be advantageous to increase the precision that applies.

8.9 Seek to avoid damage

Damage to either change parts or fixed machine elements can mean that the changeover – the rapid, accurate achievement of new product manufacturing conditions – will be compromised. Damage can be apparent as sudden condition change or as wear. This can make existing changeover tasks more difficult to complete or can mean that additional adjustment steps have to be undertaken. In some cases damage can prevent a changeover being concluded satisfactorily. Examples have been given previously in this book.

Seeking to avoid component damage means handling/manipulating components during changeover in such a way that the likelihood of damage is minimized. Procedural changes can often be employed to satisfy this objective. A simple example would be to ensure that two plane surfaces between different changeover items are always well lubricated in advance of sliding them across one another. Numerous other possibilities will exist.

8.10 Identifying, agreeing and implementing standard procedures

This improvement technique seeks to ensure that everyone who conducts a changeover does so in an identical way.[11] As far as it can be determined, this needs to be the best way possible. In this manner time-consuming poor practice can be eliminated.

A reference changeover model is a model of best practice, within the confines of the existing manufacturing system. A reference changeover model should thus be the basis of documented standard changeover procedures, which should set out who is to do what, when and how. Standard performance times for each separate task will be written down. It will be understood what sequence these tasks are to be conducted in, and by whom. Training and practice will reinforce this standard way of working.

Strictly a standard changeover procedure should apply only to the type of changeover for which it was derived. As circumstances change, for example when equipment design changes are made, the changeover procedure should be revised accordingly. Thus in times of rapid and continual improvement it can be inadvisable to concentrate upon refining procedures and enforcing their adoption. Rather, standard changeover procedures should come more into play to cement best practice once other changes (improvements) have largely been completed.

8.11 Prior checking that required changeover items are present

A changeover will be conducted more efficiently if the items that are to be used are in their anticipated locations in advance. It might be planned to conduct these preparatory tasks in external time – but there can still be merit in subsequently checking this has been done. It makes sense that checking occurs with sufficient time available to rectify any omissions that are found.

Checking item availability can be done by moving around the different locations where these items should be placed using a checklist. A different way of checking is described in Chapter 10 as a case study, where all necessary items were deliberately placed on a trolley within eyesight of the operators. Aids such as a shadow board or careful identification can also aid the checking task, thereby helping determine that the right item is in position (for example, the correct die set, rather than a look-alike die set). Colour coding can be a useful option in this regard.

Checking should include the availability of tools. Similarly, attention should be given to the data that are to be used and, for all changeover items, if necessary, the format in which they are presented.

As intimated with reference to changeover trolleys (above), consideration will also be warranted as to where all changeover items are to be located. It is one thing to have everything neatly racked in an anteroom ready for use. It will almost certainly be more useful, however, to locate items adjacent to their point of use.

A final point can be made in respect of who is to do the checking. Those who are responsible for placing changeover items ready for use can be those who also – at the same time – confirm their availability. Check lists can be used for this purpose (which, once completed, are made available to the changeover operatives). However, using check lists in this way can be unwieldy as considerable control is typically required. As the case study in Chapter 10 will show, such use of check lists is not the only option to avoid undertaking checking as a later, separate, operation.

8.12 Prior checking of item quality standards

As explained in detail previously, variation from a sufficient standard of item quality is highly likely to detract from a changeover. Like checking that items are present, it can be highly beneficial to check, in advance of their use, that changeover items are of sufficient quality. This technique's contribution once again is to ensure smooth, rapid progress of the changeover once the line has been stopped. Many of the observations made above in respect of changeover item availability also apply when verifying changeover item quality.

8.13 Batch sequencing

The reader is referred to coverage of this topic within Chapter 3. A shopfloor team should be aware of this technique, but will not necessarily be in a position to bring about change.

8.14 Eliminating superfluous activity

If a specific task is not warranted it can simply be eliminated, without altering the changeover in any other way. Eliminating superfluous tasks will almost always be a 'no-cost' option. The objective is that operator activity that contributes nothing to the changeover should be dispensed with. What constitutes a superfluous task, however, as is described below, need not always be immediately apparent. As elsewhere, improvement options might be more easily identified when a flexible perception of a changeover task is adopted (Chapter 5).

Consider walking away to search for and retrieve a hand tool to release a clamp. This is inefficient working practice, where simply picking up the hand tool is all that needs to take place. Permanently siting this tool in a convenient, fixed location would make it possible to conduct this task in a greatly more efficient manner. Alternatively, however, looked at in another way, the 'walking' element of retrieving the hand tool can be perceived as a task in its own right. It is a task that can be eliminated.

Some 'tasks' are much easier to identify as being superfluous. Taking a coffee break is a simple example. It is activity that can occur during the changeover, but which contributes nothing to it, save to delay it. Often such breaks would be more viably conducted during periods of normal manufacturing. Similarly, we have observed that paperwork is often completed during the changeover period. Often this could be done once the line was back in full production.

Another instance of superfluous activity might be setting line elements that do not in any way affect the new product. An example might be setting a series of guide rails when, for the particular product in question, not all of these guide rails are necessary. Alternatively, a final manual cleansing step, applying a solvent and using a rag, might occur for all colours when changing ink in an ink duct in a print operation. This might be wholly unnecessary when changing over, say, from yellow to deep red.

8.15 Address problems that arise

'Problems' arise during many changeovers. By this we mean that there are events that delay the changeover that were not anticipated when the changeover started. What is important here is that these problems – these unexpected events – are recorded. Attention can then be given to ensuring that they do not recur.

As an example, engaging the services of a fork-lift truck driver can often be necessary at certain times during a changeover. Involving this facility at the time it is required can sometimes be difficult. In other words a problem is encountered – waiting for this individual's services. Chapter 10 describes how this problem was addressed in one particular instance – with the hope that it could be avoided in the future.

Other typical problems might concern parts that are missing, incomplete, incorrect or damaged. Similarly, information could be missing or unclear. Alternatively, skilled personnel might be required to break away from the changeover to address a machine breakdown elsewhere in the factory. Whatever the problem, there is typically an opportunity to ensure that it does not happen again.

8.16 Conducting individual changeover tasks efficiently

Seeking 'efficient' work practice can have significant overlap with other possible improvement techniques. As noted in Chapter 7, this is of no consequence. Rather, an opportunity exists to assess potential improvement options from a different perspective.

Working more efficiently is not the same as working harder. Quite the opposite: working more efficiently might be regarded as executing a changeover task in the easiest possible way. The question is, how is this improvement to be achieved? There are perhaps no generally applicable rules, but some guidelines can be set out.

What is being sought is refinement of the task, where all its current 'slack' is to be cut out. Attention needs to be given to the way that time is spent conducting the task. For example, what proportion of time is spent moving unnecessarily? Is time being spent interpreting information, or seeking information related to the task? Is it clear what the task involves? Have all aspects of preparation prior to the task commencing been attended to properly? Are tools to hand and in good condition? Are the optimum tools and devices being employed (see also below)? Often, many opportunities for improvement can be identified.*

Another consideration is the integration of tasks such as preheating. This is efficiency gained by careful time planning – to ensure that other tasks are able to occur on time in their planned sequence. Such tasks involve a delay, and if possible should be conducted, or at least commenced, in external time. The use of parallel facilities for this purpose might also be considered. Finally, for permanent components such as UV curing lamps, benefit might also be possible by adopting an intermediate stage temperature (turning UV lamps down when not in use, or during changeover, rather than turning them fully off).

8.17 Employing the best tools/handling/storage aids

A diverse range of tools and other aids is available to be purchased, for example handling aids or storage aids. For press tool systems, for example, devices to assist changeover will feature prominently in press ancillary exhibitions. These aids might range from simple manual clamping devices, through to die set lifters or specialist equipment to enable die presetting – prior to rapid substitution within a corresponding specialist die set.

---

* These considerations together apply when deriving reference task performance times (Chapter 7).

To be able to make the most of these opportunities the changeover practitioner needs to be aware of what aids or devices are available. This should be done in general terms, in addition to knowledge of changeover aids available specifically for the process in question.

8.18 Ensuring excellent identification

Excellent identification assists good changeover practice. In doing so it saves time. Excellent identification should apply to all items involved in a changeover. This will include items such as written procedures, as well as more obvious items such as change parts, consumables and tools. Identification can be by written labels or signs. Colour coding might also be employed. Whatever method is chosen, the improvement team should be seeking absolute clarity in what is presented. Item selection mistakes, ultimately, should become near-impossible to make.

Consider an example of a production line where operations occur at successive stations. These successive stations are identical except for the tooling that they employ. In such cases it becomes essential to mount the correct tooling into the correct station. Station 1 could be painted, for example, blue. Other colours could be chosen for the four remaining stations. All tooling that is to be used in station 1 likewise could be painted blue. A changeover trolley might be used, upon which the change parts are loaded. With five different colour sets of tools on the trolley it is immediately clear what tooling is to be picked up for use in each successive station.

This, though, does not mean that in preparing the trolley the correct blue tool has been picked up and placed upon it, in preference to alternative blue tooling (for a sister product) that is also available. To overcome this potential problem yet further thought will need to be given to distinguish the various station 1 (blue) tooling. Careful racking, labelling or other options are possible. Perhaps, for example, a secondary colour stripe could be added to distinguish 'blue station' tools from one another. What ever method is chosen, lack of ambiguity should always be striven for.

Colour coding can be used in other circumstances. A particular application can be the painting of spanners or sockets that match similarly coloured bolts on the machine. For example, 13 mm bolt heads could be painted green, as could all corresponding 13 mm sockets and spanners.

Despite care in making identification as clear and as unambiguous as possible, errors might still be found to be occurring. Particularly in these circumstances attention might also be given to 'foolproofing' the assembly of components, such that it becomes physically impossible (by design) to mount components that do not belong with one another (see Chapter 9).

8.19 Ensuring cleanliness

Reiterating what has been written previously: changeover items that are clean are easier to identify; to handle; to locate precisely; to work upon. It is good practice in any case to maintain a clean working environment. Cleanliness indicates discipline. It can make a job easier, more pleasant and quicker to complete. For example, crawling about in a mess of oily swarf might be avoided. Alternatively, a bolt that has just been dropped will probably be much more easily retrieved. Numerous small contributions to the time a changeover takes to complete can be made.

These, though, are perhaps not the major reasons for attention to cleanliness. In many environments there is a real danger of contamination of changeover items. By this mechanism both the quality and the duration of the changeover can be affected. An example would be swarf being caught below a die set when it is loaded into a press. The die set will not sit correctly and any manner of problems might arise in achieving the output quality that is desired.[12] Unexpected adjustments may well occur while an attempt is made to compensate for the problem. As another example, in the printing industry, ink contamination might occur. Ink contamination might ultimately lead to the changeover being aborted because the printed colour does not sufficiently match what the customer requires.

A requirement of high standards of cleanliness can be written into changeover procedures. Achieving cleanliness, for many components, or for much of the work space, can often be completed in external time.

In some circumstances cleanliness can be essential for safety reasons. For some food production lines, for example, very exacting standards of cleanliness are required.

8.20 Conclusions

The described techniques cover all potential organizational improvement opportunities of which we are aware. The improvement practitioner should take strict care to avoid preconceptions that some techniques are either better or more important than others. Given the diversity of changeover situations, every potential improvement opportunity should be assessed on merit. Moreover, any assessment of potential opportunities should include those with both an organization bias and a design bias. Potential design-biased techniques are described next.

References

1 Schonberger, R. J. (1990). *Building a Chain of Customers*. Free Press, New York. This is a frequently made point, particularly by consultants/academics approaching changeover improvement from a management (as opposed to an engineering) standpoint. While true, it can obscure a wider picture, in which the role of design, including low cost design, is more fully considered.

2 Zunker, G. (1995). Fifty percent reduction in changeover without capital expenditures. *PMA Technical Symposium Proceedings for the Metal Forming Industry*, pp. 465–476. Zunker's paper provides a good example.

3 Shingo, S. (1985). *A Revolution in Manufacturing: The SMED system*. Productivity Press Inc., Massachusetts.

4 Shingo, S. (1985). *Op. cit.* ref. 3. This example is made in detail in Shingo's book.

5 Mather, H. (1992). Improving your changeover performance. IMechE seminar, Birmingham, UK. Mather is one of many commentators who commends the use of parallel equipment.

6 Hay, E. J. (1989). Driving down downtime. *Manufacturing Engineering*, **103**, No. 3, 41–44. Likewise, Hay is clear that providing parallel equipment can be highly beneficial to changeover performance.

7 Shingo, S. (1985). *Op. cit.* ref. 3. Shingo distinguishes 'separating' and 'converting' changeover tasks into external time to reduce the elapsed changeover time. We do not propose an improvement technique that is directly analogous to 'converting' changeover tasks. Rather, we describe the reallocation of resources and the breaking of task interdependencies.

8 Anon. (1983). Automatic tool change system for car component production. *Sheet Metal Industry*, **60**, No. 2, 93–94. This example relates to Tallent Engineering, suppliers of automotive products, who set up two parallel lines side by side.

9 Shingo, S. (1985). *Op. cit.* ref. 3. The potential importance of this technique is noted by Shingo, who acknowledges that, in some circumstances, two people working in parallel can reduce changeover time by more than 50 per cent. It is revealing to contrast this improvement – for a technique within the 'SMED' methodology's final 'streamlining' phase – with the changeover time reduction in that it is often expected of the methodology's first two concepts ('separate' and 'convert' tasks into external time).

10 McIntosh, R. I., Culley, S. J., Mileham, A. R. and Owen, G. W. Reinterpreting Shingo's 'SMED' methodology (paper in preparation). The hypothetical example in the main text highlights features exhibited by real changeovers. Two changeovers can be consulted. The first changeover is that described by the reference changeover model of Figure 7.12. A second real changeover is described in this paper.

11 McIntosh, R. I. (1998). *The Impact of Innovative Design on Fast Tool Change methodologies*. PhD thesis, University of Bath. The 'SMED' methodology describes standardizing features of equipment but it does not emphasize standardizing work practices. Neither is this aspect strongly featured in most other changeover improvement texts. Nevertheless, one company that we have worked with was particularly emphatic in ensuring that standard changeover procedures are always maintained, displayed and used. In its internal changeover training material the company writes: 'By the end of the ["SMED"] workshop a new [standard procedure] must have been developed and agreed by people on the shopfloor. This is one of the most important points, if not the most important.'

12 Smith, D. A. (1991). *Quick Die Change*. SME, Dearborn. Smith describes this particular situation in some detail.

9

Design-led improvement techniques

Alongside organizational improvement techniques, the changeover practitioner should also be aware of a full range of techniques that emphasize equipment or product modification. The description below of potential design-led techniques again coincides with the classification made within the matrix methodology.

As before, there will be seen to be some overlap between the techniques that are described. Considered in another way, some specific improvement actions can benefit the changeover simultaneously in a number of different ways. Also, some techniques described in this chapter can require that other changes, employing other techniques, are undertaken at the same time if they are to make a worthwhile contribution.

Some specific examples will be given, to help convey what specific techniques seek to achieve. The reader should be aware (see also Chapter 7) that the few examples that are presented represent but a tiny fraction of what can be done.

9.1 Break task interdependencies (by design change)

The description below follows on from the explanation of breaking or alleviating task interdependencies in the previous chapter.

Design change can be applied in a wide variety of ways. Consider Figure 7.12, which presented a reference changeover model. Part of this overall changeover involves the printing plates, wherein new plates are automatically loaded into the machine. Loading takes approximately 1 ½ minutes per plate. This particular equipment is designed such that all the print decks are mechanically coupled to one another, including the plate loading mechanisms. The design is such that new plate loading has to occur consecutively, with the main drive rotating slightly between the loading of successive plates. Interpreting loading each plate in turn as a separate task, the loading of, for example, the third plate is dependent on the second plate having previously been loaded. Were this solid coupling to be disconnected, all plates potentially could be loaded simultaneously.[1]* Parallel loading of printing plates can save 7 ½ minutes from the total elapsed changeover time when six plates are to be

* This is not necessarily a simple retrospective mechanical change to make. If instead it had been designed as part of the original machine, this feature could perhaps have been much more easily provided.

loaded. The extent by which interdependency relating to this task has been alleviated is small: simply the plates can now be loaded in parallel. All other interdependencies relating to this plate loading task are essentially exactly as before.

Another option, which typically also requires new parallel equipment (increasing resources), is to design for presetting – in external time – to occur. For example, preset CNC tools (for example, drills or milling cutters) are widely employed. Many tool changes can be required during a typical CNC machining cycle. These tools have already been set in external time. The preset tool is quickly, accurately and securely loaded, in the larger module of which it is a part.

In another example that we have witnessed, presetting again aided the changeover, this time for a steel shearing device. Cutter elements were mounted to the top and bottom shoes of a die. Previously the die itself was permanently attached to the shearing machine – during changeover it was only the cutter elements that were exchanged. This was a difficult operation, where the cutters had to be set with considerable precision if they were both to remain undamaged and yet cut satisfactorily. An improvement was devised whereby a separate, fully prepared die set (on which the cutters were set away from the machine) was mounted instead. The design provided for this die set to be loaded, aligned and secured with ease. Here, by breaking task interdependencies, the complex and time-consuming cutter setting tasks no longer needed to occur while the line was stationary: the time at which this task could be conducted was changed.

Setting tasks, in reality, simply represent a block of changeover tasks, as part of the overall changeover. It can be assessed, for example, how making equipment more modular – as is here done in both the CNC tooling and die set examples above – might allow any block of tasks to be reallocated.* The fact that a particular assembly has always been manipulated as a complete entity in the past does not restrict future manipulation of that assembly as a series of modules.

Design to alleviate task interdependencies can also often be applied in terms of guards, or to overcome problems of access (see below).

9.2 Automate adjustment

Automating adjustment – a specific '*mechanization*' option (below) – is highlighted because it can be particularly beneficial in situations where access is a problem. Major time savings can result by eliminating extensive dismantling that otherwise would have to take place.

9.3 Address precision requirements (by design change)

Precision is an issue in respect of making adjustments during changeover. Design potentially offers significant scope for improvement. Two approaches might be adopted:

---

* This topic of breaking down the machine in different elements is picked up again in the later section 'Separate/combine items'. In this later section the focus is on making tasks easier to conduct, rather than, as here, being used to alleviate restrictions as to when tasks can be conducted.

- seek to improve existing adjustments (that is, continue to use the same individual adjustments as before but do so in a better way)
- in general terms, reappraise where, why and how adjustments occur

Improve existing adjustments

The practitioner typically will be seeking one or more of: adjusting with greater ease; adjusting with greater control; adjusting with greater speed. Attention will usually be given to how condition change is undertaken (for example, moving parts) and/or how condition monitoring is undertaken (for example, determining position).

Many different improvement options exist. For example, relative movement between parts might be aided by employing separate runners, or, perhaps, a screw thread might be used to enhance positional control. Similarly, a scale might be introduced to indicate position. Again, the desirability of knowing real applications at the 'examples' level of the matrix of Figure 7.4 is reinforced – something our proposed catalogue of improvement ideas (see Preface) would address.

Reappraising where, why and how adjustments occur

Alternatively the practitioner might move from considering individual adjustments. A more radical assessment can be made, where the design of the existing system as a whole is questioned. For example, are more adjustments available than are strictly necessary to set changeover items into their final position? How do separate adjustments interact with one another? Are numerous smaller changeover components being positioned, when just one or two large components (for example, large preset modules) might be substituted and adjusted instead? Is the best use being made of datums, to facilitate swift and precise adjustment? Can adjustment steps be eliminated, instead being replaced by installation of a foolproof 'fit-and-forget' assembly? Is there an avoidable build-up of positional tolerance? Can pads or other special features be incorporated that will assist in either making or monitoring adjustment? Can sensitivity to variable product quality (damaged or otherwise out-of-specification) be reduced? Often, by implementing such changes, some previously required adjustment stages can be eliminated altogether from the changeover.

Further thoughts

It can be particularly useful to substitute an infinitely variable mechanism with one that incorporates preset positions. For example, a notched locking bar might be incorporated, allowing adjustment only between fixed positions. If carefully designed, high precision can be achieved comparatively cheaply. Alternatively, fixed-length location pieces might be incorporated. Usually, adopting such an approach, the easiest mechanisms of all are those that switch between just two preset positions. Another option can be to use fixed setting pieces (where setting pieces represent preset components that are withdrawn once used).

In addressing adjustment, it is not only the speed of the changeover that can be impacted upon. The practitioner is also wise to remember the contribution that the above improvements may make to the changeover's quality.

9.4 Consider using monitoring equipment

Continuing our discussion, it is often useful to reflect that adjustment tasks typically will comprise separate elements. As noted, two core activities are altering condition and determining what the new condition is (to signify adjustment is complete). If deficient, each of these separate elements might be improved. For example, the position of a slide may be read off an existing scale quite satisfactorily, but, because of excessive friction, the slide tends to jerk when any attempt is made to move it. In such cases, perhaps, a better lubrication system might be employed, or a threaded adjustment mechanism used. In this way just the deficient element of the overall adjustment task is being addressed. Equally, the slide might instead move freely and easily, but its position is difficult to determine when using the existing scale. In such cases a more sophisticated position monitoring system might be attached and used. Once again, design attention is given only to that part of the overall adjustment task where particular improvement is called for.

Adjusting changeover components should be viewed as a separate activity to handling or moving components roughly into position.

9.5 Make items more robust (less prone to damage or wear)

Less damage to changeover components can be sought either through improved procedures (retaining the same hardware) or, by design change, making changeover items more robust.[2–4] Design improvement might be sought by making critical areas of changeover items more resistant to damage, or by reducing exposure to potential damage situations.

One possibility is to increase a component's material specification by substituting materials, where applicable, that are more resilient. For example, a cast component might have case-hardened steel inserts added in those locations where it is known to experience wear. Alternatively, mild steel surfaces of a component that occasionally experiences swarf trapped beneath it could also be upgraded (not to prevent misalignment during a changeover in which swarf has become trapped, but rather to prevent damage that could affect all subsequent changeovers).

Tool-aligning pins can become worn, making it difficult to set tooling satisfactorily, hence losing time during the changeover. Improvement of these and other tool-aligning features might be contemplated.

In cases where precision location features, particularly when made of 'soft' material, occur in exposed positions, there is always a danger that these features will become damaged during routine component handling. Disregarding changing material specification, another option would be for these features to be relocated, by design, into less exposed areas.

Finally, consideration might sometimes be given to materials that comprise working surfaces (as opposed to location surfaces). The design of a component in general terms dictates the changeover procedure that has to be employed. This can extend to the materials that are specified for the working surfaces. A case in point is the use of carbide cutting edges as opposed to tool steel cutting edges. Although offering many potential advantages, carbide cutters are notoriously brittle, and this can be a major consideration during changeover when the cutters are set.

9.6 Upgrade changeover item quality (secondary design)

Chapter 5 highlighted that improvement can sometimes usefully be sought on other equipment, remote from the equipment undergoing changeover (secondary design). The quality, or specification, of items supplied for the changeover might be improved – which can have a considerable effect upon adjustment activity during changeover (Chapter 3).

9.7 Act upon standard change part features

Moving from micro-variability of components, there is also often considerable advantage in standardizing identified features of change parts. The potential of this technique might be appreciated by considering standard domestic electrical sockets. Were they not available, a crude, time-consuming alternative would be necessary: disabling the mains supply and manually rewiring each different appliance (change part) into position. Or take a simple automotive example. A particular vehicle might be specified with either steel road wheels or alloy wheels. Changeover between the manufacture of these different products (the vehicle with either steel or alloy wheels) will be very considerably eased if – as happens – each wheel type employs a standard number of securing bolts on a standard PCD (pitch circle diameter).

In more general industrial terms, if a substitute change part has entirely different mounting features, then additional changeover tasks will be involved in accommodating these differences. Entirely different mechanisms for securing the part might need to be employed, or substantial work might be involved in adapting existing mechanisms. If, conversely, the same securing features applied throughout, then comparatively less changeover work is likely. For example, pedestal clamps might be employed to secure a change part. This part is to be exchanged. If these same clamps can be used on equivalent features on the replacement part, all that might be necessary would be to loosen/tighten the clamp nut by half a turn, and to swing the clamp arm by 90 degrees. This is vastly less work than removing the pedestal clamps and then rebuilding them in a different location, to match alternative features of the new change part.

If seeking to apply this improvement technique, the practitioner should observe what features of each and every part are employed during the changeover. Where equivalence does not exist between these features, the practitioner should assess whether advantage might be gained, if possible, by making them so. An oft-cited example is the equivalence of key die set dimensions, to enable die sets to be exchanged much more easily and much more quickly.[5]

One particular way of standardizing features can be to standardize the equipment that is used. Although potentially expensive, in some cases this would still represent a more viable alternative to retrospectively changing features of existing, non-standard equipment. Returning to the die set example, a common die set (from one manufacturer's range) could be used as the basis for all change tooling used in a nominated press. Perhaps, as well, a series of identical presses could be used (again, from one manufacturer's range).

9.8 Act upon standard product features

Standardization need not apply only to machine components. Product standardization similarly can significantly benefit changeover performance.

We visited one factory where a mark had been introduced at a standard position on a toothpaste tube's printed surface. The same mark was applied for each of the different tubes that were being manufactured (both for different sizes and for different print designs). This was a reference mark for subsequent operations. As it was now in a standard position, changeover tasks for these subsequent operations became greatly simplified – and hence reduced in duration.

A similar example was witnessed at an automotive factory. A range of cast engine blocks was being machined. A tapped hole in certain blocks was made at a non-standard angle. Subsequent standardization upon a single angle throughout the product range wholly eliminated a significant body of tasks when changeover between different batches of engine blocks occurred.[6]

An account by Kobe[7] indicates the thought given by Honda to designing a car to facilitate assembly line changeover between left-hand and right-hand drive models. Stated design goals were for there to be as few as possible non-standard parts between these two products (the LH and RH drive cars), and to make part location as common as possible. It is reported that 'the firewall is the only unique stamping for a RH drive car'. Elsewhere a common visible example of automotive design that eliminates changeover tasks between RH and LH drive vehicles is a single, centrally mounted windscreen wiper. Although such ideas are not necessarily easy to adopt retrospectively by a site improvement team, such options – if conceived – might at least be suggested to product designers for possible future use.

The use of product 'carriers'

A potentially highly useful option can be included under this topic of product standardization. It need not be necessary to standardize product features *per se*; the same effect might be achieved by placing the product in a carrier. This carrier is specific to its product, but shares common features with other members of its family of carriers (which separately relate to other products from the product range). An example of using a product carrier is shown in Exhibit 9.2, where one of a number of possible advantages of using the carrier is also discussed.

9.9 Standardize features beyond the immediate changeover equipment

This option considers applying standardization away from the items that are physically exchanged. Information can be more clearly and unambiguously conveyed, or standard devices can be used to store or transport changeover items. A case study in Chapter 10 provides some examples.

9.10 'Foolproof' location

Foolproofing component location can sometimes yield considerable time savings in terms of otherwise necessary checking, rectification or compensation (see also Chapter 5, particularly Exhibit 5.5).

The changes that need to take place to prevent incorrect location can often be very simple. In Exhibit 5.5 the use of staggered pins and corresponding clearance holes was described. Many similar options can be contrived.[8]

9.11 Add devices to aid existing tasks

Proprietary devices can be bought, or original devices might be fabricated. Usually fabricated devices will be relatively simple, for example specialist setting pieces, or location aids. This need not always be so. As the anticipated complexity of novel devices rises it can be wise to assimilate good design practice from what others have done.

An overview of proprietary devices

Consider proprietary devices, for example quick-release handles. Use of these devices eliminates specific tasks associated with using hand tools (finding the hand tools, bringing them to their point of use, employing them and then replacing them after use). Alternatively, quick-release handles both speed up and simplify the clamping of a component. The idea's application may have arisen through practitioner knowledge of the device's use in similar circumstances elsewhere.

Numerous proprietary devices are available. These range from power devices (hydraulic or electric) to manual devices (for example, clamps and catches). Magnetic devices are available. Devices that provide feedback of condition (notably position) can be bought. Ultimately single components might be added (for example, special location pins). A few useful devices might be:

- position sensors
- hydraulic tool runners
- couplings
- alignment/location pins
- hydraulic lifters
- magnetic clamps
- overhead lifting aids

An investigation of proprietary devices – to assist in-house design of original devices

Knowing what devices are available is highly advantageous. As amplified in Exhibit 9.1, a good understanding of these devices can also be helpful if trying to develop specialist in-house designs. The subject here is a quick-release device that many people will encounter every day (but one that is not necessary appreciated as such): a motor vehicle seat belt.

Exhibit 9.1 Design for changeover lessons from a motor vehicle seat belt

The design of a seat belt makes it very simple to use, facilitating, if required, rapid driver changeover. Contrast an automotive seat belt with a possible equivalent detachable device in industry for, say, an 80 kg mass.

Automotive seat belts can be considered further, determining features of the design that allow them to function so successfully. Different locking mechanisms are in use for seat belts, including those that employ latches or magnetically retained clasps. For seat belts the commodity of strength is highly important. Yet the circumstances in which high strength is needed have been assessed. Engaging a seat belt requires little force, as does disengaging the belt. No separate tools are required to complete the task. Engaging a seat belt also requires little care or thought, partly because the device's precision requirements are achieved automatically when it is used. Inertia reel seat belts have almost entirely replaced the use of static seat belts whereby, by design, automatic belt adjustment occurs, further speeding up driver changeover. The seat belt design is also essentially foolproof, where the user will immediately be aware that fastening has not been completed satisfactorily: either the belt engages or it does not – in modern designs there is no scope for partial or false engagement. Finally, again assisting in its use, the seat belt uses a single clamp point to secure the belt portions together.

Thus the apparently simple seat belt device in fact simultaneously incorporates a number of design features that together contribute to its easy and rapid use. The example indicates some features that might be sought if special-purpose devices are being designed:

- not using tools
- consideration of where forces are applied, and what the magnitude of these forces is
- single-person operation
- 'foolproofing' location
- minimizing contact/alignment/securing points or surfaces
- incorporating the above features within a low number of parts

Design rules for changeover

Attributes of good design are described in Exhibit 9.1. However, this single example will not provide a full picture. When designing more complex novel devices, the practitioner might decide to make use of the design rules of Chapter 5. To develop a design successfully the practitioner should be clear exactly what the design is seeking to do.

9.12 Modify equipment to aid existing tasks

This is another wide-ranging improvement technique. This being so, again, knowing examples of previous modifications can be particularly advantageous (see Figure 7.4 and related discussion). It is possible here to give only a flavour of some of the things that might be done.

This particular technique is nominally differentiated from the previous technique in that components or devices need not necessarily be added. This, however, will not always be so: in practice the two techniques will often considerably overlap one another.

As with all techniques, it is being sought to reduce or eliminate 'excesses' of existing tasks (in this case 'effort'). A reduction in effort can be thought of as simplification of the existing task, or reduction in the time necessary to complete it. The current improvement technique can be focused in some potentially highly beneficial directions: for example, lightweighting components; securing; positioning. Reduction in effort is sought to the point of task elimination. To show what can be achieved, some examples of task elimination are later presented in Exhibit 9.2.

Lightweighting

Lighter components are likely to be easier to handle (for example, to move, store or position). Being easier to conduct, these handling tasks are likely to be completed more quickly.

Benefit also might be gained in other ways. A heavy object might require the use of special lifting devices. For example, a mobile overhead gantry might be employed. When required, this gantry might have to be brought into position (assuming in the first place that it is currently available to be used). To bring the gantry into position might mean that surrounding objects have to be cleared to gain access. The lifting equipment then has to be brought into use. Chains and lifting hooks probably will be needed. In all, the activity involved in removing or replacing a heavy component or assembly can be extensive. If, alternatively, the component was sufficiently light to be moved manually, all this foregoing activity essentially would be eliminated.

Securing

An objective here can be to eliminate having to employ screws – or at least to eliminate having to turn them more than, say, half a turn. By now familiar options such as the use of key-hole slots can be extremely useful.[9]

Other modifications that dispense with using screws can also be considered. These might include using sprung clips of many different types, or perhaps adaptation so that a magnetic device might be employed. Or, sometimes, the physical restraint that is employed can be completely rethought: it might be possible, for example, simply to rest a component on pins. Similarly, interlocking components together in other ways entirely without the use of screws might be possible.

Positioning

Changeovers can be aided by relying on simple, consistently used reference datums. Exhibit 5.6 explored this issue. Exhibit 5.5 presented an example of the considered use of datums, where the manufacturing cost of mould components was higher, but where this initial outlay would be more than paid back by the simplification that the design imposes on the mould changeover.

Exhibit 9.2 Examples of eliminating changeover tasks for transportation mechanisms

Most industrial processes require transportation of the product between equipment as it undergoes successive manufacturing operations. Transportation mechanisms frequently require to be changed to enable different sized products to be passed. As highlighted by the following examples, options can exist to eliminate changeover of transportation mechanisms.

Example 1

A plastic bottle printing factory used inks that needed to be thermally cured. The bottles thus had to be transported through an oven. It was important that nothing contacted the ink before it was cured. Plastic bottles of significantly varying shape and size were being printed.

To save changing over the transportation system, the plastic bottles, each of which were only printed in the centre of their front and rear faces, were conveyed from the print station and through the oven in perforated shallow convex trays. No matter what bottle shape was under production, contact with the tray could only occur away from the wet ink surfaces. The situation is illustrated in Figure 9.1.

Example 2

Aluminium bottle caps similarly needed to be transported through an oven for their ink to cure. As shown in Figure 9.2, transportation was effected on a pin chain, on which different sized caps could be located and which, like the perforated trays of the previous example, allowed hot air to circulate relatively unimpeded.

Other 'non-changeover' transportation mechanisms are also possible. Cylindrical components that can be contacted on their outer surfaces might be transported by Vee belts as shown in Figure 9.3. Like all 'universal' solutions, an ability to retain the desired level of control of the product needs to be considered.

Pockets, as shown in Figure 9.4, also might be used to eliminate changeover of the transportation medium.

Pockets, like the other solutions cited here, might not be applicable if precise orientation of the product is required (for example, to allow it to be removed and positioned with ease, ready for the next processing operation), or if the product range consists of a wide variety of sizes and/or irregular shapes. In this case it might be appropriate to mount the products on a carrier. The carrier's 'working' surface, in which the product sits, would suit the product that it is to carry. Elsewhere the carrier would incorporate standard features, to allow it to be used on the machine in an identical way to all its sister carriers – thus eliminating a body of changeover tasks that would otherwise be necessary were a different product to be transported.*

* This technique of employing a carrier can have other advantages, as described in an example in Chapter 10. It might also be applied, for example, to limit on-line activity (rather than completely eliminating a series of changeover tasks).

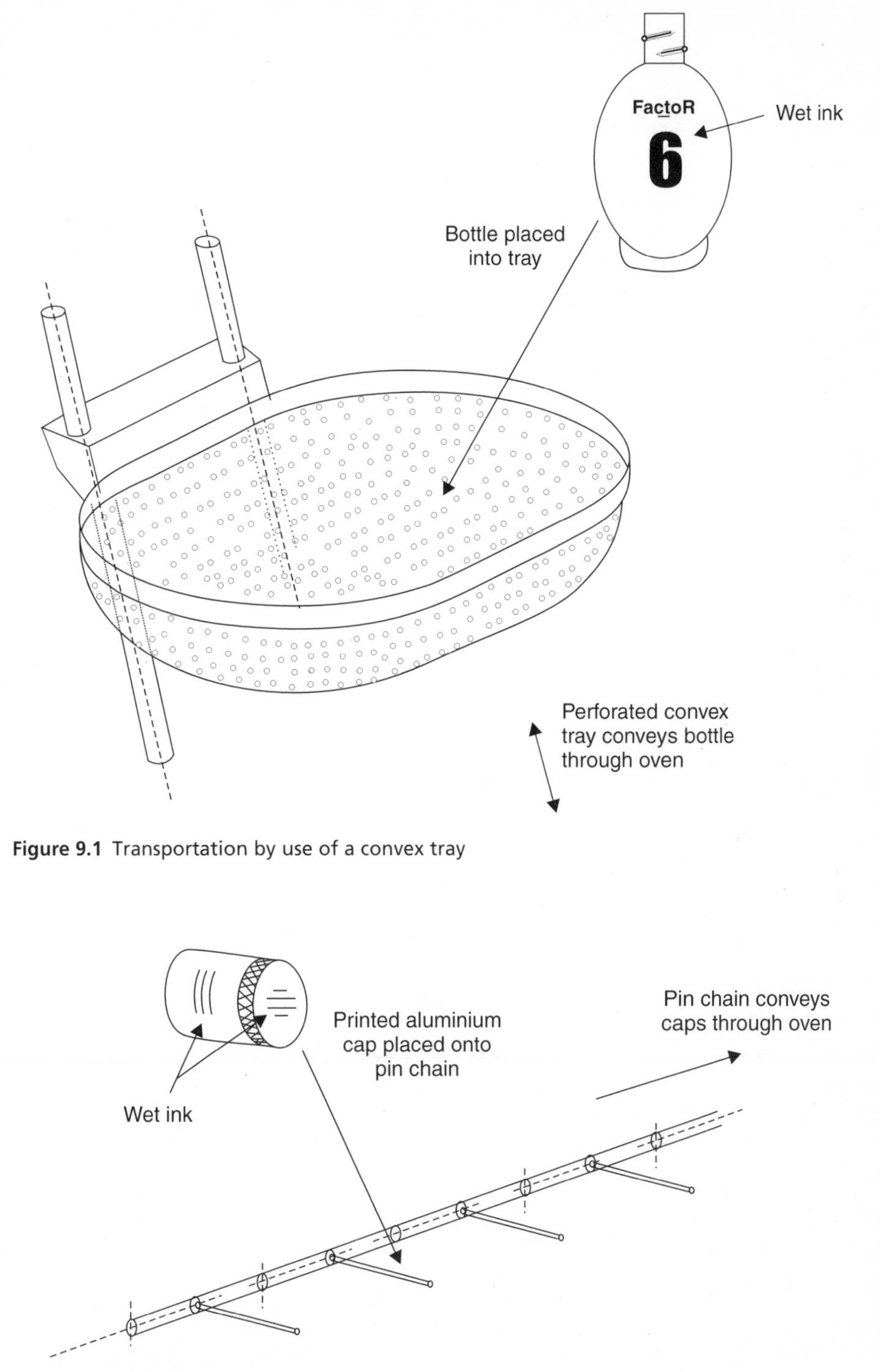

Figure 9.1 Transportation by use of a convex tray

Figure 9.2 Transportation by use of a pin chain (cap not to scale)

Exhibit 9.2 (*Continued*)

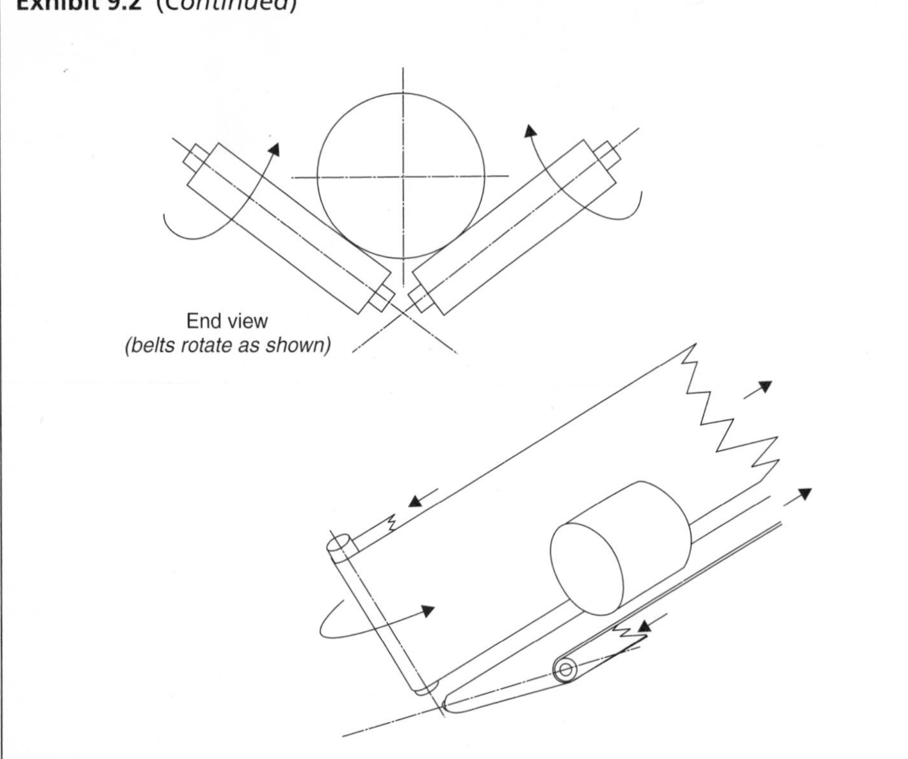

Figure 9.3 Transportation by use of Vee belts

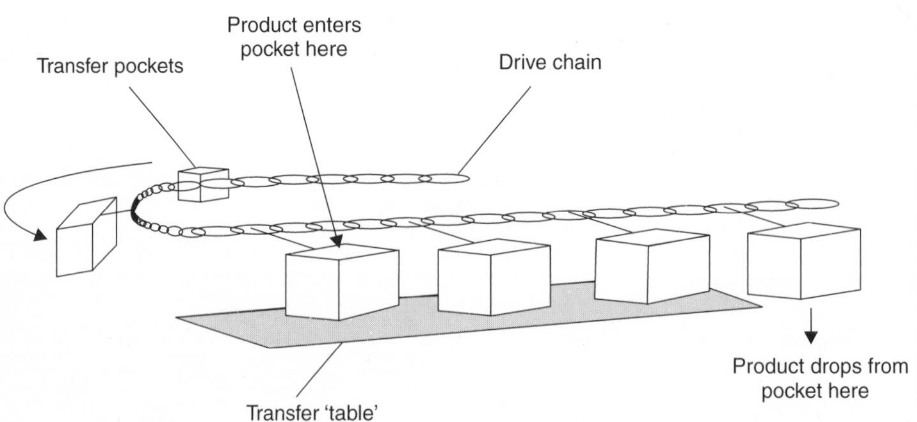

Figure 9.4 Transportation by use of pockets

Note that all these designs might be particularly attractive to original equipment manufacturers – where the designs might be little different in their complexity and cost to other designs that do not share the capability that these designs exhibit.

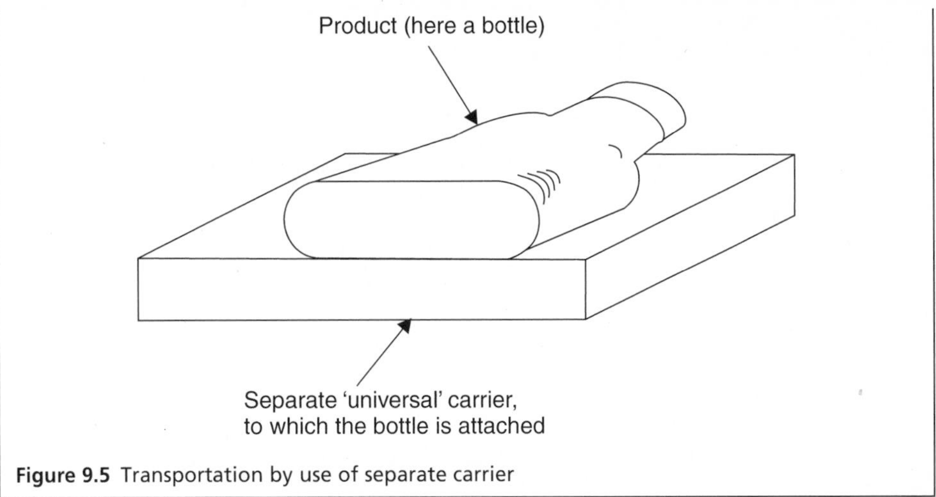

Figure 9.5 Transportation by use of separate carrier

9.13 Improve access

Access problems can sometimes be substantially overcome without resorting to using design – simply by better organizing how tasks are conducted. For example, we have sometimes witnessed wholly inappropriate hand tools persistently being used. On more than one occasion using a ratchet and extension bar in place of a difficult-to-use spanner would represent a simple and highly effective solution.

On other occasions design change does have to be undertaken if access problems are to be alleviated. For example, inaccessibility is sometimes apparent in terms of tasks needing to be conducted in, say, an elevated position. In this case the permanent construction of a separate platform might be considered (see also Exhibit 5.3).

Another option can be to apply medical 'keyhole surgery' holes to mechanical systems. An example is shown in Figure 9.6 – which perhaps is used to gain access to a particular bolt.

Options that allow obstructive components or assemblies to be left in place might not always be possible. Better access can also be facilitated by making obstructive parts easier to move (in some cases moving parts even when this was not done previously). Or, by design, an obstructive assembly might be reappraised as separate modules – where only a small part of what was previously a whole assembly is detached.

Safety guards will exist for most manufacturing processes. Guards are sometimes given little attention in respect of changeover improvement. They might often be redesigned to be quickly and easily removed entirely from a work area – rather than, as might currently happen, being left partially in place and getting in the way. One useful option can sometimes be to lift guards whole into the factory's roof area.

Finally, access problems can be usefully considered alongside the option of mechanization – whereby access problems can often be wholly overcome.

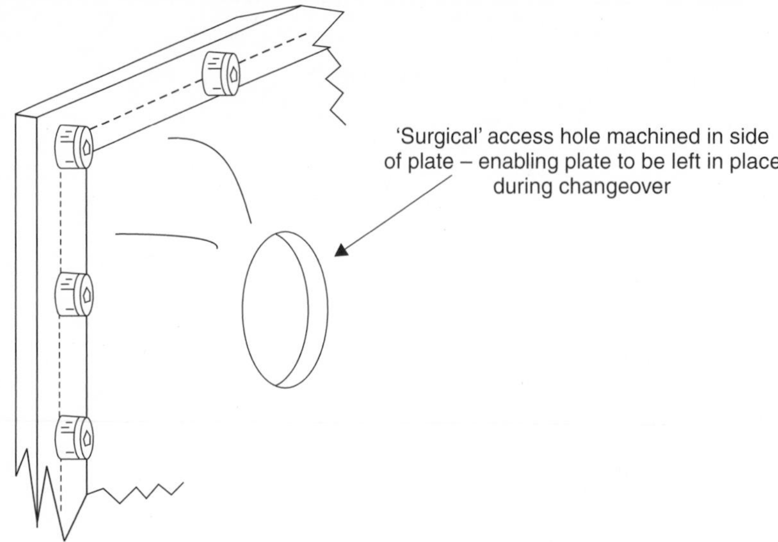

Figure 9.6 'Surgical' access hole

9.14 Mechanize activity

Mechanization is a specific case of adding devices to aid changeover, and mechanization of just a single task can sometimes have a substantial wider impact upon the changeover as a whole. For example, consider Exhibit 5.3 once again, which describes mechanization of a key changeover task (the adjustment of inaccessible roof-mounted tracking). Mechanization of this one task was not identified for the purpose of easing this one task, but rather was proposed as a means greatly to simplify (and speed up) the changeover as a whole. Numerous previous tasks were eliminated.

Such a situation can be contrasted with mechanization that only impacts upon a single task. In one example, prior to improvement, a heavy machine crosshead was lifted by manually turning four threaded shafts (each shaft was coupled together with a chain, and thus turned in unison). The improvement constituted substituting an electric motor to drive the chain and hence lift the crosshead.

The practitioner might consider how motive power is to be provided. The above example employs an electric motor. As noted elsewhere (for example, see Exhibits 5.1 and 5.3), more specialized servo motors might also be employed. Similarly, there is often scope for hydraulically activated devices to be employed.

There are occasions where it is prudent to introduce a fail-safe feature into a mechanized operation. For example, it would be wise to ensure that heavy masses that are hydraulically clamped (upper die shoes, for instance) cannot become insecure – or at least, if this is possible, that line shutdown occurs immediately. Alternatively, failure to complete a mechanized operation successfully can sometimes result in physical damage.[10]

Mechanization beyond 'powered devices'

Mechanization might be thought of more broadly, and also be applied to information exchange. Machine settings might currently be entered by hand. A possible improvement could be for settings or information to be stored electronically, for example on a floppy disk. We have witnessed this on a complex print installation (see Chapter 10), and are aware that it is also used for plastics moulding equipment. As well as allowing information to be exchanged more quickly, it also greatly restricts the chances of data entry errors being made.

9.15 Seek single-person working

A changeover task that is conducted by two people will often be inefficient. For example, one task we observed required that operators stood on either side of a process line. For much of the time one operator was waiting while the other operator was conducting the task alone. This is not the only way that losses can be incurred: arranging for people to be available when required can sometimes be even more significant[11] (also see Exhibit 5.1).

Many opportunities can exist to convert multi-person tasks to single-person working. Often these opportunities will involve adding devices or employing mechanization. The objective will be to bring task elements within control of one individual, within his or her working space. Or, by design, task elements might be eliminated.

9.16 Reduce task content

Reducing task content refers to reducing repetitive operations when executing that task. This is sought without otherwise fundamentally altering the task.

The technique of reducing task content is perhaps more easily explained by means of an example. A 300 mm diameter circular cover hatch might need to be removed from a machine. The hatch might be unstressed and of low mass. It does not need to be airtight. The hatch is secured by a series of twelve screws about its perimeter. In these circumstances, quite possibly, the hatch could equally be secured by just four screws, without in any way altering its functionality.

An equivalent motoring situation can be narrated. Some motor vehicle road wheels are secured on just three studs (some Renault motor cars, for example). More normally, four studs are used. Other vehicles of similar mass and with similar sized wheels might have their wheels secured on five or more studs. While there might be good reasons for increasing the number of studs that are used, the time taken to change over a wheel in the event of a puncture would not be one of them. As is widely known, a Formula 1 racing car has its wheels secured in just one place – which represents the best arrangement for rapid wheel changing.

While the use of screw fasteners is generally poor in respect of changeovers,[12] there are occasions when they might need to be used, particularly where high clamping forces are required in a confined area. Another option can be to limit (shorten) the length of thread engagement to that which is functionally necessary.

Perhaps a more useful option can be to combine all service connections (perhaps, for example, for compressed air, water or electricity) using a multi-coupling. The same coupling/decoupling of the services still has to take place, but the degree of repetitive coupling or decoupling that occurs might be drastically reduced. Proprietary multi-coupling units are available to do this. Or, for example, for a die set that has multiple air-activated ejectors, branching the compressed air fully within the die set (the change part) would mean that only one external connection need be made.[13–15]

9.17 Conduct automated tasks more quickly

From the operator's perspective, fully automated tasks will require no input whatsoever once commenced. There is thus no opportunity for organizational improvement to such tasks. This is not to say, however, that these tasks cannot be completed more quickly.

Often there will be scope simply to speed up an existing automated task, completing the automated sequence more quickly, without changing it in any other way. For example, using a faster drive motor might be an option. Alternatively, in a computer-controlled sequence, for example a washing cycle, task duration might be changed within the control program (perhaps, in this particular example, in conjunction with the use of alternative solvents).

9.18 Separate/combine items

Changing how machine assemblies/components are manipulated generally can be important. The improvement technique of separating or combining items specifically considers how these assemblies or components are broken down, or split up into modules, to be handled. A single large assembly, for example, could instead be split into a number of smaller constituent parts. These constituent parts are then manipulated entirely separately, only being brought together when installed in the machine. Conversely, advantage might sometimes be possible by manipulating a large single assembly that previously was a series of smaller assemblies or components.

Once again, some specific examples can show how his improvement technique can yield advantage. One witnessed example involved the manipulation of a parallel (a precision-machined block of metal that is used as a spacing piece) in a large double-acting press. The parallel overall was approximately 1.3 m long by 75 mm square. As this component was solid metal, and thus physically difficult to manipulate manually, it was split into four separate pieces. Thus, four identical parallels were being used. Each piece separately had to be carefully brought into position during the changeover. Due to these components' mass, and because of the precision with which they had to be located, their manipulation constituted a significant part of the overall changeover. Improvement was possible by replacing these four separate parallels by a single lighter parallel (achieved by machining large pockets within it). The component's functionality was not affected in any way. Once available as a single component, the previous manipulation tasks effectively were reduced by a factor of four. Moreover, there was overall less danger of incorrect assembly.

Conversely, the splitting of what were larger components or assemblies into smaller parts can also yield significant benefit. Benefit might be derived, for example, in situations where only a small element of a much larger assembly needs to be changed. An example was a paint-spraying machine. Only the spray head needed to be changed. Before improvement the large assembly (of which the spray head was an integral part) had to be removed whole. After improvement the spray head alone was exchanged – with the remainder of the assembly being left undisturbed.

Finally, on this theme of changing components/assemblies, advantage might also be gained by arranging for constituent parts of assemblies simply to be broken down differently. This could later facilitate a much better distribution of changeover tasks (upon the separate, dismantled elements) between available resources. Alternatively, better use of datums might be afforded.

Minimize the number of parts that comprise an assembly

An alternative perspective can be taken of this current technique. Sometimes different components are present within an assembly because they each serve one particular function. For example, a component might be present only to resist forces that the assembly experiences, while separate components are used to locate the assembly, or secure the assembly. If these separate components can be combined to be multi-functional, such that fewer components overall comprise the final assembly, then fewer changeover tasks may to be required. Aspects of this technique are therefore similar to what can be involved in DFA (design for assembly).[16]

9.19 Avoid using hand tools

It can be surprising how much time is lost because hand tools need to be used. First, the time required to pick up, use and replace tools will contribute towards the total changeover time. Second, there can be numerous opportunities for the required tool to be unavailable. There is then delay while this tool is found.

Many of the devices noted earlier eliminate the need for separate hand tools. Their use might be sought for this reason alone.

9.20 Final thoughts

Ultimately the practitioner should always believe that nearly any task can be eliminated: what is more critical is whether the task can be viably eliminated.*

This is one reason why it is important for team members to be aware of the targets they are working to achieve (time reduction sought; expenditure; when improvement is sought by). The extent of retrospective design work and the disruption caused to scheduled production should be reflected upon. The improvement practitioner should

* The original design of both the process equipment and the product – rather than retrospective design – should also be appreciated in this light.[17]

always be conscious of the possibility of installing alternative, more changeover-proficient equipment – if the changes that are being contemplated are becoming too extensive.

Although the issue of equipment replacement should previously have been addressed by senior managers, any decision earlier in the initiative will necessarily have been made without the practitioner's current experience. As noted elsewhere in this book, equipment replacement is starting to become more viable in many industries, where highly changeover-proficient machinery is now under development or becoming commercially available.[18–22]

9.21 Conclusions

This chapter is intended to give further insight into the application of design. Possible improvement techniques presented both in this and in the previous chapter can significantly overlap one another. This is of little consequence: what these techniques do, in unison, is to provide a wide insight into ways that improvement might be gained.

References

1 Hayes, R. (1996). Totally shaftless. *Printing World*, October, 24–27. A description is given of an equivalent printing press where a solid-coupled primary drive is eliminated.

2 Taguchi, G. and Clausing, D. (1990). Robust quality. *Harvard Business Review*, January–February, 65–75. Taguchi and Clausing differentiate between what better product design and better control of manufacturing processes can contribute to product quality: 'Quality is a virtue of design. The robustness of products is more a function of good design than of on-line control, however stringent, of manufacturing processes.' Although this is not a true like-for-like situation, including the interpretation of 'quality', an inference is clear: that procedures, however rigorous, cannot be expected to compensate for indifferent design.

3 Remich, N. C. (1997). From 15 min. to 90 sec. die changeover. *Appliance Manufacturer*, **45**, No. 7, 44–46. Remich describes that during changeover: 'Use of fork trucks caused expensive wear and tear on dies.'

4 Forth, K. D. (1994). Quick die change helps auto stamper produce to order. *Modern Metals*, **50**, No. 9, 30–35. Forth describes that by incurring less handling, less tool damage and better-quality products result.

5 Shingo, S. (1985). A *Revolution in Manufacturing: The SMED system*. Productivity Press Inc., Massachusetts. Shingo is one author who chooses to use this example.

6 Lee, D. L. (1987). Set-up time reduction: making JIT work. *Proc. 2nd Int. Conf. on JIT manufacturing*, Ch. 45, pp. 167–176. A similar situation is reported.

7 Kobe, G. (1992). Engineer for right hand steer: how Honda does it. *Automotive Industries*, **172**, March, 34–37.

8 Shingo, S. (1986). *Zero Quality Control – source inspection and the poka-yoke system*. Productivity Press Inc., Massachusetts. Further advice can be found in books on this topic, such as that by Shingo.

9 Shingo, S. (1985). *Op. cit.* ref. 5. Shingo's work in respect of securing and positioning components is well worth referencing.

10 McIntosh, R. I., Culley, S. J., Mileham, A. R. and Owen, G. W. Reinterpreting Shingo's 'SMED' methodology (paper in preparation).

11 Kings, R. (1999). Discussion of changeover on farm equipment, Worcestershire, March. The importance of reducing the number of people that conduct a changeover can be underestimated. Like manufacturing industry, farmers use equipment that needs to be changed over. An example here was potato-planting machinery – changing the spacing for different potato varieties (Marfona = 8½ inches, Wilja = 12½ inches). The machine was designed such that a single-person changeover could not be done: two people were needed. Finding help – where staff were dispersed across the farm – could be difficult, leading to lengthy delays.

12 Mather, H. (1992). Reducing your changeover times. IMechE seminar, Birmingham, UK. Mather is one of many authors who strongly advise eliminating the use of screw fixings wherever this is possible, noting that using screw threads, for all but their last half turn or so, represent wasted effort.

13 Anon. (1988). Nicht nur, um zeit zu sparen. *Plastverarbeiter*, **39**, No. 5, 148–149. As we have described previously, the plastics moulding industry is relatively advanced in terms of changeover proficiency (largely because the market has dictated that it needed to be): many different examples of the techniques that we describe can be witnessed on moulding equipment.

14 Anon. (1998). Quick change debuts for four-level stack molds. *Modern Plastics*, **75**, No. 4, A8.

15 Rozema, H. and Travaglini, V. (1995). Quick mold change systems for high volume stack molds. *Proc. Annual tech. conf. ANTEC 1995*, Soc. of Plastics Engineers, Vol. 1, pp. 1011–1015.

16 Boothroyd, G. (1992). *Assembly, Automation and Product Design*. Marcel Decker, New York. Boothroyd is one of the principal exponents of DFA.

17 Kobe, G. (1992). *Op. cit.* ref. 7. Kobe describes changes that Honda have made to facilitate changeover between RH and LH drive cars. The relative simplicity of what has been done – by employing design at the outset – is set against the benefits that the company accrues. Kobe observes: 'Its nothing that every other US manufacturer couldn't do if they wanted to – but they have to want to do it first.'

18 Mason, F. (1996). Prototype stampings? No problem! *Manufacturing Engineering*, **117**, No. 1, 2. Mason describes developments in metal stamping machinery, where a high changeover capability has been designed into the equipment. The design is such that effectively all previous manual changeover tasks have been replaced by an automated, computer-controlled system.

19 Fultz, P. (1999). One-to-one marketing – for real. *Direct Marketing*, **62**, No. 4, 63–67. Significant developments are occurring in the way that printing can be undertaken. Digital print machines are being developed that do not require manual changeover in the way of conventional printing presses. Changeover is rendered as changing the digital (computer) information that is sent to be printed. The changeover capability of such equipment potentially can allow economically viable printing in unit batch sizes.

There can still be problems associated with the commercial use of such equipment – for example, its speed of operation, reliability and cost – but it can still open up new opportunities. Opportunities for individually tailored marketing are discussed in the literature. Other opportunities exist. At the time of writing, for example, recorded music is commonly available on CD-ROM format. Notwithstanding that other parallel developments have to occur in respect of CD-ROM manufacture and distribution, responsive stock-free manufacture to order of low-volume obsolete CD sleeves might be facilitated – for example, sleeves for reissues of Freddie King's *Texas Cannonball* album (with Jim Franklin's original artwork), originally available in the USA between 15 May 1972 and 3 August 1974.

20 Ueno, K. (1998). New guideless CNC shaper for helical gears. *Gear Technology*, **15**, No. 2, 17–19. Changeover by means of reprogramming is again described.

21 Renda, J., Nauss, E. and O'Neill, D. (1997). Five steps to enhanced stencil printing. *SMT Surface Mount Technology Magazine*, **11**, No. 5, 92–96.

22 Covell, P. (1997). Sleek look for Pretty Polly. *Packaging Today*, **18**, No. 4, 63–67.

10

Case studies

A number of case studies are presented in this chapter. In particular these case studies investigate the application of design. Wider management and maintenance issues are also reported, supporting our assertion that many issues away from the shopfloor can affect the success of the initiative on it. The case studies also show other aspects of the approach that this book advocates.

Case study 1 highlights the prominence that design can have, even when there is no explicit focus on applying design. **Case study 2a** assesses improvement to a plastic bottle conveying mechanism. **Case study 2b** considers the importance of datums in making improvement. **Case study 2c** again assesses the use of datums. Continuing an explanation that was commenced in Exhibit 5.8, the case study investigates the need to use genuinely fixed datums when seeking to minimize adjustment. **Case study 3a** considers a situation where a large investment in new equipment was made. A lithographic print line was bought with high flexibility in mind, where a sub-15-minute changeover time ultimately was being sought. A reference changeover model showed the extent of the work that was going to be necessary to achieve this level of performance, and demonstrated that this improvement was not going to be possible by organizational improvement alone. **Case study 3b** considers the importance of good communication in respect of changeover logistics.

10.1 Case study 1 The use of design

This study indicates the extent to which design solutions can be present in changeover practice; perhaps over and above what a typical improvement team might expect. It suggests that design changes (often minor design changes) can be common.

We undertook an assessment of improvement ideas that were proposed during a major site initiative. These ideas were contributed by factory personnel who had been extensively trained – by an external consultancy – on use of the 'SMED' methodology. In general, except issues such as the use of 'keyhole' slots, the application of design had not been promoted as part of this training.

Our study was undertaken in collaboration with a shopfloor team. The research brief was not to act as consultants, but rather to understand the thinking and procedures of the factory improvement team and, if necessary, to act as a catalyst to develop and progress particularly good ideas. The attachment sought to record and

Table 10.1 Different types of resource required for 26 improvement ideas

Resource needed for the 26 proposed ideas	*Number of occurrences*
Components to be made or modified	19
Drawings required	16
Design input required*	13
Parts or equipment to be purchased	10
Organizational changes	5

* Refers to non-trivial design where involvement of professional design or process engineers could be required

Table 10.2 Cost breakdown for 26 improvement ideas

Cost	*Number of occurrences*
£0–100	14
£101–200	5
£201–500	4
£501–2000	1
>£2001	2

foster the development of the team's own ideas. There was an explicit policy not to promote or impose external solutions.

In total, 26 improvement ideas were identified and investigated. These ideas were analysed at the conclusion of the attachment, and a summary of the resources required to implement the proposed changes was made, as shown in Table 10.1.

As Table 10.1 relates to just 26 improvement ideas, it shows that in general there is considerable overlap between the resources needed. However, the category of 'organizational changes' warrants special mention. This category applies to change without altering the equipment in any way. For example, a proposed organizational change was to ensure that scrap was removed before the changeover started. This represents a change, requiring new discipline, to the existing changeover procedure. Alternatively, where design changes are made, the existing changeover procedure will be structurally altered.

A further analysis was made on the cost of implementing the improvement ideas, the results of which are shown in Table 10.2.

Taken together, Tables 10.1 and 10.2 confirm that there was a high incidence of ideas employing design change, but that these design ideas were often of a simple, low-cost nature. This study thus indicated that the application of design – albeit often of a non-substantive nature – can be common. It indicates that a deliberate design focus can be beneficial.

10.2 Case study 2a Improvement of a plastic bottle conveying mechanism

This case study describes improvement in stages to a plastic bottle conveying mechanism. The mechanism was part of a silk-screen printing machine, printing successive colour images on blow-moulded plastic bottles.

It was interesting to note the variety in the design of different manufacturers' equipment. As demonstrated by alternative equipment in use at the factory, some available designs are more changeover-proficient than others. Equally, some designs are rated as providing a higher-quality print. Other equipment attributes can also be considered, like the speed of operation, and the cost of manufacturing new change parts (for a new bottle shape). Changeover performance is but a part of the total capability of a machine.

A brief description of the bottle transportation requirements

The bottle conveying mechanism is quite complex. The bottle needs to be passed between successive print stations and ultraviolet (UV) curing stations in a closely controlled manner, including controlling the bottle's orientation. During transportation the bottle is held in shaped, fabricated mild steel carriers by means of a partial vacuum. During printing cycles air pressure is applied to the inside of the bottle as the ink is applied, to provide a rigid surface to print against.

Figure 10.1 schematically describes the print line. The mechanism being described applies between the bottle unloading station (from the infeed conveyor) to the inspection camera. The conveying mechanism is fully enclosed by protective screens during running, both for isolation of the moving parts and to shield the operators from UV light.

This print equipment was bought largely for its claimed changeover performance. Like later described equipment, the expected level of changeover performance was rarely met – sometimes with a very considerable deficit being apparent. The equipment was designed to print either round or 'oval' (irregular) plastic bottles. The difference was noted in changeover type when changing production from 'oval' to 'oval', as opposed to 'round' to 'round', or from 'oval' to 'round', or from 'round' to 'oval'. Wholly different changeover tasks were involved dependent on the type of changeover that was to take place. These different types of changeover took markedly different times to complete. The poor changeovers that were being witnessed applied for the comparatively simple changeover from 'oval' bottles to alternative 'oval' bottles.

The original design and initial improvement

Figure 10.2 schematically shows the original design of one of the bottle carriers, as supplied by the equipment manufacturer.* The welded steel carrier is essentially a shaped, hollow chamber, to which a partial vacuum is applied to hold the bottle in position. The edges of the carrier need to conform closely to the bottle shape.

The original method of attaching these carriers to the machine is highlighted by Figure 10.2. A short 5 mm socket head screw was used, that passed through a clearance hole at the base of the carrier's hollow rectangular leg. It was difficult to remove and replace this screw. In short, access was a problem. There was also an ever-present danger of the screw dropping into the vacuum chamber. The changeover time for 37 carriers (mainly of this type) averaged at about 25 minutes. Alignment of the carrier type shown was taken care of by the precision machined recess into which the carrier leg fitted.

---

* One carrier type is focused upon here. Other bottle carrier designs were used in certain other sections of the machine (see the next case study).

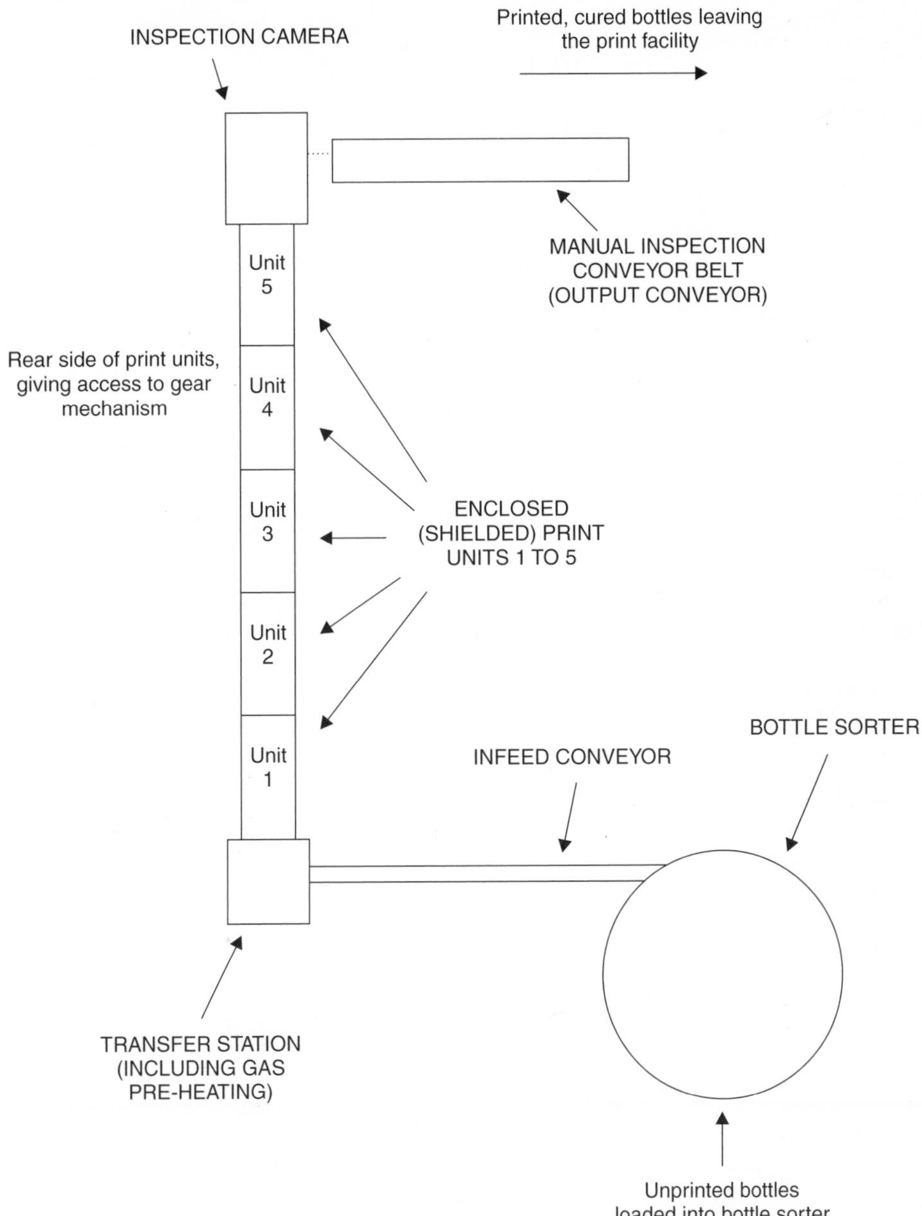

Figure 10.1 Schematic of UV print installation

An initial improvement, shown in Figure 10.2, was made that had two elements to it. First, the socket head screw was altered in the way shown, by welding a metal sleeve over a longer screw. This raised its hexagon drive into a much more accessible position. Second, as is almost inevitable when a screw thread is used in a changeover application, undue effort was involved in fully releasing (and subsequently fully

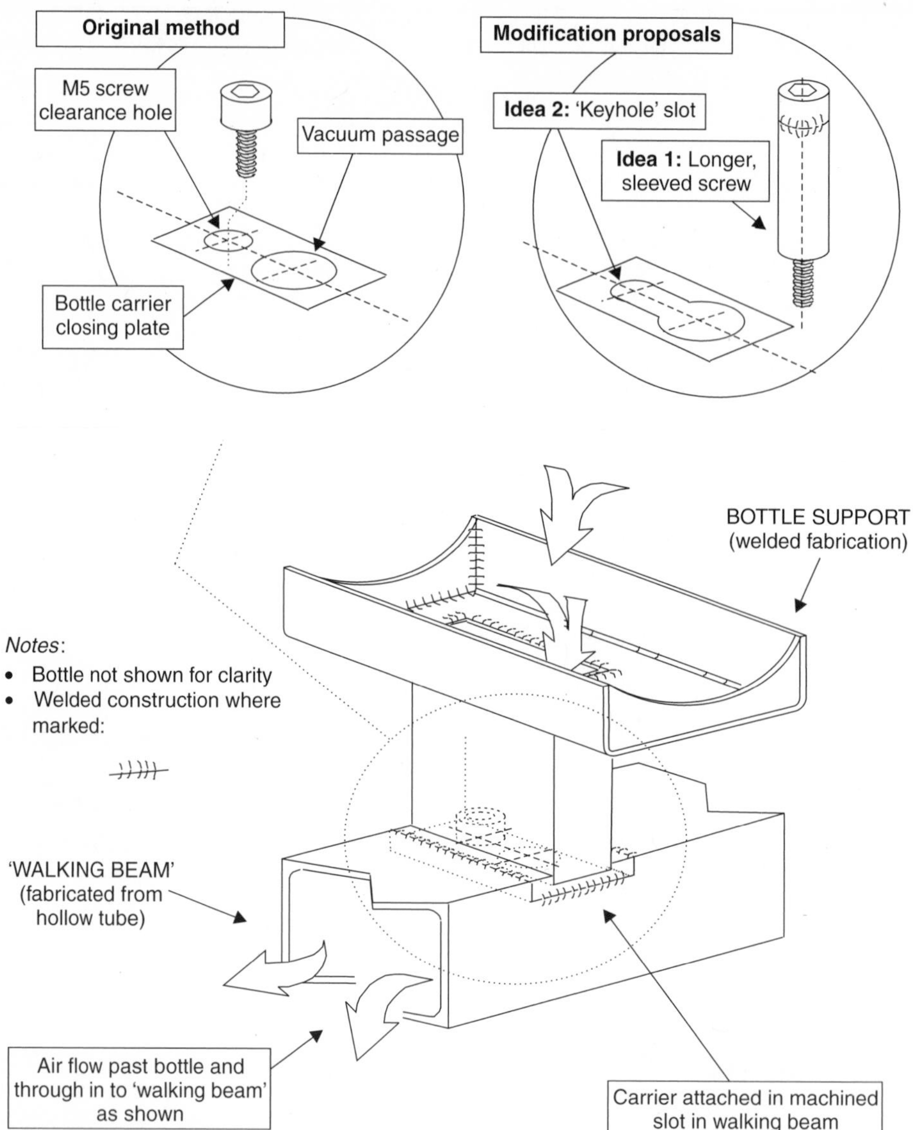

Figure 10.2 Early modification proposals to assist the bottle carrier changeover

fastening) the screw.* It was also difficult to get the screw started into its inaccessible threaded hole. The use of a keyhole slot overcame both these difficulties, wherein the screw is simply loosened or tightened a couple of turns, but is otherwise left in place. The improvements shown brought the changeover time for this type of carrier down from approximately 40 seconds each to approximately 15 seconds each.

* Repeated turns to engage/disengage the full thread length is necessary, when it is only the last turn of the screw that performs the clamping operation.

In considering the improvements detailed in Figure 10.2, a design change was sought specifically to overcome known problems with this changeover task. In particular problems with access and effort were highlighted. The solution was simple in its conception and implementation. Also, the improvement was cheap and permanent.

Further improvement

In time further design improvement was conceived. This subsequent improvement again allowed a step change in changeover performance to be made. Also, as will be described, the new design yielded other important benefits.

Whereas the change described above in Figure 10.2 might be considered as relatively straightforward, the next design improvement arose when greater attention still was focused on shortcomings of the design. Although improvement had already been achieved (Figure 10.2), exchange of the fabricated bottle carriers was still relatively awkward. The use of a hand tool was still necessary. Also, it was noted that a greater portion of the change part than was necessary was being exchanged: strictly it is only the upper working surface that needs to be replaced.

It was appreciated that there was scope to split the existing one-piece bottle carrier into two or more smaller parts. Specifically, given the range of bottles that were to be printed, there was no need to detach the bottle carriers at their point of contact with the walking beam (see Figure 10.2). Instead a wholly different design for rapid detachment and alignment could be considered. Figure 10.3 shows the changes that were made.

Different method of bottle support manufacture

This later design improved changeover performance. Some other highly important attributes of the new design were also realized.

The original mild steel bottle support was hand finished. As such it was very difficult to maintain exacting dimensional control. Variation existed between nominally identical individual bottle carriers. Adjustment often had to be undertaken elsewhere on the machine during changeover to compensate for this lack of repeatability between individual carriers. The bottle profile in the new design shown in Figure 10.3 was CNC machined in nylon. This nylon component was permanently attached to a 1 mm thick mild steel base. With the greatly enhanced precision available via this manufacturing route, the requirement to conduct any possible post-mounting adjustment was dispensed with. Assessed another way, the quality of the bottle supports was improved, and this improvement had a direct impact on changeover performance (as well as an impact on the security and positional accuracy with which the bottle was held).

Features that enhance changeover performance

The new bottle supports shown in Figure 10.3 could be changed over as a set in approximately 4 minutes (in comparison to the original time of 25 minutes). With

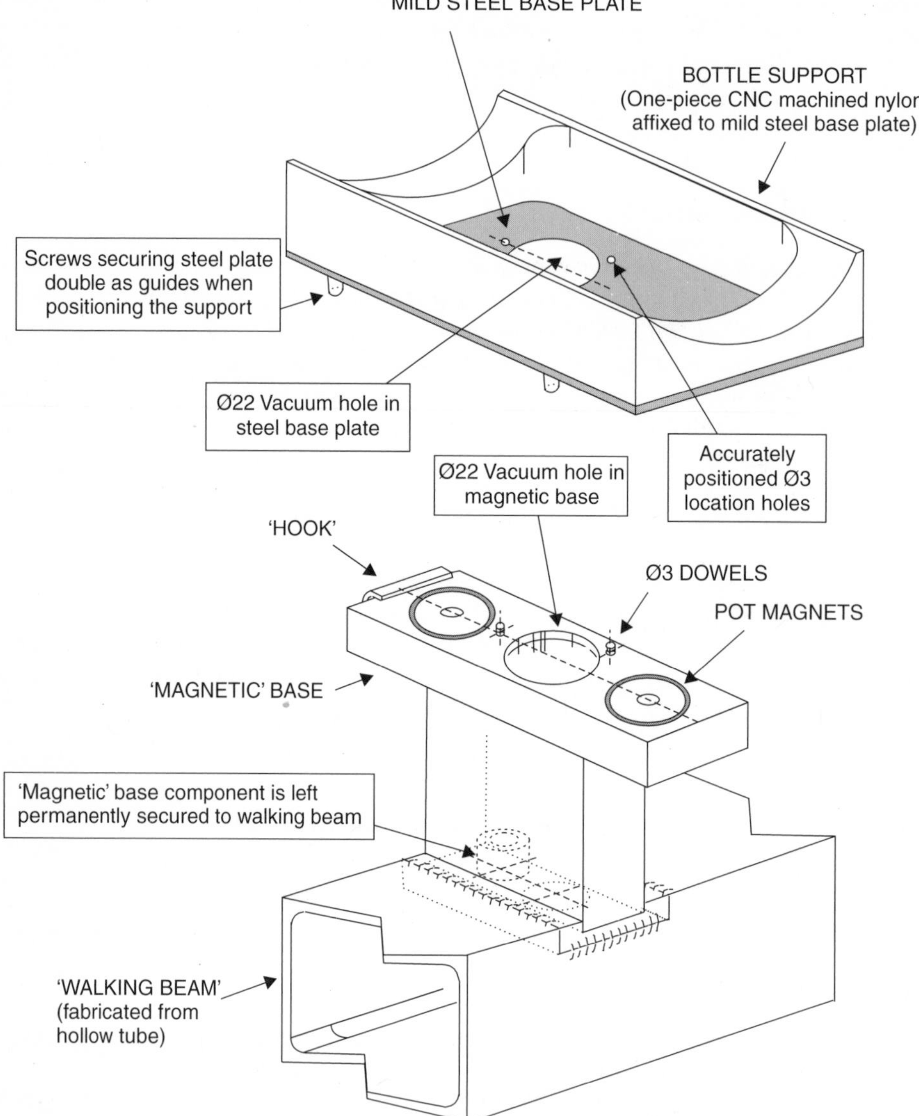

Figure 10.3 Substantive redesign to assist bottle carrier changeover

reference to Figure 10.3, some design features that enhance changeover performance can be identified. These features include:

- improving component quality – thus eliminating a significant adjustment task
- eliminating the use of hand tools
- 'foolproof' positioning
- maintaining a standard reference dimension, across all bottle carrier types (see also Case study 2b)
- rapid fastening method – eliminating the use of threaded fastener

- excellent access
- forces reconsidered
- only the functional element of a change assembly/component is detached
- guiding component into position, but with final location achieved by separate feature – both on same component

Other benefit arising from the new bottle carrier design

The design involved splitting the original fabricated carrier into two separate components. As only the upper component with the working surface was changed, and because of the completely different method of manufacture, the cost of change parts fell from approximately DM 20 000 per set to DM 5000 per set. With pressure on the chosen supplier, the delivery lead time fell from 3 weeks to 1 week. Also, the component was more amenable for manufacture by a wide range of general machine shops, and a local fabricator could be used if desired. This was not previously the case. As noted, the quality of the components rose because of the revised method of manufacture.

10.3 Case study 2b The importance of datums/1

Case studies 2b and 2c both describe the contribution that fixed datums can make in achieving rapid, high-quality changeovers. Considered use of datums would normally occur to reduce adjustment activity during changeover. The work described here additionally shows that extreme care sometimes has to be taken that the datums employed are reliable.

This next case study effectively is a continuation of the case study presented immediately above. It applies to the same bottle printing equipment. Once again, the description is of bottle carriers, transporting the plastic bottles through the equipment (see Figure 10.1). In this case a slightly different bottle carrier type is under scrutiny where, as shown in Figure 10.4, the carrier is mounted at the end of swinging arms (rather than on the walking beam shown in Figure 10.2). These bottle carriers are again accurately CNC machined in nylon. Indeed, for many bottle designs that pass through the machine, these detachable top portions can be identical, irrespective of the magnetic base to which they are to be mounted. Only the top component is separated, whereas previously the bottle support was changed in its entirety as a mild steel welded fabrication. Again, the changed design incorporates in-built precision, allowing different parts to be exchanged without undertaking some of the previously needed adjustment steps to compensate for lack of repeatability between parts.

Figure 10.4 presents a view of the bottle carriers attached to the end of swinging transfer arms. These arms function in unison, first to load the bottle to the print position on the 'bottle rotation shaft', and subsequently, using the next of the two arms, to remove it from this position. The 86.6 mm dimension shown is a datum dimension, the achievement of which allows the bottles to be manipulated correctly. The 86.6 mm dimension applies in turn to each of the two transfer arms.

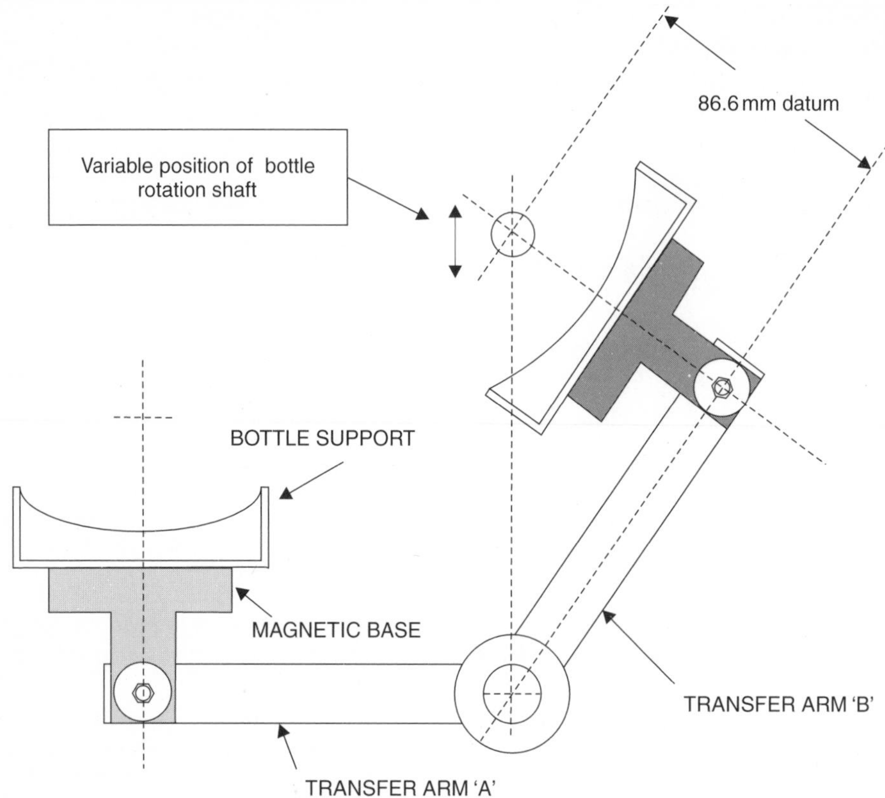

Figure 10.4 Transfer arms and the 86.6 mm datum

With this new design of bottle carrier, achieving the 86.6 mm dimension should be straightforward. The magnetic base portion of the support is permanently attached to the transfer arm, and is not moved at all during changeover. Using specially made setting pieces, this magnetic base was carefully positioned when first installed. Machining of the detachable nylon portion of the bottle support can be precisely controlled. Once affixed, the position of the bottle carriers on the transfer arms, at the end of their stroke, was anticipated to be consistent at 86.6 mm relative to the 'low' position of the bottle rotation shaft.

In all cases where a fixed relative dimension or position is to be achieved, attention should be given to all elements that define the relative dimension or position. This means, in this instance, that the lowest position of the bottle rotation shaft (to which the 86.6 mm datum dimension applies) needs to be consistent for both transfer arms and from one bottle design to the next. It was indicated that a consistent 'low' position would always be achieved for this shaft, and this information was initially accepted at face value. This was an error. Despite the care that had been given to designing precise, quick-release bottle carriers, the changeover was found still to experience considerable difficulty. In particular the bottles were found not to transfer satisfactorily through this portion of the machine. Adjustments needed to be made that were expected have not been necessary.

The cause of these problems was found to lie within the mechanism that drove the bottle rotation shaft. The mechanism that defined the movement of this shaft typically needed to be adjusted during changeover. When an 86.6 position was set for transfer arm 'B' (as illustrated in Figure 10.4), it was found that this position was not automatically, as expected, repeated when transfer arm 'A' was elevated.

The solution to this problem was comparatively complex. The mechanism that imparted vertical movement to the 'bottle rotation shaft' had to be modelled. This model confirmed that an adjustable length tie-bar was needed within this mechanism, in place of a previously fixed-length bar. The length of this adjustable length bar had to be calculated prior to the changeover commencing. This length was set on a duplicate tie-bar, prior to the changeover commencing. The tie-bar was then installed as an additional task during the changeover.* A rig was made to aid setting the adjustable bar to its required length both quickly and precisely. A quick-release feature was designed for these bars, to allow them to be changed over as quickly as possible.

10.4 Case study 2c The importance of datums/2

The above case study describes a perceived fixed datum that was in fact subject to variation. Additional changeover work had to be introduced to re-establish this datum.

The need to understand selected datums was also demonstrated elsewhere on this machine. The machine's silk screens have been described, including illustrations, in Chapter 5 (see Exhibit 5.8). By making use of well-chosen datums it was intended that many previously needed adjustments would be eliminated – being replaced by the rapid, precise location of the screens straight into their working positions.

This case study is presented to show again that attention to datums is important, yet benefit often will only be achieved when all possible variables have been taken into consideration. Machine components should be considered in unison, alongside all other components with which they can interact or otherwise be related to.

The effectiveness of the improvements described in Chapter 5 was tested. A new set of screens was made and these improved screens were installed. At this point the problem of non-repeatable screen images in relation to the screen frame dowels – and hence the scales by which the frames were set – was understood and had been corrected (see Exhibit 5.8). Nevertheless, upon setting all scales to zero and printing some bottles, it was found that the five images still did not line up with one another. Except the advantageous use of toggle clamps to secure the screens, the changeover time was nearly as before, taking about an hour to complete: the scales proved to be of no use.

The problem, in fact, was very simple and the situation is shown by Figure 10.5. Improvement had been made to individual screens. The screen images were now precisely controlled relative to the dowels on which the screen frames were mounted.

---

* Although an additional task, its inclusion eradicated otherwise protracted and difficult adjustment tasks that would have been necessary elsewhere on the machine.

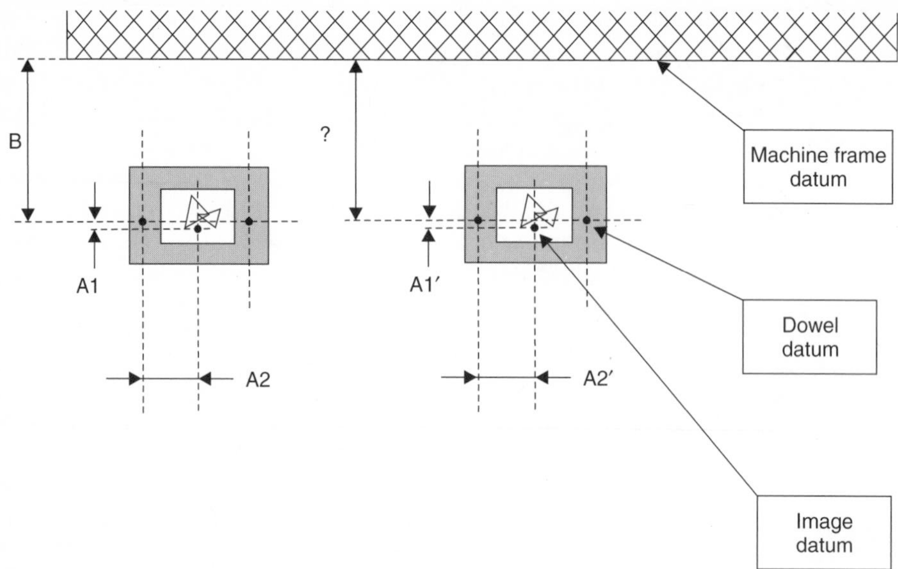

Figure 10.5 Schematic describing relevant datums

Yet these dowels (and the scales alongside them) were not set to each other via a common datum on the machine. In other words, dimension A1 is the same as A1′. Likewise, dimension A2 is the same as A2′. However, no common dimension applied in respect of dimension B. Once this had been addressed all scales could be set to a common value and the screen changeover time fell to be consistently below 5 minutes.

10.5 Case studies 2a, 2b and 2c General notes

Each of the three case studies in this series have shown how exploiting design change can improve changeover performance. These changes can be assessed in relation to the design rules of Chapter 5 or to the matrix methodology of Chapter 7. Changes have been made to overcome identified problems.

This print equipment demonstrated considerable potential for design change (including many opportunities that are not reported here). Retrospective improvement was undertaken. These same opportunities were also originally available to the OEM who, had these improvements been incorporated, would have had a more competitive machine available for sale, better able to meet likely market demands (see Chapter 2).

Machine specification

A final point in respect of this equipment is worth making. Chapter 3 discussed machine specification for changeover. This particular equipment was bought largely for its advertised changeover capability. A target changeover time was set, but this target time took no apparent account of the type of changeover that was to be undertaken. Sometimes the target time could be achieved. More often the changeover lasted far longer. There were still notable problems with both the changeover procedure and the quality of items delivered for use during the changeover, yet the machine design itself also imposed limitations on changeover performance. On one occasion three of the machine manufacturer's employees were witnessed conducting a changeover. This changeover took in excess of 8 hours to complete. The factory had already purchased the machine by this stage. The factory's changeover target was just 2 hours – and was based on just one skilled fitter and one assistant conducting the work.

10.6 Case study 3a Improvement limitations

With reference once more to print equipment, this next case study again highlights that a manufacturer's claimed changeover time needs to be scrutinized. A prospective purchaser of capital equipment ideally should have all aspects of a machine's capability demonstrated (and thus confirmed) before any payment is made. In respect of changeover, care should be taken that a representative type of changeover is being demonstrated (ideally, the most difficult type of changeover that will be experienced), and that no 'short cuts' are being taken that would not be possible in a factory situation.

Like the previous silk-screen print machine, the lithographic print equipment of the current case study was bought largely because of its claimed short-batch manufacturing capability. However, no evidence of this capability having been demonstrated could be determined.* The machine was bought, representing the largest ever capital outlay on a single piece of manufacturing equipment by the purchasing company. Once in use it was found that the expected changeover time was being far exceeded.

This print line has been the subject of previous exhibits in this book. Changeover performance was crucial if the line was to be financially viable. At the level of performance that was being experienced, shortly after first installation, it was calculated that heavy losses were being incurred in the line's operation.

A major initiative was commenced to improve changeover time (and, although not explicitly demanded, changeover quality). A changeover time of 30 minutes was sought. In due course it was expected that this time could be reduced to 15 minutes. The business expected Shingo's 'SMED' methodology to be employed to achieve this improvement, and that no significant expenditure or design change was needed.

* This statement needs to be qualified. Many of the individual changeover tasks were automated, and these had been witnessed taking place prior to purchase. Instead, no evidence could be determined that a changeover as a whole had been witnessed, of which these automated elements are only one part.

The destacker, the loader and the unloader (see Figure 10.9) were all new pieces of equipment (to a new design, that had never been built before). These line elements were highly sensitive to sheet quality variation – during both normal running and changeover. This was only properly determined once the line had been installed, indicating as well that it is desirable to test sensitivity to item quality variation when evaluating equipment's changeover capability.

Organizational improvement – eliminating the 'slack' of current changeover practice – was anticipated largely to yield the target level of performance.

Study of existing changeover practice commenced and researchers were on-site for a 6-month period. Three separate factory improvement teams were initiated, covering each of the three shifts that the factory worked. Company specialists were also assigned to address the changeover problem.

An overview of changeover difficulties

Significant deficiencies in the changeover operation were assessed. Unexpectedly, it was determined early on that systematic false recording of changeover performance was occurring (see Exhibit 3.7). The true changeover time in fact was far worse than senior managers were being led to believe. Major difficulties were assessed in changeover logistics (getting items into position for the coming changeover) and in the quality of these items. Changeover practice also had ample scope for improvement; in getting personnel to work in a much more focused, efficient manner.

These points have been explained earlier in this book. The purpose of this case study is threefold. First, it is presented to show that organizational refinement alone will yield improvement only up to a certain point. Second, it demonstrates that management sometimes need to be able to adapt their expectations of where, and how, improvement can be found. Third, a wider assessment is made of how management of the machine's development, installation and operation compromised the changeover time that was experienced.

Determining improvement available by more efficient working practice

Consider the first of these issues, determination of the likely changeover time arising by organizational improvement alone. The researchers had been involved at the factory for approximately 2 months. The improvement teams were deliberating the changeover problem, seeking improvements that could be made to their working practice. Some progress was being made. Importantly, key printers were involved, and an atmosphere of mutual trust and cooperation was judged to have developed. The decision was taken to define a 'reference changeover' (see Chapter 7). This exercise demands operators' honest participation if it is to be of real use. Building blocks for the reference changeover model are the individual changeover tasks, and self-recording of a number of changeovers on video occurred to gather information in respect of these tasks. From these video recordings the time for the individual tasks was written down. These are the times for the existing way of working. The next step was to assess what the 'reference performance' times for each task should be (again, see Chapter 7). Sessions were conducted to determine these times. The printers led the discussion. Some reductions were acknowledged as being possible.

This was useful in itself, but was not the only benefit of deriving the model. The model demands that all slack working practice be eliminated. The work noted above addresses slack on an individual task basis, but slack – poor practice – exists also in terms of, for example, work distribution and tasks that need not be part of the changeover at all. Defining the model was an exercise therefore in identifying where all slack currently occurred – by relating existing practice (from the video data) with an agreed model of ideal practice.

The model also served other important purposes. It was highly useful, as is explained below, in identifying limitations in management's improvement expectations. It also served as the basis of a written standard changeover procedure (for this specific type of changeover). Further, the model indicated, in the nature of the difference between conceptual and actual performance, that significant difficulties were experienced because changeover items were of poor quality, and thus not permitting the changeover to proceed smoothly according to plan.

A reference changeover model for this line – for a four-colour/four-plate changeover – is shown as Figure 7.12. The extent of the work that was going to be necessary to achieve the target level of performance (initially a 30-minute changeover) is shown. Further, the model demonstrates that improvement to this extent was not going to be possible by organizational improvement alone.

Management expectations

The print line changeover imposed pressure on the factory managers. Some unrealistic perceptions of the problems that were being faced were observed, and questionable decisions were taken as to how the initiative should be taken forward. Besides the problems of testing changeover performance during commissioning, management dictated:

- In order to reduce its duration, the changeover should have elements withdrawn from it that were not 'real changeover activities'. One example was the remaking of print plates when these were found, during the changeover, to be defective. Thus, if remanufacture added 3 hours to the changeover, it was deemed that this 3-hour period should be deducted from the overall changeover time. As argued previously, this is wrong. A changeover must be conceived as presented by Figure 1.2, where all activity that occurs during the elapsed changeover time is changeover activity, no matter what the cause. Some factory managers were also highly reticent to accept, erroneously, that the line's run-up phase also represented changeover activity.
- That unrealistic demands were made of the printer's performance (see also below).
- That application of Shingo's 'SMED' methodology, at little cost, would yield the expected levels of improvement.

Expectations of the print operators' performance

The second of these points, where the printer's performance was being questioned, was particularly of concern in respect of the number of adjustment iterations that the print line typically experienced. Iterations occur to produce finished (printed) components of sufficient quality. Figure 10.6 schematically illustrates how these iterations occur during the run-up phase of the changeover.

The figure shows that line start-up involved a series of very short runs. During this period both colour matching and print alignment takes place. The product from each of these trial runs is inspected and, as necessary, fine adjustments are made. Research showed that the number of iterations that needed to occur was highly dependent both

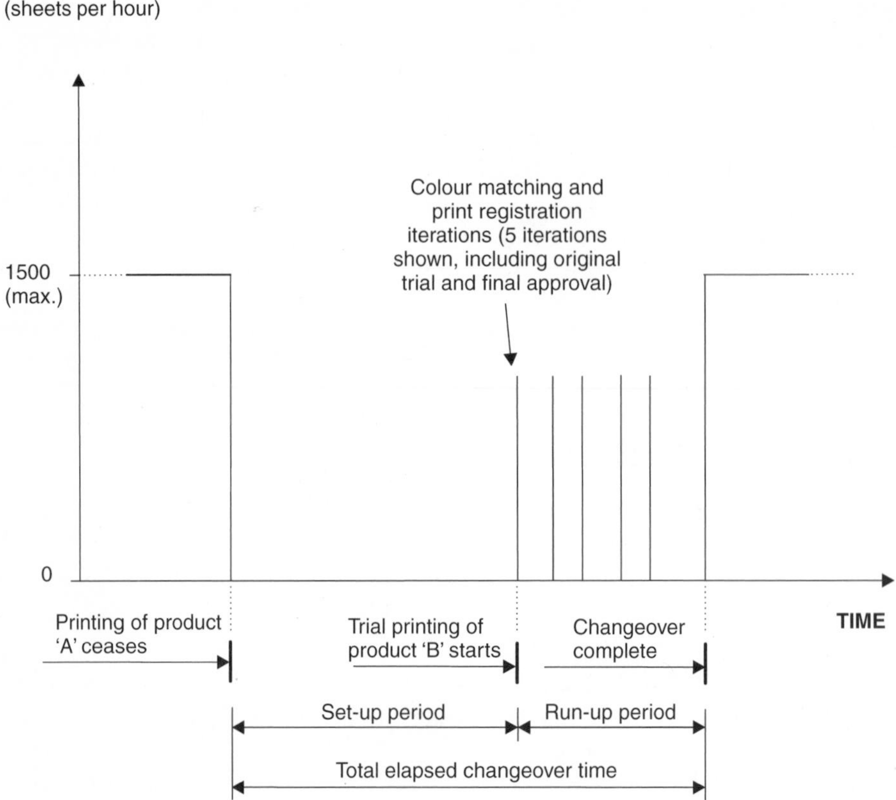

Figure 10.6 Colour matching/print registration iterations during run-up

on the quality of the inks and the quality of the plates that were used in the changeover. Other factors could also contribute, such as the quality of the white base coat that was being printed upon, or of the predefined ink key settings that determined, within limits, the amount of ink that was laid down at different locations across the print surface. With the equipment as it currently existed the researchers independently concluded that perhaps four or five adjustment iterations represented the minimum that should be present in the reference changeover model. Observations over a long period, including recorded evidence, supported these assertions. The trouble was taken to visit a comparable state-of-the-art European print shop, to assess their changeover practice. Their working practice further supported that four or five adjustment iterations would normally be expected.

Senior managers, however, strongly disagreed with this judgement. It was ruled that just two iterations should suffice. Further, it was stated that this improved performance should be achieved by the printers sharpening their working practice. This contrasted with what others associated with the project concluded: that two iterations would only be possible, if at all, once dramatically improved component quality had been achieved (most notably in respect of the printing plates, ink and diskette-held ink duct setting data). Whereas management could argue a right to direct improvement effort, this was contrary to the devolved authority and autonomy usually associated with successful shopfloor improvement exercises.

Perhaps the driving reasoning behind management's position was the contribution that having just two adjustment iterations made to a faster changeover time. The difference between two and five adjustment iterations can be evaluated from a reference changeover model. Figure 7.12 illustrates the reference changeover model for this line for a four-colour/four-plate changeover. It shows just two adjustment iterations occurring. A complete adjustment iteration is modelled as taking approximately 10 minutes to complete. Were five adjustment iterations to take place, the changeover time would extend to approximately 75 minutes, instead of the modelled 45 minutes of Figure 7.12.

How should the next improvement be sought?

Figure 7.12 assesses the best possible changeover time that could be expected on this line, for the modelled changeover type, before physical changes (design changes) to improve changeover performance were made. At the time of the study the best possible changeover time was rarely being achieved. Significant work had been done to refine the printer's work practice. Other important work also contributed to better changeover logistics, which was now occurring where possible in external time and in a better identified, better communicated manner (see also the following case study). Some early attention was being given to the quality of the ink supplied for the changeover and the quality of the diskette recorded ink duct setting data. In each regard the changeover was being improved. Improvements were largely costing little to implement although, when strictly applying the 'SMED' methodology, some improvement opportunities were not necessarily readily highlighted.

Although helpful, this did not mean that sufficient work was in progress to achieve the target sub-30-minute changeover, let alone the later hoped for 15-minute changeover.* Design improvements could be contemplated, like decoupling the print decks when changing the printing plates, thus allowing parallel loading (see the previous chapter), but all design improvements being considered were of a substantive (specialist; expensive; disruptive) nature. In this regard design improvement opportunities were different to many of those that were apparent on silk screen printing line of Case study series 2a to 2c – which on-site teams could far more easily progress.

Wider management considerations: a comparative assessment

This case study so far has discussed on-site research that we conducted. More recently an opportunity has been taken to speak to one of the company's managers, who was then heavily involved in supporting the installation and operation of this equipment. These discussions add a further perspective to the line's changeover problems. A second near-identical line also featured in the conversations that were held. This

* A four-colour/four-plate reference changeover model was assumed to represent the 'average' of the changeover types that were being undertaken. Both objectively and based on the factory's installation justification models, this premise probably underestimated the real situation, where many changeovers taking place were of a type that took longer to complete.

second line was installed at a slightly later date elsewhere in Europe. The mean changeover time for this other line very quickly stabilized at approximately 45 minutes. The researchers had only ever been permitted very limited access to this second site, and the manager's comments clarified possible reasons for the substantial discrepancy in performance between the two lines.

The points below are concerned wholly with the manufacturing environment in which the equipment was sited: comparison is being made between installations where the hardware itself was essentially identical and remained unchanged. It should be remembered also that this is a substantial, expensive, technically complex piece of equipment. Points raised during this conversation can be grouped in a number of categories:

1 Business plan versus installation/operational capability
2 Studio support
3 Integrating responsibilities
4 Local management involvement/installation team involvement
5 Maintenance issues
6 Printer morale
7 'Scientific' line operation (replacing printing as a 'black art')
8 Overoptimistic internal 'selling' of the machine

1 Business plan versus installation/operational capability

Four possible installation sites across Europe were originally considered. The company largely chose the first print line's location based on the quality of local management's business plan. In conversation it was suggested that a fundamental discrepancy existed between the business plan and the capability of this local business team in terms of installing and operating the plant. The former aspect was prominent in the company's location decision; the latter aspect was said to have been insufficiently considered.

It was further stated that this same oversight also applied when the decision to install the second line was taken. In practice this second site (as is reported further below) proved to be far more adept at operating the line as intended – including changeover performance.

2 Studio support

It was decided early on to have a well-equipped studio on each premises to support the print lines – instead of relying on external resources. This was the first time that the company had located a studio at one of its factories. The intention was that better quality, more responsive support could be afforded (including for changeover purposes).

At the first factory many of the potential advantages of this decision were negated when, for political reasons, studio personnel were subsequently denied access to the shopfloor. At the second factory it was observed that the studio had always worked much more closely with their planners and line personnel.

3 Integrating responsibilities

The equipment was wholly new to each site, from the destacker through to the unloader (see Figure 10.9). Training of personnel to operate the lines was therefore essential. The opinion was passed that the second site's operational success was partly due to the better integration of engineering and print functions. Not only did these personnel work better together, it was also stated that cross-functional capability was also engendered: that the printers acquired a new level of engineering capability and responsibility, while the engineers equally could assume certain print responsibilities.

4 Local management involvement/installation team involvement

The installation at each respective site was supported by a corporate team. The machine builder also provided considerable on-site expertise during commissioning.

It was found that there were notable differences between the involvement of the local management teams at the respective sites. One site always strictly used this external resource only to provide support when and if called upon, if the site themselves failed to resolve the immediate difficulties being experienced. The site thus approached the installation and handover in an involved, participative manner. The contrary was true at the original site, where a very high reliance was placed upon external support, where it was suggested that local management, engineers and printers were 'living in the pockets of the installation people'. Thus one site demonstrated an active local management involvement throughout, far beyond that at the site where satisfactory operation was never experienced. It was judged that, through their commitment, one site much more quickly gained an good understanding of the equipment. Also, the successful site employed (and still employs) the same key management personnel throughout the installation and commissioning period. This was not true at the original site, where frequent changes in key personnel occurred.

5 Maintenance issues

The machine is highly complex. In addition it has a small 'operational window' – by which we mean that parameters, including parameters of incoming or consumable items, require to be very tightly controlled. If parameters are not tightly controlled the line will not print wholly satisfactorily. Maintenance conceptually should assume a high priority under these circumstances; to ensure that small parameter changes because of wear or damage do not take the machine out of its operational window.

Consider, for example, the control/feed rollers within the destacking module. It was known that significantly different maintenance effort was expended at each respective site. At one site the rollers were regularly inspected, and cleaned or replaced. At the other site far less attention was given to this component. The printers consequently had to make adjustments elsewhere upon the machine to try to compensate as the rollers' performance degraded. In time control became ever more difficult to achieve, particularly when more and more variables were adjusted. Similarly, it became increasingly difficult to establish the root causes of the problems that were being experienced.

6 Printer morale

It was observed that printer morale was very different between the two sites. At one site printers were considered to be motivated, inquisitive, active; refining their new skills and on top of their tasks. The printers at the other site worked more on a reluctant 'must-do' basis, and were less inclined to extend their participation – a pattern of behaviour that can become increasingly difficult to break.

7 'Scientific' line operation (replacing printing as a 'black art')

With reference also to the points made above under the heading 'Maintenance issues', the machine's complexity can be further considered. It was shown earlier that losing control of one operational parameter, particularly in the context of equipment with a narrow operational window, can rapidly compromise performance. Re-establishing optimum conditions can become increasingly difficult when interrelated parameters all stray from their ideal condition.

The company as a whole was keen to establish a 'scientific' print operation, in contrast to the current 'black art' operation of its existing print lines. By this the company meant moving printing from a specialist and subjective operation, where there was inherently a high reliance on printers' understanding of how the machine could be compensated to be made to run satisfactorily. Instead rigid control of conditions was sought where, for each job, known settings would always apply. The company experienced difficulties imposing a 'scientific' operation, including changeover, in part due to the complexity of the equipment, particularly in terms of the number of parameters/variables associated with it (and their complex interrelation).

In these terms, as was observed, a difference in changeover performance can be expected if there are also discrepancies in the respective sites' maintenance effort. If operational parameters are allowed to get out of control (out of limits, or into an unknown state) simply because of poor work practice, then the same failure to achieve 'scientific' operation, including conducting changeovers, is likely. At one site the printers rigorously ensured that critical parameters were always reset at the conclusion of a run. The other site – where the poor changeovers were witnessed – failed to do this.

8 Overoptimistic internal 'selling' of the machine

The line's operational problems are noted in the current series of case studies. These problems were compounded by overoptimistic internal 'selling' of the machine within the company – claiming performance that it could not immediately achieve (or was never likely to achieve). Changeover performance is a case in point. Claims of a 15 changeover time being established were misplaced – particularly when seeking to apply the 'SMED' methodology and low-cost organizational improvement.

Problems with the line led to a number of high-profile management casualties. The poorly performing line has recently been withdrawn from its original location, and has been reinstalled in another country. Its changeover performance is now apparently much improved, but still does not match that of its sister installation.

Step-change improvement over former designs of print equipment

The company previously exclusively used printing presses whose design had not significantly altered over a period of approximately 40 years. Changeover time for an elderly two-deck line was, at best, 30 minutes per deck. Six-colour printing necessitated employing three consecutive two-deck line passes. Specified changeover times (for equivalent colour/plate changeovers) on the new line were at least three times faster. Also, operational speed was approximately double that of a two-deck line (where the need for multi-passing when three or more colours are used should also be remembered). Installing this new line was deliberately selected as a way to significantly improve performance, in preference to upgrading existing equipment (Chapter 3).

Changing market demands

Changeover performance was important not least because of the smaller, more frequent orders that were being placed by the company's customers (in which regard the company believed it was no longer competing satisfactorily). In a period of one year average batch sizes fell from 4200 units to 3200 units. Three years later customers were demanding batches that averaged at just 2000 units. Despite these changing conditions, one of the new six-colour lines was still not being fed with appropriate work (multi-colour, small-batch orders that it was originally bought to process) – largely because of the operational problems it was experiencing. In turn, because of these operational problems, some of the elderly lines that the new line was intended to supersede were still being kept and used – thus incurring still further unplanned costs.

Planned working time versus changeover time

With reference to Chapter 4, the mean ratio of planned production time/planned changeover time of 57/43 per cent indicates the extent that this business values manufacturing flexibility above outright productivity.

Line OEE (overall equipment effectiveness)

It was noted that both the new lines' OEE figures (Chapter 3) were less than that of the ancient print lines they replaced. Pressure to maintain high OEE performance equated to pressure not to print in small batches (again, see Chapter 3).

Revising the internal business structure to operate the new equipment

Chapter 3 highlighted that any new equipment should be matched to a business capability to operate that equipment. If necessary the existing plant structure (skills profiles; work demarcation; planning; maintenance; etc.) will need to be revised to allow its successful operation.

Two further – relatively minor – oversights show the thought that needs to be given to this issue. First, the company did not foresee the need for a competent electronics engineer. Second, if small batches were to be run, ink mixing and supply needed to be revised accordingly (for which changes to the number of ink personnel, work space and equipment all subsequently had to be assessed). These issues further inhibited installing and operating the line as intended. Each took time to resolve.

'Hexachrome' printing

Hexachrome printing has been described previously in Chapter 3 (Exhibit 3.10). Dependent on the product mix,* hexachrome printing can allow ink changes to be substantially eliminated from a changeover.

It was discussed that the use of hexachrome inks also raises other possibilities in terms of scheduling work through these new six-colour lines. Multi-image/single-design plates are currently used (the same image is repeated as many times as can be fitted into the available space). As well as eliminating or reducing ink changes, hexachrome could also allow multi-image/multi-design plates to be used, thereby enabling different designs to be printed side by side on the same sheet. The relative proportions of each design on a series of plates (one to six plates) could be matched to the upcoming batch sizes for these designs that were to be concurrently printed. In this way changeover intervals could be usefully extended.†

In conversation it was agreed that in principle this was a very attractive idea. This tactic is equivalent to using a single high-speed print line as a series of parallel low–medium speed lines. Whether the company had planning systems that could plan work in such a way was questioned. It was also appreciated that other significant logistics/manufacturing problems might result if the line were to be operated in this way. For example, the multi-product sheets have to be slit and separated accordingly (which could be difficult for different sized images).

'Reduction-In' effort by organizational change: an example

Since the line has been in use there have not been any significant design modifications to improve changeover performance – largely because the equipment did not lend itself to design change by an internal improvement team (see earlier text). Organizational improvements have dominated, varying considerably in their cost, effectiveness and ease of implementation. One organizational improvement involved change to the way that old ink was removed from the ink ducts. This change is described here by way of example, showing how the matrix methodology (Chapter 7) might be applied.

Originally the old ink was both removed and transferred into waste ink tins by use of spatulas. The ink duct surfaces were then washed with rags and solvents. As shown by the reference changeover model (Figure 7.12), this is a comparatively time-consuming task.

At the time of our research the improvement team members were all conscious that this was a changeover task where, potentially, the effort of conducting the task could be reduced. A design change – involving the use of sleeves – was briefly tried and rejected. More recently it has been assessed whether the spatulas are in fact the best tool to conduct this task. It was proposed that a suitable industrial vacuum cleaner could be used instead. This was tried, and has proved to be successful, shaving approximately 1 minute from each separate ink duct wash.

* Some product mixes will include products employing 'house' colours that cannot be replicated by the hexachrome process.

† With a slight – negligible – penalty in terms of manufacturing flexibility (Chapter 3).

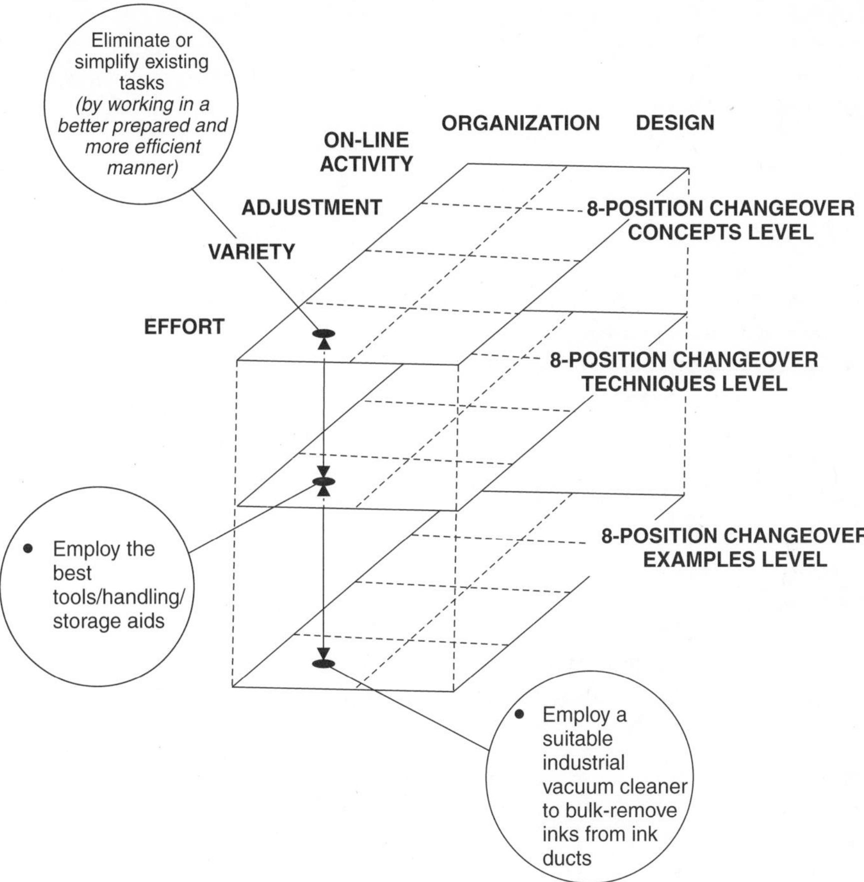

Figure 10.7 Matrix entries for the ink duct cleaning example

The desirability of addressing this task was highlighted by the reference changeover model (based on video recordings). The matrix methodology then assists in identifying possible solutions. Once 'excess effort' has been identified, the team can consider either design- or organization-led improvement concepts, alongside their corresponding techniques (Figures 7.7 and 7.8). The matrix entries for this particular example are shown in Figure 10.7. Although we describe entering the matrix in this instance from its concepts level, this need not always be so.

Although not yet available, this particular case also indicates the benefit that might be had by access to a suitably referenced catalogue of improvement examples (level 3).

10.7 Case study 3b Improving communication

It was noted in Chapter 3 that improved communication can aid changeover performance. This case study is a continuation of that presented immediately above.

Attention is here focused on the logistics aspect of the changeover, describing steps that were taken to deliver changeover items correctly to their assigned locations in preparation for the upcoming changeover. As part of this logistics improvement effort, attention is also given to 'foolproofing' the delivery and use of items – as far as was practicable – to overcome this site's poor record of commencing a changeover with items that were subsequently found to belong to other print jobs. Foolproofing in the context of this case study relates to employing nominally correct items (for the changeover in hand). Foolproofing here does not mean guaranteeing that the items provided for use are of requisite quality.

Unusually, this site used certain consumables, notably silver inks, that had a short shelf life once mixed. Although instances of their use were rare, their potential use placed additional pressure on the scheduling/delivery system that was in operation.

Early improvement effort: the use of a checklist

Major items for each changeover included:

- the base-coated sheets to be colour printed
- the inks
- the print plates
- the ink duct setting data (on diskette)
- a print proof (to colour match to)
- a design information folder (that gives details of line parameter settings for the design to be printed)

Complication was added by the need to introduce sheets into the machine in the correct orientation and, particularly, by the need to ensure that both inks and plates were mounted into their correct print stations (or decks) – from the six that are available for use.

An initial improvement was to bring all changeover items into a store area in a small anteroom at one side of the production facility. Prior to the changeover commencing it was intended that an operator would take the time to check that all items were present. This procedure was found to be less than satisfactory in practice. Principally difficulties arose because the printers had to make a conscious effort to be absent from the line to conduct the check shortly before the next changeover was due. Sometimes production demands, particularly if problems were being encountered, made this difficult. Also, if it was found during checking that an item was not present as required, there was usually insufficient time available to rectify the problem before the changeover was due to commence. The anteroom was of limited size, and it was difficult to store more than one set of changeover items within it. Also, because the anteroom was relatively poorly placed, additional time was added to the changeover in transferring these items to their point of use.

Seeking further improvement

The delivery of changeover items (above) showed scope for improvement. It was decided to improve these logistics issues as a factory-wide initiative. Better control was sought, including control of production scheduling. Ultimately it was sought to ensure that all items would be present for use on time, in preordained locations. Further, because of prior problems, particular attention was given to ensuring that the *correct* items were delivered to these locations.

Steps were taken to significantly improve communication between the print operators and other members of the support staff. The use of visual communication was increased, making use of prominent, clear images in preference, for example, to handwritten print codes on loose sheets of paper. Some examples of this change are described below. At the same time it was decided that changeover items, except the sheets to be printed, should be newly located on trolleys that could be lined up alongside the print line itself, ready for use (see below, including Figure 10.9). Disks were used with these trolleys, indicating that the trolleys were fully loaded and ready for use. Checking became a part of the trolley loading. The disk was painted amber on one face, and green on its reverse face. The slot into which the disk was inserted was painted red. Only when the last item was in position could a disk be placed over this red mark, signifying completion, including checking, of trolley loading. Responsibility for checking that all items were present was thereby transferred to the changeover item provider – who were given the responsibility always to check the state of completion of the trolley when they placed their items upon it.

A further problem addressed by this use of 'flags' was that of involving fork-lift truck operators. These drivers were needed to bring the new size base-coated sheets into position to be loaded onto the line. At a suitable interval before the changeover was due the amber/green disk was reversed by the lead printer, signifying that a driver was to halt and receive instructions. Hence, red (disk not inserted) indicated 'not ready'; amber (disk inserted) indicated 'prepared'; green (disk inserted) indicated 'action'. Previously the print operator typically had to leave the line to locate a fork-lift truck driver and negotiate for their help. The drivers were under new instructions to give high priority to this line changeover. The issuing of radio communication sets was also under consideration, pending the success of the system outlined above.

Some other details of improved visual communication to aid changeover

Other visible communication techniques were employed in improving the logistics aspect of this changeover. Some features, which are now explained, concerned:

- lining up new jobs
- confirming ink specification and print station (print deck 1 to print deck 6)
- presenting design information
- presenting the correct floppy disk (with ink duct setting data for the new print job)
- ensuring print proof visibility

These are what might be regarded as primary changeover issues. As with most changeovers, there were also numerous smaller components that also needed to be in place (and confirmed as being in place) if the changeover was to proceed smoothly. These items, among many others, included:

- spatulas
- torque wrench

- wash solvent
- gloves
- wipes
- ink tin opener
- print blankets
- densitometer
- telephone
- sidelay setting pieces

Better ink tin labelling

Indicative of the changes that were made, Figure 10.8 shows how a revised print label, affixed to the tin(s) containing the ink, presents both essential and potentially useful

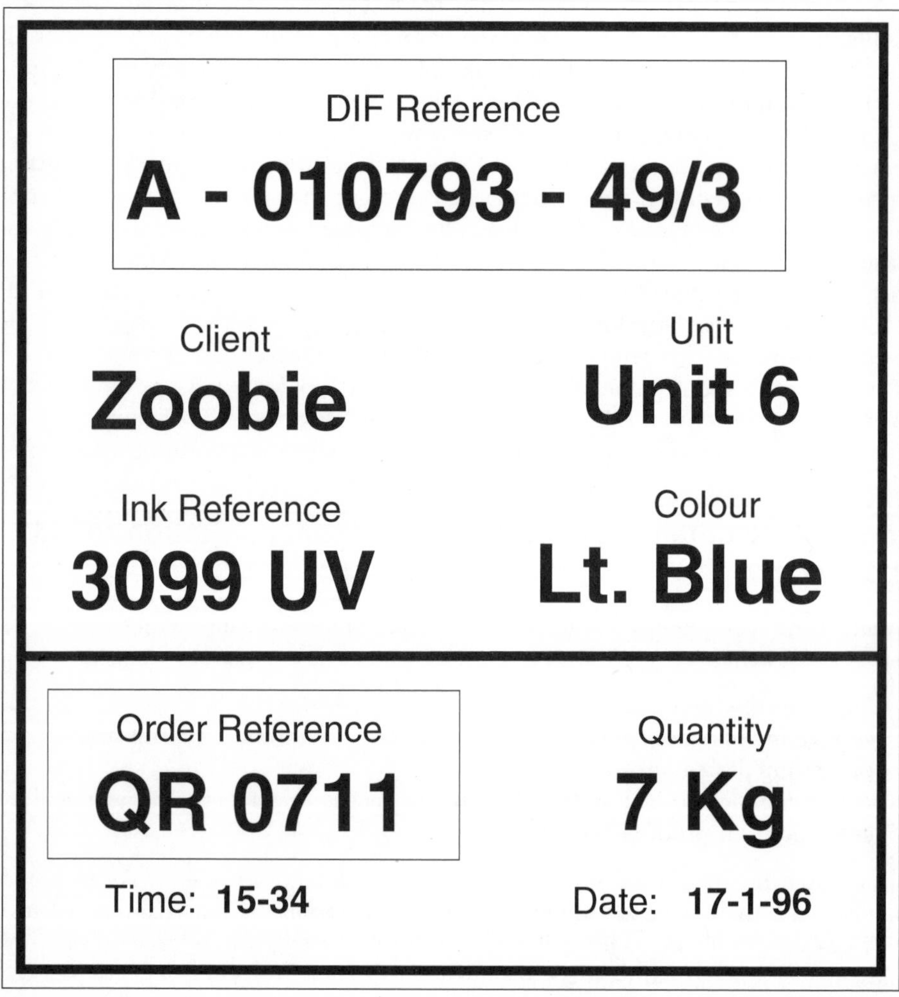

Figure 10.8 Improved ink tin labelling, presenting both essential and potentially useful information to the printer

information for the printers. Previously just the ink reference number was written, by hand, onto a tin. This then needed to be separately cross-referred to the print job in hand via information in the design information folder (see below).

Changed workplace layout, including the use of changeover trolleys

Other changes concerning the presentation of pending print jobs were made. Figure 10.9 shows how the use of trolleys aided the changeover. As indicated, the floor was carefully marked out to ensure consistency in the trolley positions. The trolleys' overall positioning was such that it was clear to the lead printer which jobs, in order, were next in line to be printed. Further trolleys that served other purposes were also strategically placed about the line:

Trolley purpose	*Number used*	*Colour identification*
Changeover trolleys	18 (max.)	Cream
Sample trolley	1	Blue
Aborted changeover trolley	1	Red
Transfer trolley (prepared off-line)	1	Green
Scrap trolley (for a print line 'crash')	1	Black

The careful layout of secondary items also took place. Shadow positions where these items were to be located were marked out. An item that was not present was readily apparent to the printers. It was a task of the second printer to walk around the line immediately prior to a changeover commencing, checking that these secondary items were all present as required.

Trolley layout

As shown in Figure 10.10, the trolley layout was carefully considered. The print line has six print stations. The inks are to be used at a predefined station, as are the plates. Sometimes six new plates and six different inks are used (a full six-plate/six-colour changeover). More usually fewer plates and fewer inks will be changed over.

To minimize the danger of loading either the wrong plates or the wrong ink into a specific print station, use was made of clearly recognized dummy components. In the case of the inks, this was an empty tin, boldly numbered with the station that it represented. For example, a dummy tin marked '5' on the trolley indicated that no ink change was to occur for print station 5. As well as use being made of dummy tins, the trolley was also compartmentalized. For every changeover there should thus be six tins present, be these dummy tins or 'real' tins.

The same procedure was applied for the plates. Dummy plates were clearly marked. The plates that are to be used, like the ink tins, are very clearly labelled. Again, both dummy plates and the plates that are to be used are loaded into assigned compartments.

A further feature of the trolley is the way that it clearly presents to the printer each of: the DIF (design information folder); the print proof; the floppy disk; the batch order details. Each of these items is marked with an image of the design that is to be

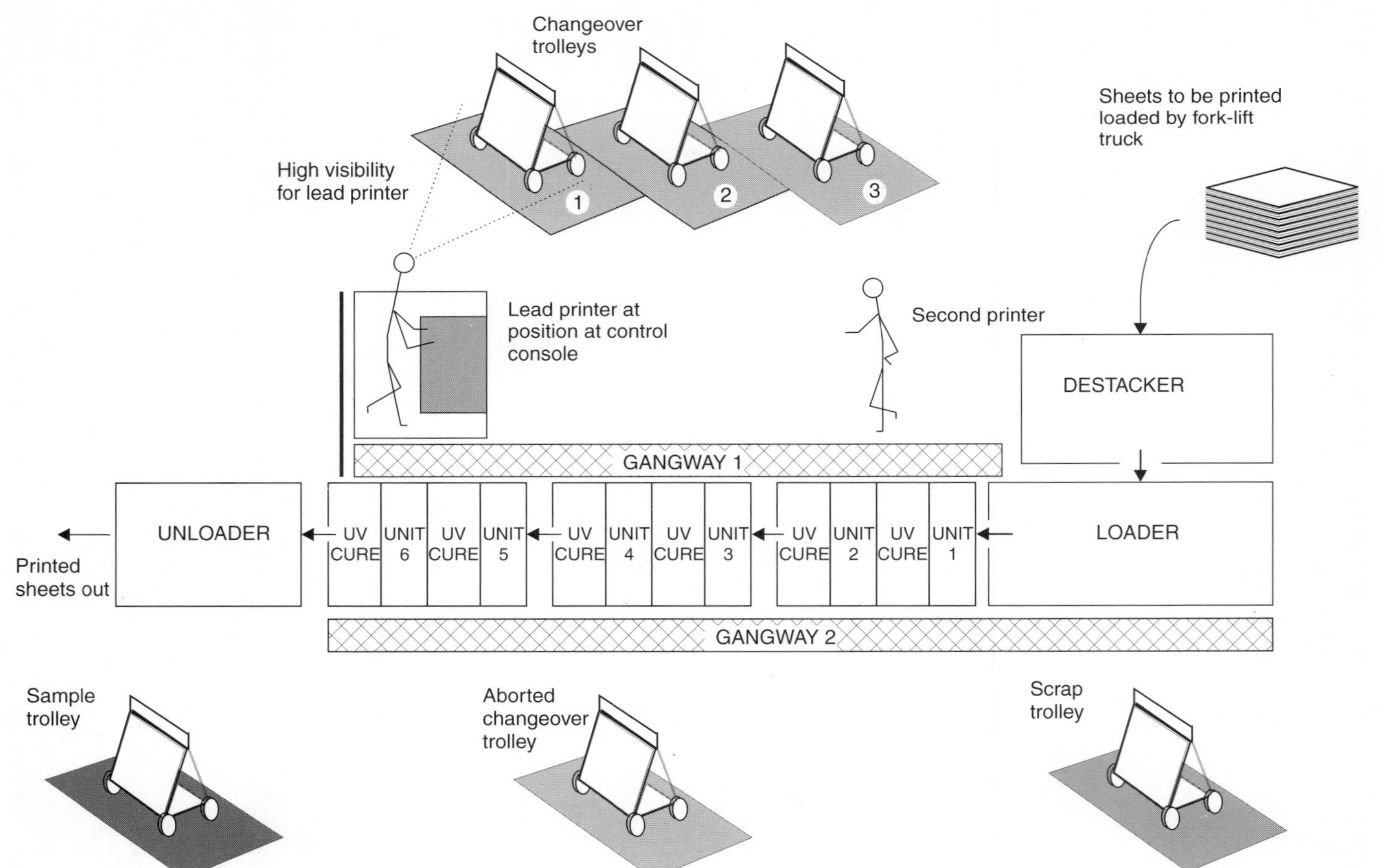

Figure 10.9 Schematic (not to scale) showing marked out trolley positions around the print line

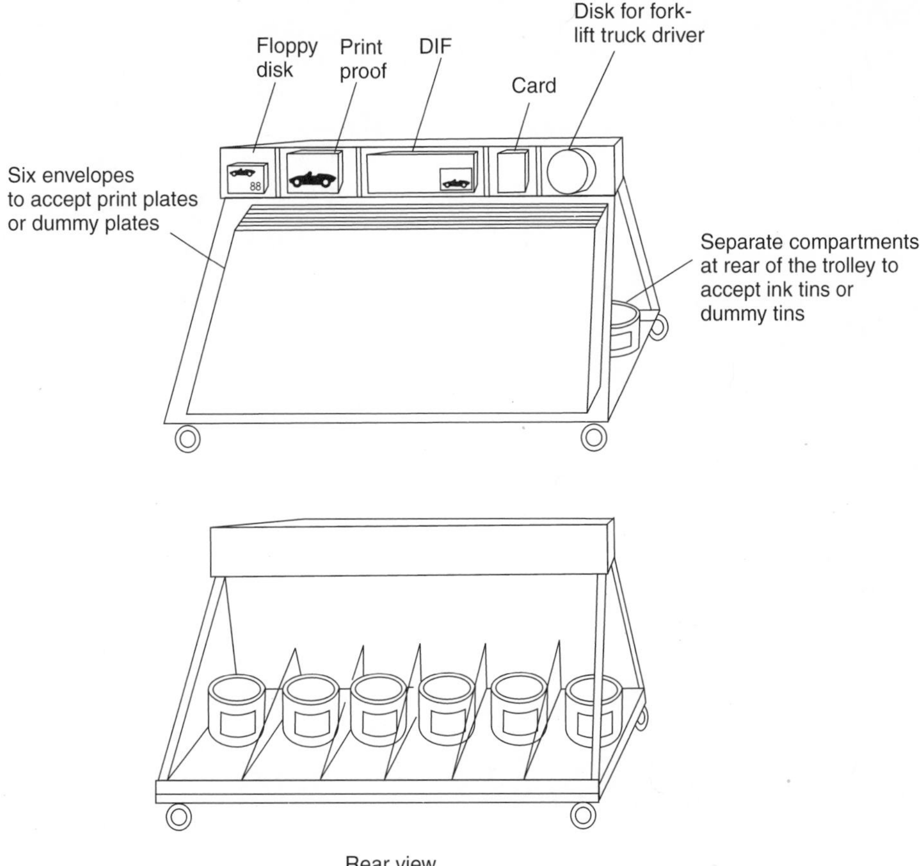

Figure 10.10 The changeover trolleys

run (the print proof is nothing more than a master image in any case: other items have images affixed to them). In this way the presence of an incorrect item is much more clearly distinguishable. As marked on Figure 10.10, these items are all located into compartments at the top of the trolley, behind a Perspex screen. They are immediately visible to the lead printer prior to the changeover. As noted, the trolley also incorporates a flag system, both for the printer and for the fork-lift truck driver.

The design information folder (DIF)

The DIF provides essential information concerning the print design it relates to. It incorporates, for example, details of the inks that are to be used, and the print stations they are to be used on. It also contains varied additional information to allow the print line to be set up (parameters set) to allow the design to be run. Like the floppy disks and print proofs, DIFs are all located, when not issued on the shopfloor, in a central store area. There is just one master copy.

At the commencement of the study a DIF typically was a haphazard collection of information on a series of sheets of paper. Changes to improve the DIF, shown by

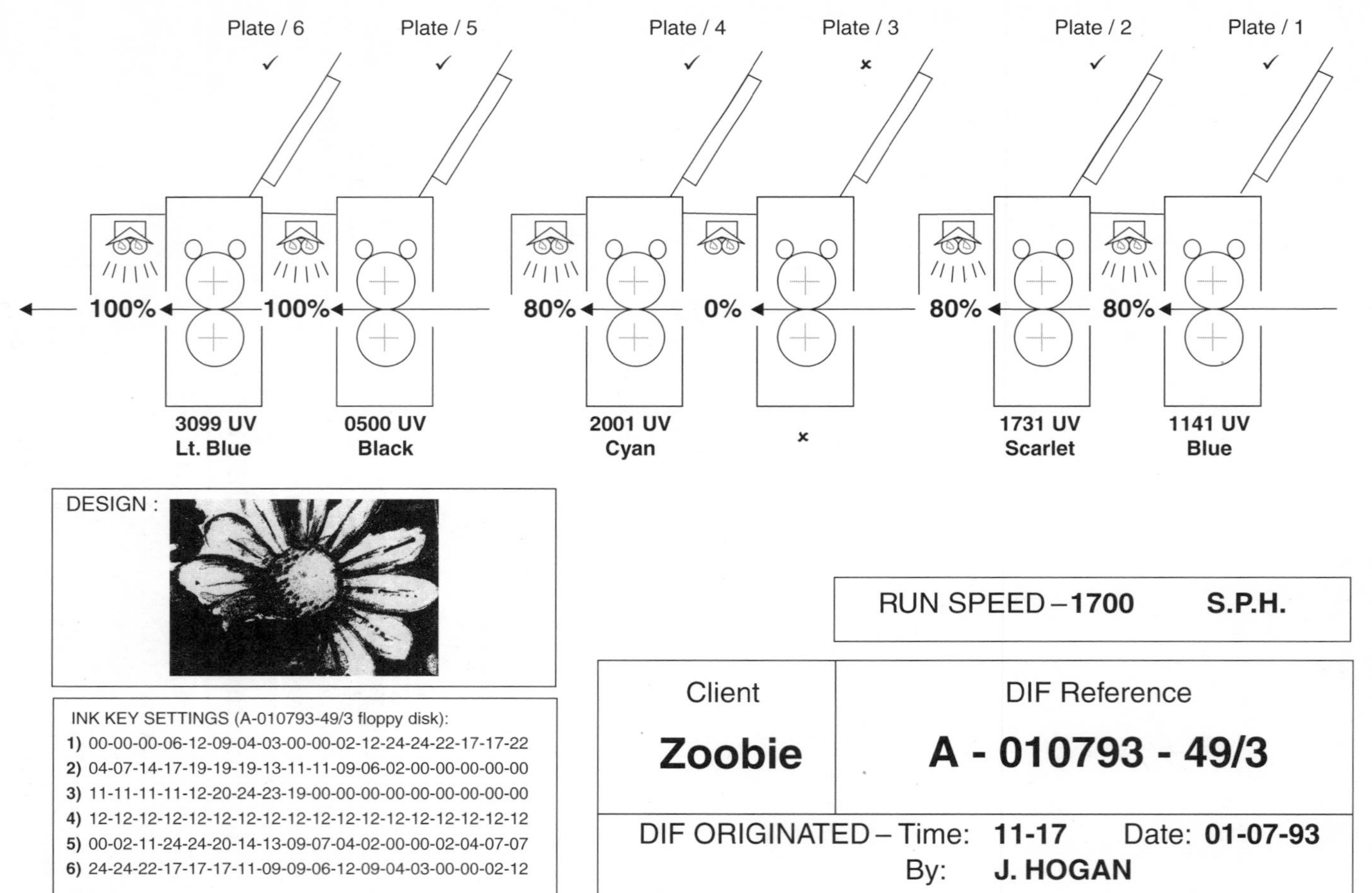

Figure 10.11 Simplified DIF (here omitting key print deck settings that appear on factory orginal, and loader/unloaded settings)

Figure 10.11, concentrated on assembling all necessary information onto a single sheet of A4 paper. A consistent format to this sheet was used. As noted above, the sheet included an image of the job that was to be run, thus doing much to eliminate any confusion as to the print design that it related to. Images also assist when presenting other information, for example writing ink code details by the image of the particular print station that the ink was to be used on.

Updating the DIFs represented a considerable body of work. Both the procedures and the hardware used to control and issue the DIFs needed to be revised (a computerized system needed to be introduced).

10.8 Conclusions

These case studies highlight wide-ranging issues. Together, along with other case studies throughout this book, they convey issues that are not always adequately recognized. They indicate that some current approaches to changeover improvement may achieve only limited results.

Of the issues raised in this chapter the considered application of design is perhaps the most important. We firmly believe that design can often make a dramatic contribution to faster, simpler, higher quality and more easily sustained changeovers. Much of our book sets out to raise design's profile in this regard – whether undertaken as part of a retrospective improvement initiative, or whether undertaken by OEMs. For example our case studies highlight that many changeovers are such that a body of core tasks cannot simply be translated into external time. In these cases changeover tasks require to be structural altered if, say, sub-ten minute performance is to be achieved. Our book identifies how design might be beneficially applied. Design need not necessarily be substantive, and our approach at all stages of an initiative is to consider the potential application of design alongside seeking organization-biased improvement.

The potential of design is not the only important issue that has been studied. Also of particular note are run-up issues, changeover item quality, validating equipment manufacturer's performance claims and the role of shopfloor improvement activity as part of a wider, structured, overall initiative. Management perceptions of changeover improvement are challenged: as suggested above, we believe it is not sufficient to expect significant and lasting improvement to be achieved only by promoting low-cost organizational refinement to shopfloor procedures. Such an approach should not always be expected to drive changeover times down into single-minute figures. In some cases this approach will suffice. Often, however, it will not.

Appendix 1 Changeover terms

Part 1: Principal terms that we apply in relation to changeover activity

(Note: Only summaries of these terms are provided here. Further explanation is given within the book, often assisted by illustrations).

Elapsed changeover time **(T_c)**
The period that the changeover disrupts normal, steady-state production. In most cases the elapsed changeover time will comprise a *set-up period* and a *run-up period*.

The set-up period
The readily identified period between the manufacture of the final piece of the previous batch and the first piece of the new batch. The line is stationary during this interval. Production of the first piece of the new batch does not imply that this first piece is of production quality, nor that it has been manufactured with the line running at its full production rate. Our definition of a set-up period (or set-up time or interval) does not equate to *internal time* (see below).

The run-up period
The interval between the manufacture of the first piece of the new batch and steady-state manufacture at full production rate and at full production quality. The run-up period is typically dominated by adjustments to the manufacturing hardware to enable full-scale production to commence. The run-up period can be extensive and can take many different formats. How work is conducted during the *set-up period* typically will influence the nature and extent of the run-up period. The run-up period represents an integral part of a changeover.

Run-down period
Occasionally a line may be gradually run down at the conclusion of production of the prior batch, rather than being abruptly halted.

External time
External time in respect of the forthcoming changeover relates to the period of production of the previous batch. Typically many changeover tasks can be conducted in external time.

Changeover quality
Every changeover has a time component and a quality component. Changeover quality refers to establishing manufacturing parameters to high precision, resulting, for example, in reduced scrap; higher production volume; greater line reliability.

Changeover item quality
Reference is made to items that are either directly or indirectly involved when conducting a changeover. Changeover items can include: change parts; consumables; the product under manufacture; manufacturing system elements. The quality of each of these items (their condition relative to a standard norm) can significantly affect the outcome of the changeover.

Type of changeover
Changeover from the manufacture of product 'A' to the manufacture of product 'B' can involve markedly different tasks to those tasks involved when changing over the same line from product 'B' to product 'C'. Changeover from product 'A' to product 'B', in such instances where different tasks are conducted, is a different type of changeover to that from product 'B' to product 'C'.

Reference changeover model
A reference changeover model is a description of optimum performance within the confines of the existing manufacturing hardware (by adopting organizational improvement alone, without applying design changes). The model should be derived with wide-ranging participation from those on the shopfloor who contribute to the changeover. The model describes successive task elements, divided between those who are to conduct the changeover. Task elements are completed where possible in external time.

The 'Reduction-In' strategy
We have classified that four 'excesses' can apply when conducting changeover tasks: excess *on-line activity*; excess *adjustment*; excess *variety*; excess *effort* (see also below). The 'Reduction-In' strategy seeks to reduce or eliminate excesses associated with each changeover task.

Reducing on-line activity
A reduction in on-line activity is a reduction in a task's contribution to the *elapsed changeover time*. Improvement is sought by reallocating tasks and resources, enabling tasks to be commenced at a different time within the overall changeover procedure (including during *external time*). Reduction in on-line activity does not constitute fundamentally altering the task in any way. Thus the task will take the same time as before to complete.

A reduction in on-line activity does not directly equate to 'separating' or 'converting' tasks into *external time*, as advocated by the *'SMED' methodology*. These two *'SMED' concepts*, however, are wholly embraced by reducing on-line activity.

'Reduction-In' effort/'Reduction-In' adjustment/'Reduction-In' variety
Excesses similarly may be addressed in any of these further 'Reduction-In' categories. There should be no preconceptions that any 'Reduction-In' category, including reducing on-line activity, should be dominant.

The organization–design spectrum
Improvement may emphasize organizational change to existing changeover procedures, or may emphasize design change to the process hardware (which also typically imposes change on existing procedures). Most improvement opportunities will comprise elements of both organizational change and design change – at varying points along the organization–design spectrum between 100 per cent organizational change and 100 per cent design change.

Design rules for changeover
Rules are presented to assist the application of design to improve changeover performance. The generic design rules are intended particularly to assist original equipment manufacturers.

The matrix methodology for changeover improvement
A tool to assist identification of a full range of potential improvement options. The matrix methodology can be used in conjunction with the *'Reduction-In' strategy*.

'SMED' methodology/'SMED' concepts/'SMED' techniques
Shingo's shopfloor 'SMED' methodology (Single Minute Exchange of Die methodology) comprises three active 'conceptual stages' or 'SMED' concepts. Improvement techniques are assigned to these active conceptual stages.

Overall methodology for changeover improvement
Our proposed global framework in which to conduct a changeover improvement initiative

Kaizen
Refers to a process of incremental, fully participative improvement of equipment and work practices.

Poke-yoke
Term used to mean 'foolproof' or 'foolproofing', as applied by Shingo ('Zero Quality Control – source inspection and the *poke-yoke* system', Productivity Press, Massachusetts, 1986).

'Universability'
The ability to process different products without equipment modification/adjustment.

Part 2: Changeover terms that we do not adopt

'SMED' system
Terms can be loosely applied in the context of the acronym 'SMED', for example ' "SMED" method' or '"SMED" system'. We have taken care to use only the terms set out in the previous section of this appendix.

'SMED': Related terms
Terms such as 'IED' (Inside Exchange of Die); 'OED' (Outside Exchange of Die); 'NTED' (No-Touch Exchange of Die); 'OTED' (One-Touch Exchange of Die) can also be employed. We likewise avoid the use of these terms.

'SMED': Internal time
Caution is needed that the widely held notion of 'internal time', as applied when adopting the 'SMED' methodology, is not applicable when defining the elapsed changeover time as comprising both a *set-up period* (when the line is stationary) and a *run-up period* (when the line is once again in operation).

SUR (Set-Up Reduction)
The term SUR is sometimes used to describe changeover activity. We avoid the use of this term.

Make Ready
Similarly, changeover activity can also be described under the term 'make ready', particularly in the print industry. Again, we avoid the use of this term in our book.

Appendix 2 Acronyms

AGV	Automated guided vehicle
CD	Compact Disc
CEO	Chief executive officer
CI	Continuous improvement
CNC	Computer numerical control
DFA	Design for assembly
DFM	Design for manufacture
DFMA	Design for manufacture and assembly
DIF	Design information folder
DNC	Direct numerical control
ELSP	Economic lot scheduling problem
EOQ	Economic order quantity
FMS	Flexible manufacturing system
GT	Group technology
JIT	Just-in-time
LH	Left hand (or left handed)
M/C	Machine
NC	Numerical control
OBI	Open back inclined
OE	Original equipment
OEE	Overall equipment effectiveness
OEM	Original equipment manufacturer
OLA	On-line activity
PCD	Pitch circle diameter
PLC	Programmable logic controller
QR	Quick release
RH	Right hand (or right handed)
SMED	Single minute exchange of die
SPC	Statistical process control
SUR	Set-up reduction
SWG	Standard wire gauge
TDC	Top dead centre
TQM	Total quality management
TPM	Total productive maintenance
UV	Ultraviolet
WIP	Work-in-progress

Appendix 3 Changeover audit sheet

A blank changeover audit top sheet and continuation sheet(s) are provided. These sheets may also be located at http://www.altroconsulting.com.

CHANGEOVER AUDIT SHEET

Changeover Equipment: **Recorded By:** **Document Ref.:**		**Date / Time:** **Changeover Personnel:**
Changeover Start Time:	**First Piece Manufactured at:**	**Changeover Complete at:**
Changeover From/ To: *(Changeover type)*		

Notes:

(i) When manually timing successive changeover tasks using a stopwatch and clipboard, complete only Section 1.

(ii) Section 2 entries should be completed at a later time. These entries should be made by the improvement team together, including those who have just been recorded.

(iii) If a video recorder with an inclusive timer is being used, all analysis sheet entries may be made if desired at a later time, when replaying the video.

(iv) 'Activity Type' and '"Reduction-In" Opportunity' entries might be made using the key presented below.

Activity Type Key (some suggestions):

Prob *= Problem* ***Mvt*** *= Movement*
Sec *= Securing* ***Wait*** *= Waiting*
Adj *= Adjustment* ***Interrupt*** *= Interruption*
Loc *= Location* ***Rem*** *= Removal*
Tr *= Trial* ***Multi*** *= Multi-person*
SS *= Special Skill* ***Access*** *= Access*
Tool *= Tool employed* ***Clean*** *= Clean*
Data *= Seeking/using data*

'Reduction-In' Excess Classification:

On-Line Activity
Adjustment
Variety
Effort

CHANGEOVER AUDIT SHEET *(cont.)*

Document Ref.: **Sheet No.:**

Section 1					*Section 2*		
Task ('major')	Task ('detail')	Time Complete	Who?	Notes	Activity Type	Duration	'Reduction-In' Opportunity

Appendix 4 Suggested classifications

Suggested classifications for use in the 'Activity type' column of the changeover audit sheet (Appendix 3).

Mvt = *Movement*
Is a tool, for example, being moved too far/too much while the line is stationary? If an item is being removed and subsequently replaced, is it still being moved too far? Are there features that can be modified that make the item/assembly in question difficult to move (for example, is it heavier than it could be)? Are there devices such as runners that could be incorporated to ease movement? Can items/components be moved in unison rather than separately – perhaps being built up more off-line?

Access
Are access problems apparent? If so, would the use of alternative hand tools, perhaps, help? Can other assemblies be changed to improve access? Is there scope to split any assemblies so that they do not have to be fully removed to gain improved access?

Sec = *Securing*
Securing of changeover items can be done by many different means. Is the item under review being oversecured? Are threaded fasteners being employed, in which case can these fasteners be substituted by quick-release devices? Or can 'keyhole slots' or other such mechanisms be employed to reduce the manual turning of threaded fasteners? Are there any opportunities to use power accessories – either to turn threaded fasteners, or to replace them (for example, remotely activated hydraulic clamping units)?

Wait = *Waiting*
When one or more of those conducting the changeover is not engaged – doing no task, even though available to do one – then that person is waiting. Can tasks be better distributed (sequenced) to eliminate occurrences of waiting?

Adj = *Adjustment*
Adjustment is apparent in virtually every changeover, but does adjustment always have to occur? Can items be mounted so that they achieve the desired processing parameters 'out-of-the-box'? Can adjustment be made easier by adjusting to known, easily identified positions? Can design modifications be made, to accommodate all change parts/products universally without adjustment? Changeover item quality can also have a considerable impact on adjustment activity. Are the items being supplied for the changeover of requisite quality?

Interrupt = *Interruption*
Interruption refers either to voluntary or involuntary distraction from 'real' changeover tasks (but does not refer to doing these tasks slowly or inefficiently). By implication, therefore, all such interruption could be eliminated.

Loc = *Location*
How is location being effected? Are too many constraints being imposed, or, perhaps, is the item being insufficiently constrained? Is location occurring to sufficiently high precision and, if not, might advantage be gained by improving this aspect? Are location features prone to wear, damage or contamination?

Rem = *Removal*
Is removal necessary? Can a way be conceived to prevent removing an item if it is not a change part? If a change part is involved, can the removal task be made easier? Is there somewhere for the removed item to be well stored (close by, adequately marked/orientated and prevented from damage/deterioration)?

Tr = *Trial/inspection*
Is this trial really necessary? Can the changeover be improved, particularly by ensuring that necessary changeover parameters are correctly met, to eliminate its need?

Multi = *Multi-person*
Can the need for multi-person execution of a task be removed? For example can handling devices be used or, perhaps, lightweighting of large assemblies/components take place? Can mechanization be employed?

Data = *Data Reference*
Is the data/information that is being used available in full, clearly presented and unambiguous?

SS = *Special Skill*
Can the operation be deskilled – or can other personnel be trained/authorized to enable them to conduct this task?

Tool = *Tool Employed*
Does a tool need to be employed for this operation? Can the need to use a tool be eliminated?

Prob = *Problem*
(Problems are difficult to provide guidance on, given that by our definition they are not expected, repeatable events during a changeover. Focus should be on taking steps to try to ensure that the witnessed problem cannot recur.)

Clean = *Cleaning*
Can provision be made for cleaning to be more easily conducted? For example, can smaller critical contact surfaces (that require to be clean) be employed? Can better cleaning materials/devices be made available? Can cleaning be automated, or can chutes, for example, be used to redirect swarf so that cleaning effort is reduced or eliminated? Is it possible to instigate new procedures, to preclean items in external time and then to handle these items better (in respect of avoiding contamination) when bringing them to the changeover machine?

Appendix 5 Improvement option assessment sheet

A blank improvement option assessment sheet is provided. This sheet may also be located at http://www.altroconsulting.com.

Assessment: ✓✓ = Strongly positive; ✓ = Positive; × = Negative; ×× = Strongly negative **External involvement**: ⌖

Changeover task	Improvement idea	Estimated time saving	Cost–benefit assessment	Other impact	Implementation issues	FINAL RANK

Appendix 6 Changeover improvement work plan

A blank changeover improvement work plan (PDCA chart) is provided. These sheets may also be located at http://www.altroconsulting.com.

CHANGEOVER IMPROVEMENT WORK PLAN:

Date:

(PDCA Chart)

DESCRIPTION OF IMPROVEMENT BEING MADE / BENEFIT SOUGHT	RESPONSIBILITY	START	TARGET COMPLETION	STATUS	NOTES
				D C P A	
				D C P A	
				D C P A	
				D C P A	

DESCRIPTION OF IMPROVEMENT BEING MADE / BENEFIT SOUGHT *(Continued)*	RESPONSIBILITY	START	TARGET COMPLETION	STATUS	NOTES
				D C P A	
				D C P A	
				D C P A	
				D C P A	
				D C P A	
				D C P A	

Index